WIRELESS AND EMPIRE

Wireless and Empire

Geopolitics, Radio Industry, and Ionosphere in the British Empire, 1918–1939

AITOR ANDUAGA

OXFORD
UNIVERSITY PRESS

Great Clarendon Street, Oxford OX2 6DP

Oxford University Press is a department of the University of Oxford.
It furthers the University's objective of excellence in research, scholarship,
and education by publishing worldwide in

Oxford New York

Auckland Cape Town Dar es Salaam Hong Kong Karachi
Kuala Lumpur Madrid Melbourne Mexico City Nairobi
New Delhi Shanghai Taipei Toronto

With offices in

Argentina Austria Brazil Chile Czech Republic France Greece
Guatemala Hungary Italy Japan Poland Portugal Singapore
South Korea Switzerland Thailand Turkey Ukraine Vietnam

Oxford is a registered trade mark of Oxford University Press
in the UK and in certain other countries

Published in the United States
by Oxford University Press Inc., New York

First published 2009

British Library Cataloguing in Publication Data

Data available

Library of Congress Cataloging in Publication Data
Anduaga, Aitor.
Wireless and empire : geopolitics, radio industry, and ionosphere in
the British Empire, 1918-1939 / Aitor Anduaga.
p. cm.
Includes bibliographical references and index.
ISBN 978–0–19–956272–5
1. Ionospheric radio wave propagation—Research—Government
policy—Great Britain—History—20th century. 2. Atmospheric
physics—Research—Government policy—Great Britain—History—20th
century. 3. Radio supplies industry—Great Britain—History—20th
century. 4. Radio broadcasting—Political aspects—Great
Britain—History—20th century. 5. Geopolitics—Great
Britain—History—20th century. 6. Great Britain—Colonies—
Social life and customs. I. Title.
QC973.4.I6A526 2009
384.5409171'24109041—dc22 2008049036

Typeset by Newgen Imaging Systems (P) Ltd., Chennai, India
Printed in Great Britain
on acid-free paper by
the MPG Group in the UK

ISBN 978–0–19–956272–5 (Hbk.)

1 3 5 7 9 10 8 6 4 2

Gure amari

CONTENTS

LIST OF ILLUSTRATIONS

Cover shows pictorial diagram of the heights of various phenomena. From G.M.B. Dobson (1926)

LIST OF TABLES

LIST OF ABBREVIATIONS

AAAS	Archives of Australian Academy of Science, Basser Library, Canberra, Australia
ACRR	Associate Committee on Radio Research
ADB	*Australian Dictionary of Biography,* ed. Bede Nairn and Geoffrey Serle, Melbourne: Melbourne University Press, 1981, vol. 8; 1986, vol. 10; John Ritchie (ed.), 1993, vol. 13; 2000, vol. 15; John Ritchie and Diane Langmore (eds.), 2002, vol. 16
Ann. Sci.	*Annals of Science*
App. J. Ho. Rep. of N.Z.	*Appendix to the Journals of the House of Representatives of New Zealand*
ARRB	Australian Radio Research Board, CSIRO
AUA	Auckland University Archives, New Zealand
Aus. Cul. Hist.	*Australian Cultural History*
Aus. Jour. Pol. Hist.	*The Australian Journal of Politics and History*
Aus. Phys.	*The Australian Physicist*
AWA	Amalgamated Wireless Australasia, Ltd.
AWA Tech. Rev.	*AWA Technical Review*
Bell Sys. Tech. Rev.	*Bell System Technical Review*
Biog. M. Fell. Roy. Soc.	*Biographical Memoirs of Fellows of the Royal Society*
BLOA	Bodleian Library, Oxford, Archives, UK
Brit. Jour. Phil. Sc.	*British Journal for the Philosophy of Science*
BRRB	British Radio Research Board
Bull. Am. Met. Soc.	*Bulletin of the American Meteorological Society*
Bull. BVWS	*Bulletin of the British Vintage Wireless Society*
Bull. Lon. Math. Soc.	*Bulletin of the London Mathematical Society*
Bur. Stan. Jour. Res.	*Bureau of Standards Journal of Research*
Can. His. Rev.	*The Canadian Historical Review*
Can. Jour. Res.	*Canadian Journal of Research*
CSIR	Council for Scientific and Industrial Research
CSIR Bull.	*Bulletin of the Council for Scientific and Industrial Research*
CUP	Cambridge University Press, Cambridge, UK

DNZB	*Dictionary of New Zealand Biography*
DPA	Ditton Park Archive, Slough, England
DSIR	Department of Scientific and Industrial Research
DTM	Department of Terrestrial Magnetism, Carnegie Institution of Washington, USA.
EOS: Tr. Am. Geo. Un.	*EOS: Transactions of the American Geophysical Union*
EUA	Edinburgh University Library, Archives
EW & WE	*Experimental Wireless and the Wireless Engineer,*
Geo. Jour. Roy. Astr. Soc.	*Geophysical Journal of the Royal Astronomical Society*
HRAS	*Historical Records of Australian Science,*
HSPS	*Historical Studies in the Physical and Biological Sciences*
HUP	*Harvard University Press*
IEEA	Institute of Electrical Engineers, Archives, London
Jahr. drath. Tel.	*Jahrbuch der drathlosen Telegraphie*
JC	*Journal of Careers*
Jour. Atm. Terr. Phys.	*Journal of Atmospheric and Terrestrial Physics*
Jour. Can. St.	*Journal of Canadian Studies*
Jour. Fran. Inst.	*Journal of the Franklin Institute*
Jour. Geo. Res.	*Journal of Geophysical Research*
Jour. IEE	*Journal of the Institution of Electrical Engineers*
Jour. Inst. Elec. Eng.	*Journal of the Institution of Electrical Engineers*
Jour. Inst. Eng. Aus	*Journal of the Institution of Engineers Australia*
Jour. Res. NBS	*Journal of Research of the National Bureau of Standards*
Jour. Roy. Soc. Arts	*Journal of the Royal Society of Arts*
Jour. RRL	*Journal of the Radio Research Laboratory*
KCLA	King's College London Archives, London, England
Mem. Roy. Met. Soc.	*Memoirs of the Royal Meteorological Society, London*
Met. Mag.	*Meteorological Magazine*
Mon. Not. Roy. Astr. Soc.	*Monthly Notes of the Royal Astronomical Society*
MUA	Melbourne University, Archives, Melbourne, Australia
NAA	National Archives of Australia, Canberra, Australia
NAC	National Archives of Canada, Ottawa, Ontario
NAUK	National Archives of England, Wales, and the United Kingdom

Notes Rec. Roy. Soc. Lond.	*Notes and Records of the Royal Society of London*
NPL	National Physical Laboratory, UK
NRC	National Research Council
NSWA	State Library of New South Wales Archives
N.Z. Jour. Sc. Tech.	*New Zealand Journal of Science and Technology*
NZNA	New Zealand National Archives, Wellington
Ob. Not. Fell. Roy. Soc.	*Obituary Notices of Fellows of the Royal Society*
Onde Elec.	*L'Onde Électrique*
Papers Ass. URSI	*Papers of the General Assembly of the International Scientific Radio Union*
Phil. Trans. Roy. Soc. Lond.	*Philosophical Transactions of the Royal Society of London*
Phys. Rev.	*Physical Review*
Plan. Sp. Sc.	*Planetary and Space Science*
PMG	Postmaster General
Poznan Stud. Phil. Sc. Hum.	*Poznan Studies in the Philosophy of the Sciences and the Humanities*
Proc. Am. Phil. Soc.	*Proceedings of the American Philosophical Society*
Proc. Cam. Phil. Soc.	*Proceedings of the Cambridge Philosophical Society*
Proc. IEEE	*Proceedings of the American Institute of Electrical Engineers*
Proc. IRE	*Proceedings of the Institute of Radio Engineers, New York*
Proc. Phys. Soc. Lond.	*Proceedings of the Physical Society of London*
Proc. Roy. Soc. Edi.	*Proceedings of the Royal Society of Edinburgh*
Proc. Roy. Soc. Lond.	*Proceedings of the Royal Society of London*
Proc. Wor. Rad. Con.	*Proceedings of the World Radio Convention*
Proc. URSI	*Proceedings of the General Assembly of the International Scientific Radio Union (URSI)*
PRO/ED	Public Record Office, Education Files, London
Quart. Jour. Roy. Astr. Soc.	*Quarterly Journal of the Royal Astronomical Society*
Quart. Jour. Roy. Met. Soc.	*Quarterly Journal of the Royal Meteorological Society*
RAAS	*Records of the Australian Academy of Science*
Rad. Res. Spec. Rep.	*Radio Research Special Report*
RRB	Radio Research Board
RRB Rep.	*Radio Research Board Report*
RRC	Radio Research Committee, New Zealand
R.R.O. Newsletter	*Radio Research Organization Newsletter*
SHPMP	*Studies in the History and Philosophy of Modern Physics*

SHPS	*Studies in History and Philosophy of Sciences*
SMLAWA	Sydney, Mitchell Library, AWA Limited Records
Soc. Stud. Sc.	*Social Studies of Science*
SUA	Sydney University, Archives, Sydney, Australia
Terr. Mag. & Atm. Elect.	*Terrestrial Magnetism and Atmospheric Electricity*
Trans. Am. Geo. Un.	*Transactions of the American Geophysical Union*
Tran. Am. Ins. Ele. Eng.	*Transactions of the American Institute of Electrical Engineers*
Trans. Newcomen Soc.	*Transactions of the Newcomen Society*
TUP	Toronto University Press
Unit. Emp. RES. J.	*United Empire: Journal of the Royal Empire Society*
URSI	International Union of Radio Science
UTA	University of Toronto, Archives, Canada
Zeitschr. f. hochfr.	Zeitschrift für hochfrequenztechnik
Zeitschr. f. tech. Phys.,	Zeitschrift für technische Physik

PREFACE

This book deals with radio development and with physical research on the ionosphere and their relations with empire, industry, and higher education in the most industrialized countries of the British Empire between the two World Wars. Significantly, by the early 1920s, most countries with imperial interests focused attention on the power of radio, in particular of shortwave, for communications, chiefly for the benefit of the military and of both radio amateur and commercial sectors. But what was striking in the British Empire was the emergence of a good number of organizations and individuals (many pioneers) that decisively contributed, by using radio, the physical knowledge of the upper atmosphere. In the interwar years, it would be erroneous (as well as deceptive) to separate the marked realism of the stratified structure of the ionosphere from the imperial, military, and commercial contexts of ionospheric physics: the nodal points of interconnection are simply too numerous. Here, therefore, it is the latter that thrusts our attention towards the pursuit of reasons behind ionospheric physics.

Essentially, the reflections on *Geopolitics, Radio Industry, and Ionosphere in the British Empire, 1918–1939* incarnate that pursuit. Our thesis is that certain structural reasons intrinsic to the radio industry, to technical education and to the geopolitical reality of the time, not only actively fomented, but were determinant for, the emergence and sustenance of physical research of the ionosphere. To prove it, we mistrusted the easy prescription of regarding the interaction of all factors at stake as explicable by direct causal effects or by simple causal relationships. Instead, we endeavoured to examine in detail the complexity of the links which interlaced the ionosphere, empire, radio industry, and education and to underline the hitherto unnoticed importance of business strategies, government entities, and national identities.

The picture drawn here may surprise those of us who have viewed radio with much the same achromatic tincture as those who devoted themselves to diffuse accounts of discovery. We, like them, regarded radio expansion as benevolent, neutral, apolitical; its relationship with the ionosphere, merely through the maternal technological link. We, like them, viewed the enormous research endeavour in ionospheric physics in the interwar years as good in itself, a laudable diversion of lucid enthusiasts in quest of transcendental goals. Physicists, it was argued, sounded the upper atmosphere and radio improved the prospects of exploration, but this imposed no responsibility upon radio. The intellectual benefits of this new knowledge of the atmosphere, in scientific and technical progress, could not be questioned. But its cognitive, distinctive features, such as the stratified character of the ionosphere, appeared uprooted from their cultural and socioeconomic habitat.[1]

For well over the last quarter of a century, historiographical studies in upper atmospheric sciences have placed the predominant emphasis on the content of their various disciplines.[2] Where the distinctiveness of diverse national patterns in ionospheric research was stressed, the objective pursued was often to identify the causes of theoretical advances. The socio-economic and geopolitical context was thus frequently used as a mere backdrop. Although cannot be questioned that in this period geophysics experienced profound internal developments, or that the analysis of disciplinary considerations is essential for its historical understanding, we think nonetheless that there is a manifest one-sidedness in this mode of examining the issue. To counterbalance it, we have presented here what we believe is a novel approach. Firstly, the task of rethinking pre-existent analyses, of evaluating the ways in which industrial, political, and military interests interfered with scientific and engineering practices has been a bounden duty. And secondly, we have given priority to the similarities in the circumstances in which radio evolved in the setting of enormous comparative potential that the British Empire was. In this respect, the emphasis on the features of similitude is by no means irreconcilable with the unmistakable plurality of factors that we have identified in the various countries. Quite the contrary; this plurality is only explicable by reference to a polychromatic diffraction of a myriad of contexts, rather than to a single 'internalist' monochrome beam.

We might develop our thesis, predominantly through a sociologist prism and apply it to all chapters, but it probably would leave us a distorted and unbalanced

[1]Apart from the discovery literature, other sources have also contributed to emphasize the neutral nature of radio. Authors such as Ch. A. Ziegler, R.P. Multhauf, and O.G. Villard have paid much attention to the role of radio technology in the exploration of the upper atmosphere, and less to the diverse contexts in which it was developed. See J.L. Dubois, R.P. Multhauf, and Ch.A. Ziegler, *The Invention and Development of the Radiosonde* (Washington: Smithsonian I.P., 2002), 26–48; Oswald G. Villard, 'The Ionospheric Sounder and its Place in the History of Radio Science', *Radio Science*, 1976, 11(11): 847–60.

[2]See, for instance, C. Stewart Gillmor, 'Wilhelm Altar, Edward Appleton, and the Magneto-Ionic Theory', *Proc. Am. Phil. Soc.*, 1982, 126: 395–423; C. S. Gillmor, 'The Formation and Early Evolution of Studies of the Magnetosphere', in *Discovery of the Magnetosphere* (Washington, DC: American Geophysical Union, 1997), 1–12.

picture. In order to explain the momentous peculiarities of some physicists and entrepreneurs, the marked idiosyncrasy of certain institutions and firms, and the cultural identities of the nations involved we should invoke precisely those components which we would deliberately leave aside with a sociological approach—such as personal virtues and ambitions, institutional style, individual personality, and national character. Hence we have sought for the bulk of the book a scheme in which certain categories and their fields of applicability at a given time and place are considered as evoking such components.

The metaphor of a thread of five pieces which represent the categories 'science', 'industry', 'government', 'military', and 'education', we believe, may serve well to define and depict our subject. Through it, we claim to show how inconclusive and insufficient Gillmor's historiographic strategy was in trying to explain the progress of ionospheric physics. Through it, we also go on to demonstrate how determinant the interplay between universities, radio companies, and government bodies was for the sustenance of radio investigation practice. Furthermore, the metaphor permits us to disclose camouflaged business strategies and the importance of large-scale industrial research for ventures such as point-to-point communication and ionospheric prediction, without neglecting the appearance of cases on the ambiguous, porous frontier between fundamental physical research and radio engineering.

In this respect, we understand that readers who are interested in simple accounts of success or failure in radio development and its relations with the set of categories designated may find this a somewhat disappointing approach. For while we allow that such factors as industry, technical education, the Empire, and the military were essential components in radio R&D for the interwar years, we have emphasized that each of these factors should not be regarded as anything but a strand of a much more intricate thread of interconnected categories. Each category has a limited explanatory scope when analysed separately; each broaches partially the complexity of the problem when treated in isolation. And it is *only* by attempting to interlace all of them along a single thread that readers will find consistence and significance in the work. With its intertwined strands, the persuasiveness of the metaphor rests upon the imbrication and overlapping of its fibres, rather than upon the expressiveness of a single, strong strand. It is certainly on the condition of the thread, rather than on that of its integral strands, that major advances in ionospheric physics and in radio depended.

In placing such a marked emphasis on the importance of specific categories which transit in parallel along an historical path, we are aware of the dangers that this kind of methodology can entail. Categorization can lead easily to the anarchic, amorphous succession of unconnected sections. Yet, we believe that such a cross-categorial approach is the most appropriate way to demonstrate the second of our theses concerning the existence of structural factors. Moreover, it provides the only way of attaining another objective, which is to proffer both causal analyses and definite explanations before pervasive common practices—for instance, the clearly defined demarcation between scientific and engineering practices, the frequent identification of ionospheric exploration with radar programmes, or radio development removed from any industrial and geopolitical context.

Despite the destruction and ruin that World War I spread everywhere, the conditions for 'a major revolution' were too propitious to preclude atmospheric geophysics, and in particular ionospheric physics, from maturing after the war. However, as the four case studies of Britain, Australia, New Zealand, and Canada corroborate, the ways in which maturity came into fruition were many and varied. Owing to the multiple factors at stake, it has been an enormous task to ascertain how the conditions of matureness crystallized with such different degrees of success and at such a different pace in the four countries which we have taken into consideration. In broaching the issue, we have placed special emphasis upon the perusal of commercial, geopolitical, and socio-cultural contexts which, we believe, were decisive in moulding not only ionospheric research but also the evolution of the radio industry and technical education. It has served us to demolish numerous established clichés. Thus, the notion that broadcasting could have been the single seed propitiatory for the exploration of the upper atmosphere loses much of its credibility. Likewise, the idea that industrial-technological innovations such as vacuum tubes and thermionic valves were the principal driving forces in the 'discovery' of the structure of the ionosphere is not too plausible. In similar vein, it becomes increasingly hard to conceive of radio technical education as playing no causal role in the materialization of ionospheric research, both academic and industrial. Instead, both investigation and radio industry and technical training emerge as the fruits of the same imperial and densely ramified economic tree that simple accounts of discoveries have often tended to ignore or at best to generalize in a simple way.

A clear illustration of the complex effects that the Empire, industry, and education had on radio development emerges from the contrast between the progress attained in three countries that were characterized by strong political interest in radio communication: Australia, Canada, and New Zealand. Their histories have much in common for the three dominions had overcome the colonial stage and undertook their march to nation-building within a larger framework, the British Commonwealth. All lived similar organizational experiences, in that they tried to emulate and transplant a model of research organization from the mother country. Furthermore, all of them witnessed the emergence of radio amateurism in the early 1920s and the commencement of state-funded radio research *circa* 1930.

However, despite apparently comparable geopolitical situations, their trajectories are not convergent. The presence of business strategies contributed to the hegemony of the Australian radio companies (especially AWA) over the whole Southern Hemisphere. It fostered not only competition and emulation but also a constructive spirit for technical education within the national radio industry. This pattern was quite unlike that in New Zealand, where the smallness of market and geographical and demographical limitations fostered the proliferation of subsidiaries of American and Australian companies. In the vast Canadian territory, too, specific circumstances related to geography, population, cultural fragmentation, and the American proximity created the conditions for the existence of a rich industrial fabric concentrated in urban areas such as Montreal and Toronto. This industry was characterized by the 'invasion' of subsidiaries of the American communications giants and by the oligopolistic monopoly on patents. Here, as in New Zealand, subsidiaries did little to promote independent innovation in their plants. Consequently, the growth of the radio industry in the late 1920s was supported in the manufacture and assembly of radio sets (often from imported components), rather that in research and development. In this respect, this was so controlled by foreign, in particular American, corporations that the incentive to R&D, which might have multiplied the openings for qualified personnel (as it did in Australia), was simply conspicuous by its absence. To explain this contrast between Australia on the one hand and Canada and New Zealand on the other, a thorough analysis of the conditions in which the radio industry burgeoned and in which higher education responded to industrial needs in the three Dominions is essential. As will be seen, for reasons of technological dependence, among other causes, the Canadian and New Zealand companies were not powerful enough to house industrial laboratories and thereby to enable the absorption of a relatively copious supply of trained manpower that was a fund of expertise and richness for Australia or Britain.

In focusing attention on the fundamental premise that our five categories cannot be construed as five worlds in an isolated relationship with each other, it is not our purpose to neglect the influence that national scientific traditions wielded on the cultivation of atmospheric sciences. In fact, often superimposed on long-established educational and industrial traditions, we have distinguished, in the four countries (though in very different degrees), recognizable scientific traditions. Obviously, these left a deeper mark in the metropolis than in the dominions, where institutional structures were usually weaker. In Britain, where they seem to have been particularly marked (probably more so than anywhere in the world), the cohabitation of styles, practices, and research schools had a richly productive character. Here, three traditions of long independent pedigree—mathematical-physics Cambridge school, laboratory-based experimental physics, and Humboldtian-style terrestrial physics—converged and intersected in the interwar years. In a similar tone, albeit with lesser intensity, the convergence in Australia of two of the three traditions observed in Britain—that is, Humboldt-style terrestrial physics and laboratory-based experiments on wave propagation—points to the existence of structural patterns. In both countries, these can be traced back to educational and economic circumstances of a highly local character, which were often stimulated by the support of government bodies and of centres of industrial activity. In Canada and New Zealand, in contrast, these patterns were less perceptible. Indeed, where initiatives in geomagnetic observations and in wireless research are most palpable, they owe little to the result of government undertakings and much more to the sustained patronage of international entities such as the Carnegie Institution.

By way of conclusion, let us make an apology for tendering surgical analysis instead of anatomic dissections, reflections in preference to descriptions. In reality, there were few options. If the history of geophysics—and in particular of ionospheric physics—still lies in a phase of growth, where narratives overcome constructive criticism, the history of technology and its role in the formation and maturation of atmospheric sciences is even better. This is in part because there is confusion on the definition of concepts, and a certain disagreement on the boundaries of the discipline. Familiar terms unreflectively used acquire several meanings, and resultant definitions, as necessary as they are, are excessively ambiguous. Such is the case with the term 'radio'. Here, it is important not to confound 'radio' in its sense of industry, commercial sector, and communications, with 'radio' in its technological sense, a term which implies a process of innovation, with an application of science implicit in that process. To confound is to conflate an innovative development with the economic and socio-cultural application of its experience.[3]

[3]Similarly, radio, wireless, and broadcasting were, and are, frequently used as synonyms. Radio—derived from the Latin *radium* for rays—was the formal, scientific term; wireless—the abridged form of its predecessor *wireless telegraphy*—was the lay, common word (preferred in Britain, not in the USA). Broadcasting, on the other hand, denoted any kind of signal transmission and reception. Although the two former were interchangeable, radio was increasingly gaining the advantage before 1939. See M. Pegg, *Broadcasting and Society, 1918–1939* (London: 1983), 2–3.

Actually, radio in its former sense is a set of commercial applications, business strategies, and legal terms, taking various connotations and acquiring different social meanings at different times. In the imperfect glossary that we shall use here, *radio* means not a simple entity but an utterance that comprises three interrelated applications: navigation, point-to-point communication and broadcasting (as it was used in the 1920s). These usually involve three generally interrelated processes: assembly, manufacture, and innovation. Each application results from specific procedures and developments, and the industrial incarnation of applications and of processes is the *radio industry*. Broadcasting was thus one single aspect of radio industry and not its entirety; the same occurred about long-distance communication. In practice, these three applications proceed at different paces within the sector of radio industry, and affected ionospheric research in various degrees and ways.

Finally, the historiographies of science and of technology, dedicated to the sociology of scientific and engineering knowledge, have tended to overlook, or even to neglect, the existence of *structural factors*. As such, they have tended to dissociate the exploration of the ionosphere from industry and education. Nowadays, diversity and complexity are being regarded as legitimate vectors for historical analyses. Science is begotten diversely, it is often said, and then it matures intricately. But between nativity and maturity runs a long avenue. After all, science grows thanks to the *racinage* of the trees at both edges of the path. However, for us the reference to follow has been (and is) the interweaving of multiple roots and shoots in the complex rhizome, rather than the single root cause. Today, with the fresh breeze of the new millennium, we see in the analysis of upper atmospheric sciences and its relationship to the various economic, political, and social contexts new modes of appreciating science in 'inter-action'—or rather in 'inter-relation'—as a highly vertebrate, tentacled enterprise, serving to reflect the rich polychromatism of traditions, practices, and motivation between the two wars.

Aitor Anduaga

ACKNOWLEDGEMENTS

Very special thanks are owed to several people who have guided and inspired the writing of this volume, who have made the experience of scholarship always challenging and fulfilling. Most particularly, I am grateful to Professor Paul Forman at the National Museum of American History, who has pushed me to think about science in the broadest terms. For the stimulus of discussion teaching me to assume the burden of criticism, I thank Professor Roy MacLeod of the University of Sydney, who kindly—and patiently!—read the bulk of the work and gave valuable criticism. Also, I owe great thanks to Professor Rod Home of the University of Melbourne. He has not only helped to guide me through the abundant material of Australia's radio but he has also led me to the path of publication. My most sincere gratitude, Rod.

I am also greatly indebted to Professor Robert Fox of the Modern History Faculty at the University of Oxford. For those of us who were fortunate enough to have spent a little time under his tutelage, he has proved that his generosity towards budding scholars was as great as his personal charm and energy. I am also grateful to Dr Gregory Good, John Heilbron, Edward Jones-Imhotep, Albert Presas, and Skúli Sigurdsson, who have been encouraging in reading a part of the manuscript and making suggestions.

I must acknowledge institutions and colleagues that have made this research possible. Above all, I am grateful to the Basque and Spanish Governments, who have provided two generous grants to cover the costs of research in Britain, Germany, Australia, and Canada. Special thanks are owed to Dr Chris David, head of the Ionospheric Monitoring Group of the Rutherford Appleton Laboratory, for his courteous help at Ditton Park Archive; Professor David Andrews, head of the Sub-Department of Atmospheric, Oceanic and Planetary Physics of the University of Oxford, and Professor C.D. Walshaw of the Cavendish Laboratory, for their insights into the scholastic background of interwar physics and atmospheric sciences; Dr H. Otto Sibum of the Max Planck Institute of Berlin, for giving me the opportunity to use the facilities of its Library and Research Centre; Dr Rachel A. Ankeny, Director of the Unit for History and Philosophy of Science at the University of Sydney, who kindly allowed me explore the ins and outs of radio science and industry in Australia;

Professor Trevor Levere of the University of Toronto, and Professor Yves Gingras, Director of the Centre Interuniversitaire de Rechesche sur la Science et la Technologie, of the Université de Quebec à Montreal, for their generous support during my stay in Canada; Dr Richard Jarrell, professor of natural science at York University, Toronto, and a fount of knowledge on Canadian science; Sheila Noble, archivist at Edinburgh University, who tracked down valuable references among Appleton Papers; and Harold Averill, Jock Given, Ailsa Jenkins, Matti Keentok, Muna Salloum, Daniel Somers, Richard White, and Trevor Wright, for their inestimable help and unstinting cooperation.

To the editors and staff of Oxford University Press I owe an enormous debt of gratitude; to Sonke Adlung, for his gentle guidance and for making the publication possible; to Phaedra Seraphimidi, for her patient forbearance and for giving me a crash course in how to make a book; to Dr John Jenkin of La Trobe University, for his helpful hints and for showing me the way to publication; and to the unsung editorial referees whose constructive criticisms were gratefully received and used.

Finally, I want to express special thanks to Dr Agustín Nieto Galán and Xavier Roqué, who at the origin of this venture generously gave me the benefit of their own expertise. My final thought is for Dr José Llombart—along with the members of the group 'Andrés de Poza' (María Cinta Caballer, Luis Angel García Castresana, Itsaso Ibáñez, María Asunción Iglesias, Juan M. Navarro, and Inés Pellón)—of the University of the Basque Country, for his constant and untiring stimulus for delving into the confines of science.

1

GOVERNMENT, RADIO RESEARCH, AND UPPER ATMOSPHERIC SCIENCES IN BRITAIN

Introduction

At the end of World War I Britain and the US encountered a new *décor* of hegemonic forces and of spheres of influence. The conflict had redesigned the geopolitical map. While a good number of historians deem the years from 1919 to 1921 to be a period in which the balance of world power inclined irretrievably towards the US and against Britain, the truth is that the interwar years witnessed a struggle for supremacy between both in specific fields, of which the upper atmospheric sciences are an excellent example.[1] In Britain, this was a period of consolidation in the organization of scientific services, as in the application of radio engineering and ionospheric physics. As a result of the massive changes provoked, accelerated, and precipitated by World War I, there crystallized in those fields a set of ideas and practices—some exported, others imported from the US and Europe, and many encouraged from within—that set Britain on a position of leadership that would be increasingly shared with the Dominions, and ultimately discussed—and in part overcome—by the US.[2] As a rule, Britain's science, industrial economy, and defence were characterized by policies seeking for promoting Empire's unity and self-reliance. An apparent readiness to set itself up as the 'principal centre of cooperation', identified with some forms of British elitism, and distanced from the goals and ambitions of its American allies, would in general remain a distinctive feature of British science for these two decades.[3]

Historians have copiously written of British 'political strategy' in the interwar period, with its lateral annexe of imperial defence, Dominions' consciousness

[1]In the interwar period Britain has been well described as maintaining not one but three 'antagonistic' approaches to the issue of international peace and her position in the world: the advocates of Anglo-American cooperation as the key to peace and security (called Atlantics), those in favour of regarding the Empire as the basis of Britain's power (imperial isolationists), and the adherents of world leadership promoting active participation in European and world politics. See Orde (1996, pp. 70–72).

[2]McKercher (1991).

[3]Holland (1981).

of nationhood, and the battle for supremacy against the US.[4] Almost all agree that, despite the loss of trade and financial resources, the outcome of World War I reinforced confidence in national and imperial grandeur. Britain without doubt remained a great world power; and her Crown, the 'symbol of free association for the members of the *British Commonwealth of Nations*'.[5] But it was the Empire that made her a primal world force. Communications, diplomacy, and security formed the pivotal trinity of Britain's imperial political strategy, and the struggle for supremacy was to have its echoes in a science intimately wedded to *imperium* and communications.

In this struggle, the history of upper atmospheric sciences meets geopolitics, and economic history meets the privileged position of Britain in the incipient and buoyant sector of radio industry and long-distance communications. Perhaps that is why a cumulus of tragic circumstances—*sequelae* of a recently concluded war, the Depression, times of low investment in science, and auguries of a new confrontation—was not sufficient to dispel the hopes and prospects raised by the progressive acceleration of the foregoing factors. On this point, we believe it is useful to reconsider the contributions of agents such as imperial defence, industrial self-sufficiency, and academic traditions in creating an 'ambience' in which not only a national science of an international stature was materialized but, as the distinguished meteorologist G.C. Simpson predicated in 1926, Britain was to 'lead the world in this branch of knowledge and research'.[6]

Certainly, the war provoked substantial regeneration in the forms and modes of government science in Britain, with far reaching consequences for ionospheric physics. From the war and the ensuing Empire's strategic redefinition emerged the Radio Research Board, the British Broadcasting Company, and an incipient radio mass-production, all through the impulse of new necessities. Hence it is correct to see the years following the war as a critical period in the relations between atmospheric sciences and government and military in Britain, as elsewhere in the US and in Europe. But if one peruses the constituents of such relations, one may find a rich soil of solid traditions germinating with ideas which were not only to impinge upon postwar practices but also to condition later physicists' conceptions of the image and realism of the upper atmosphere. In this respect, the identification of nodal points, not

[4]Useful summaries include Drummond (1974), Ferris (1989), and Orde (1978).

[5]Grimal (1980, pp. 360–2).

[6]'Discussion on the Electrical State of the Upper Atmosphere', *Proc. Roy. Soc. Lond.*, 1926, 111, 1–13, p. 12.

merely intersections, between the formulations of science as knowledge pursuit and as socio-cultural enterprise seems here clearly de rigueur.

Confluence of traditions and disparity of practices

Analysing the assembly of geophysics from frameworks of consensus, historian Gregory A. Good invokes the kaleidoscopic nature of a discipline that, unlike others, was assembled by a process of extrication and recombination, rather than of accretion of disconnected fragments. Just as the kaleidoscope showed geophysicists how the multiple images of the Earth were related to each other, so the frameworks of consensus denote nesting loci in which multiple research specializations are juxtaposed and coordinated. In depicting the maturation of geophysics in the twentieth century, Good does not interpret it as the culmination of a teleological process but as the gradual transformation of an enduring and large framework in which its multiple levels of complexity progressively evolve.[7] This view of geophysics as an intercalated set of sub-disciplines, research schools, inter-disciplinary fields, research programmes, and traditions (bound together through complex institutional and personal links) which concur in the same direction is applicable to upper-atmosphere geophysics, and in particular to Britain's ionospheric physics.[8]

Let us illustrate this last case with a metaphor. During the first third of the century, physicists and engineers harboured a profound respect for radio techniques. They searched for reliability in waves rather than contingency of natural manifestations, and preciseness in pulse technique rather than vagaries of celestial phenomena. This regard had its institutional expression, albeit variable in accordance with the country and its traditions. In the US, for instance, departments of electrical engineering and applied physics together with military and industrial laboratories practically absorbed upper-atmosphere geophysics, reducing it to a large extent to radio physics, while geophysics was relegated to geology departments.[9] Gregory Breit and Merle Tuve were just two of many American radio devotees who cultivated a practice blessed by the fertile union of science and industry, and who followed DTM's tradition of geophysical

[7]These imaginative formulations are included in Good's excellent study (Good, 2000, p. 284) on the formation of geophysics. See also Good (2002, pp. 229–39) and Doel (1997, pp. 391–416).

[8]A valuable summary containing a postscript on C.S. Gillmor's career and work is Gillmor (1997, pp. 1–12).

[9]For the institutional frameworks for ionospheric research in different countries before World War I, see Gillmor (1986, pp. 112–13). For the different traditions, see also Gillmor (1994, p. 134).

observational studies.[10] In Germany, as in France and Japan, upper-atmosphere geophysics was pursued in physics and geomagnetism centres (predominantly in universities and generally from an experimental viewpoint).[11] In Britain, the same kaleidoscopic metaphor now cast in the image of three pieces of thread serves to define its singularity. The threads represent three traditions of long-independent pedigree: the mathematical-physics Cambridge school, laboratory-based experimental physics, and a Humboldtian-style terrestrial physics, each of which is made up of fibres which incarnate sub-disciplines, styles, and research programmes. With its intertwined strands, the strength of Britain's ionospheric physics does not reside in the resistance of a single, strong thread but in the profusion and in the overlapping of fibres. Traditions, practices, and research schools did not run in parallel with the formation of the conceptual representation of the atmosphere: they converged and intersected in the interwar years.[12]

The first thread pertaining to the mathematical-physics Cambridge tradition was both flexible and capacious. It includes over a dozen graduates of the Cambridge Mathematical Tripos (*inter alia*, J. Larmor, S. Chapman, E.H. Milne, V. Ferraro, D.R. Hartree, S. Goldstein, R.H. Fowler, G.N. Watson, A.E.H. Love, H.M. Macdonald, and J.W. Nicholson), whose physical interests predominately concentrated on geomagnetism and on radio waves propagation (in diffraction rather than reflection theories), and to a lesser degree on the ionization of gases. All of them possessed an important, if not primary commitment to the mathematization of the physical sciences, which they pursued through their insistence on models and physical imagery.[13] For them, mathematical technique was coextensive with the physical problem to which it was applied, and the atmosphere was a medium in which pure and applied mathematics coalesced into one single discipline.[14] However, the graduates were not

[10]Good (1994, pp. 29–36).

[11]Here it is important to note that Försterling and Lassen led a theoretical group of ionospheric physics at the University of Cologne (apart from Jonathan Zenneck's experimental team at the Technical University in Munich). For Germany's radio physics, see Dieminger (1948, 1974, pp. 2085–93).

[12]Russell Moseley, in his classic study on British physics, distinguishes only two traditions (mathematical and experimental). Illuminating though it is, his work is concerned with the 'new physics', and throws little light on the influence of British observational tradition in physics. See Moseley (1977, pp. 423–46). For the two traditions, see also J.A. Fleming, 1939. 'Physics and the Physicists of the Eighteen Seventies', *Nature*, 143, 99–102.

[13]D.B. Wilson (1982, p. 333).

[14]Yet after 1907 the Cambridge Mathematical Tripos turned away from the physico-geometrical approach towards pure analytical mathematics. See Rouse Ball (1912, p. 322).

all 'theoreticians' in the sense that such categorization might prefigure, or by any means demonstrated an absolute dedication. Although all deemed mathematical physics as the route to the understanding of atmospheric nature, few pursued it exclusively (among them, Chapman was exemplar), and some, like Eckersley and Chree, consummated their careers in the realms of industry and public service (Eckersley at Marconi's and Chree as Kew Observatory's director).[15] Hence, this collective title is perfectly compatible with the variety of activities and functions often extending far beyond the geophysical or even scholastic.[16]

Moreover, the close affiliation of some of them with research schools did much to prolong Britain's strong, venerable heritage of mathematical physics. Of the exemplars of this distinguished lineage, the most representative is Arthur Schuster's geomagnetism school at the Victoria University, Manchester, which, through the New Physical Laboratory and Whitworth Meteorological Observatory, transcended the most routine forms of recording and collecting distinctive of magnetic observatories.[17] Throughout the nineteenth century, the terrestrial magnetic investigations of Edward Sabine and Balfour Stewart had formed a distinctly geophysical trend competing with that of George Darwin at Cambridge, a physico-mathematical geology group specialized in the Earth's structure and crust.[18] Though never too far advanced from its rival, their successor, Schuster, was at the forefront of Manchester's group—at least in geophysics. His broad-mindedness in research topics (from geomagnetism to atmospheric dynamics and cosmic rays) and its physico-mathematical character made it something of a British paragon, meriting stripes of respectability.[19] Of the virtuosos of this elevation, the most eminent were Sydney Chapman, who in 1919 would succeed his professor of mathematics Horace Lamb, and George Clarke Simpson, the future director of the Meteorological Office.[20]

It is completely consistent with our categorization that in this first thread there were pure mathematicians, especially diffraction theorists. They possessed the attributes distinguishing mathematical physicists—the use of mathematical tools for the pursuit of physical models—and their investigations were oriented firmly towards physical optics problems—the possibility of long-distance radio transmission in the atmosphere—that by any means were strange to

[15]Ratcliffe (1959, pp. 69–74) and Simpson (1928, pp. vii–xiv).

[16]For a comparison between Cambridge traditions, see Warwick (1993).

[17]R.H. Kargon (1977, pp. 220–37).

[18]Kushner (1993, p. 200) and Good (2000, p. 274).

[19]Simpson (1935) and Lightman (2004, pp. 1780–4).

[20]Gold (1965, pp. 157–75) and Pedgley (1995, pp. 347–9).

the core of geophysical topics which atmospheric reflection experimentalists encompassed.[21] However, unlike these, they constructed mathematical representations with extremely simplified atmospheric models. In 1903, Macdonald explained Marconi's transatlantic transmission with a surface diffraction theory being predicated on an over-idealized atmosphere (a free space with zero conductivity and uniform dielectric constant) and a quasi-imagery Earth (as a perfect conductor).[22] Others, like Nicholson and Love, basically subscribed to his assumptions, while being at variance with how to approximate the analytical solution of diffracted field intensity.[23] So competently (if fruitlessly) did Cambridge mathematicians (along with Henri Poincaré and a German group led by Arnold Sommerfield and Jonathan Zenneck) explore this enigmatic phenomenon that, for the first two decades of the twentieth century, their diffraction theories became the sole theoretical alternative to Eccles' ionic refraction.[24] But since 1918, as G.N. Watson—yet another Cambridge-trained mathematician—had demonstrated that the surface diffraction alone could not explain empirical observations (in particular those deriving from the Austin-Cohen formula), the leverage exerted by diffraction proponents upon the atmospheric realm has waned.[25] The diffraction theory proved mathematically consistent, but empirically invalid. In a last attempt, Watson applied the complex-variable techniques to the physical model of a homogeneous and conducting boundary with the aim of eliciting quantitative predictions consistent with the Austin-Cohen empirical formula.[26] Incidentally, the last of the diffraction saga advantaged the advent of reflection theories in the 1920s.[27]

The variety of research programmes and methods is one of the most startling, enriching, and distinctive characteristics of our second thread, the experimental physics tradition. The most systematic studies of this tradition, belonging to Romualdas Sviedrys, mention no fewer than 24 physics laboratories[28] and 15 academic engineering laboratories at the end of the nineteenth

[21]E.A. Milne (1939–41, p. 473) and W. Wilson (1956, p. 209).

[22]Macdonald (1903) and Whittaker (1935, pp. 553–5).

[23]Nicholson (1910) and Love (1915).

[24]To be sure, the diffraction and reflection theories did not vie directly with each other; their modeleers had complementary rather than competitive worldviews, as C.P. Yeang has recently proved. For a convincing study of both theories and communities, see Yeang (2003).

[25]G.N. Watson (1918).

[26]G.N. Watson (1918–19).

[27]Whittaker (1966, p. 523).

[28]University of Glasgow (1866; W. Thomson); University College, London (1866; G.C. Foster); University of Edinburgh (1868; P.G. Tait); King's College, London (1868; W.G. Adams); Owens College, Manchester (1870; B. Stewart); Clarendon Laboratory, University of Oxford (1870; R.B.

century (most of them electrical engineering)[29]—a figure that was to grow considerably with the NPL and the electrical and electronic firms in the interwar years. Although, as time went by, the share of responsibility and prestige would permeate unmistakably through industry and the military, the academic physics laboratories (predominantly the Cavendish), rather than electrical engineering's (unlike the US), were the centrepiece and exemplar of the laboratory-based radio research. It is in these centres, many of them Cavendish-minded in their practices and aspirations, in which a constellation of researchers (among others, W.H. Eccles, J.A. Fleming, O. Lodge, R.J.S. Rayleigh, E.V. Appleton and his pupils, B. van der Pol, J.A. Ratcliffe and his team, R.A. Watson-Watt, C.T.R. Wilson, G.M.B. Dobson, F.A. Lindemann, R.L. Smith-Rose and his team) encompassed well over two thirds of the research mentioned.[30]

Of several of the respects in which the experimental tradition differed from the mathematical physics, two are worthy of mention. First, by their training (over half of them were Natural Sciences Tripos graduates)[31] most experimentalists knew relatively little advanced mathematical physics, in particular electro-magnetic theory, by comparison with their colleagues in the

Clifton); Royal School of Mines (later Royal College of Science, 1872; F. Guthrie); Royal School of Science, Dublin (1873; W. Barrett); Queen's College, Belfast (1873; J.D. Everett); Cavendish laboratory, Cambridge (1874; J.C. Maxwell); Armstrong College of Science, Newcastle-on-Tyne (1875; A.S. Herschel); University College, Bristol (1876; S.P. Thompson); City and Guilds Technical College, Finsbury (1879; W.E. Ayrton); University of Aberdeen (1880; C. Niven); Mason College of Science, Birmingham (1880; J.H. Poynting); Queen's College, Galway (1880; J. Larmor); Trinity college, Dublin (1881; G.F. Fitzgerald); University College, Liverpool (1881; O. Lodge); University College, Nottingham (1882; W. Garnett); Firth College, Sheffield (1883; W.H. Hicks); University College, Cardiff (1884; E.H. Griffiths); University College of North Wales, Bangor (1884; A. Gray); City and Guilds Technical College, London (1885; W.E. Ayrton); and Yorkshire College, Leeds (1885; W. Stroud). See: R. Sviedrys, 1976. 'The Rise of Physics Laboratories in Britain', *HSPS,* 7, 405–36, pp. 416 and 432.

[29]University College, London (1878); Finsbury Technical College, London (1881); Mason Science College, Birmingham (1882); King's College, London (1882); Royal Indian Engineering College, Cooper's Hill (1883); University College, Bristol (1883); Central Institute, London (1884); Firth College, Sheffield (1885); Yorkshire College, Leeds (1886); Owens College, Manchester (1886); University College, Liverpool (1887); University College, Dundee (1889); University of Edinburgh (1890); University of Cambridge (1893); and University of Glasgow (1900). Ibid., p. 431.

[30]'The directors of the physics laboratories formed a homogeneous group. Most of them had taken the mathematical tripos at Cambridge (e.g. J.J. Thomson, Larmor, Stroud, and Griffiths), others had been partially trained at the Cavendish (Schuster and J.A. Fleming).

[31]Of those mentioned, C.T.R. Wilson, Appleton, Ratcliffe, Lord Rayleigh, Dobson, and Van der Pol read for the Natural Science Tripos at Cambridge and familiarized themselves with the Cavendish.

Mathematical Tripos. This is illustrated by examples such as B. van der Pol's, who, towards 1920, in an unsuccessful attempt to smooth over the discrepancies between diffraction and reflection theories (and thereby to conflate both *modus operandi* into one) was forced to solicit the mathematician G.N. Watson for help. The point could be made equally well for Appleton, whose magneto-ionic theory was mathematically modelled by Douglas R. Hartree in 1932 (the well-known Appleton-Hartree equations for the refractive index of a wave propagating in a magneto-plasma),[32] and by (mysteriously) his Austrian pupil Wilhelm Altar.[33] In dynamic and laboratory teams like Appleton's and Ratcliffe's, their members demarcated a defined intellectual territory (clearly experimental) and circumscribed their practices within it, leaving it up to mathematicians to pursue their sacred ideal of atmospheric mathematization. In this respect, the confluence of both in the 1920s allowed the edifice of physical models to unite with the sophistication of mathematical techniques.

A second distinctive feature of the experimental physicists was their faculty of stimulating new avenues. The diffraction theorists, like the theoretical geomagneticians, often formulated rigorous mathematical solutions of problems deriving from straightforward, highly speculative physical models. The lack of verisimilitude led them to stagnation and a blind alley. The experimentalists, by contrast, in their pursuit of the coherence between physical models and experimental data, provided a fresh fillip by analogy and replication. Towards 1900, Joseph John Thomson, in his experiments with electrical discharges, recreated physical processes in vacuum tubes and vessels, in which the mysteries of the upper atmosphere could be immaculately revealed to the profane.[34] The streams of electrical particles, supposedly irradiated from the sun, impinging on the higher atmosphere (and thereby forming the aurorae) were those that hypothetically appeared in the experiments with ionized gases.[35] By studying wave propagation through ionized means, the new microphysics of the late nineteenth century revealed not only the internal molecular structure of gases

[32]Indeed, Appleton corrected his mathematical limitations thanks to the extensive advice of Fellows of St. John's College, such as E. Cunningham, S. Goldstein, and others such as M. Taylor and G. Builder. See Appleton (1932b, pp. 642–50) and Hartree (1931).

[33]The mystery resides in the fact that Appleton apparently plagiarized W. Altar (or at least appropriated his ideas to complete the magneto-ionic theory, hiding his contribution), as Gillmor has credibly demonstrated. See Gillmor (1982). For a view against Gillmor's accusation, see Wilkes (1997).

[34]*A History of the Cavendish Laboratory 1871–1910* (Cambridge: Cavendish Laboratory, 1910), 229–32; Crowther (1926, 1974, pp. 148–51, 160–9, 263).

[35]Lord Rayleigh, 1941. Joseph John Thomson. *Ob. Not. Fell. Roy. Soc.,* 3, 587–609, pp. 590–1.

but also reanimated the Scandinavian interest in magnetism and the aurorae and vitalized the emerging area of radio communications.[36] In abandoning the strictly mathematical realm in favour of empiricism and the reproduction of natural processes in laboratory, Thomson and his colleagues at Cavendish had the satisfaction of at once provoking new epistemic situations and affirming the validity of a practice in which the gift for quantification had given way to the subtler gift for intuition.[37]

Given such favourable circumstances, it is not surprising to see the dedication of an increasingly high number of experimentalists to radio propagation, especially after 1910, when wireless telegraphy teaching became relatively commonplace at the universities. Yet, by the same token, the partial abandonment of mathematization and its replacement for experimentation help to explain why the atmospheric-reflection theories remained at an explicatory level, rather than predictive, failing in their objective of predicting results consistent with the Austin-Cohen formula. Largely as a result of Eccles' investigations (and also J.A. Fleming's and Lodge's), which were epitomized in his ionic refraction theory of 1912, the realm of wireless phenomena became broader than diffraction theorists' (engaged only on long-distance transmission).[38] Now, as never before, in return for an unprecedented, if partial aperture towards experimental microphysics, physicists encompassed phenomena apparently as unrelated as diurnal variation of atmospherics, the directive antenna pattern, and wave transmission. Gradually, what before was a geometric optic problem now extended to the qualitative behaviour of any kind of wireless phenomenon. Certainly, the experimental radiophysicists were the greatest beneficiaries of this new orientation, and it was above all their want of a solid theoretical framework that led them to effect their noteworthy and, by the standards of the 1900s, atypical fraternization with mathematics in the 1920s.

The changed perspective of physics professors was unequivocal. The interest in atmospheric electricity and in the electrical field of thunderstorms, which had languished in the late nineteenth century, was revived (especially with C.T.R. Wilson's work at Cavendish);[39] studies on fading and atmospherics

[36]Gillmor (1997, p. 4).

[37]See the introduction in Jed Buchwald, 1985. *From Maxwell to Microphysics: Aspects of Electromagnetic Theory in the Last Quarter of the Nineteenth Century*. Chicago: UCP.

[38]Eccles' theory explained three phenomena: the diurnal variations of atmospherics, the wave transmission along the Earth's surface and questions of directivity in Marconi's aerial. See: Eccles (1912).

[39]On Wilson's work in atmospheric electricity and thunderstorms, see Blackett (1960, pp. 282–7).

proliferated; the involvement of physicists in the field of radio communications became common; and, perhaps most important, the upper atmosphere became also an object of experimentation, and not only of observation.

Mathematical physicists and experimentalists, however, were only a part, albeit fundamental, of the success. Fortunately for British science, there existed an observatory-based data recording tradition—which conforms to our third thread—that had allowed geophysicists such as Edward Sabine and Balfour Stewart, especially in the mid-nineteenth century, to contribute substantially to the advancement of geomagnetic knowledge.[40] Then, Humboldt's requests for the expansion of terrestrial physics were taken into consideration by the Royal Society and the British Association for the Advancement of Science, which established observatories throughout the Empire and promoted magnetic crusades in ways that endorsed the most mundane styles of observation and analysis of geophysical variables appropriate for the Humboldtian sciences.[41]

The cases of the Observatories of Greenwich and Kew (together with Lerwick and Eskdalemuir in the early twentieth century) illustrate the girth and scale of the tradition. Greenwich, the main focus of observational astronomy from its foundation in 1675, also shared a modest magnetic section which was transferred to Abinger in 1925.[42] Kew, for its part, was the centre of geomagnetic attention and training par excellence.[43] It is true that its main task was the verification of scientific instruments (over 180,000, most of them clinical, had been tested and calibrated between 1886 and 1895), followed by regular and systematic recording (e.g. of magnetic field and electrical potential gradient).[44] But the emphasis on investigation (almost entirely by its director, C. Chree) was by any means insignificant by the standards of that time. With the Scandinavian school and Greenwich (in the figure of Walter Maunder) as main rivals, Chree gained respect not only for his varied magnetic work (over 80 papers and monographs) but also for combating speculation and hypothesis, so abundant in terrestrial magnetism, with a zeal and a scrupulousness

[40] 'Sabine, Sir Edward', *Proc. Roy. Soc. Lond.*, 1892, 51, xliii–li; P.G.T., 'Stewart, Balfour', *Proc. Roy. Soc. Lond.*, 1889, 46, ix–xi; Lightman (2004, pp. 1909–13).

[41] Morrell & Thackray (1981, pp. 523–31).

[42] Malin (1996, p. 71); Tarplee (1996, pp. 1–10).

[43] See 'Kew Observatory Bicentenary, 1769–1969', *Met. Mag.*, 1969, 98, 161–90; 'Some Aspects of the Early History of Kew Observatory', *Quart. Jour. Roy. Met. Soc.,* 1937, 63, 127–35.

[44] Simpson (1928, on p. viii); H. Barrell, 'Kew Observatory and the National Physical Laboratory', *Met. Mag.*, 1969, 98:171–80, p. 171.

quite unknown in his contemporary colleagues.[45] Again, the presence of Eskdalemuir and Lerwick in the Shetland Islands owed much to the position and pattern of electric currents associated with magnetic disturbances reaching their greatest concentration at those latitudes.[46] The records of atmospheric electricity, one of the most prestigious and successful fields of observation in Britain, initiated at Kew in 1843, Eskdalemuir (1908), and Lerwick (1926), and the sections for the study of aurorae (photographic since 1921 at Lerwick), ozone (with Dobson spectrograph since 1926 at Lerwick), and meteorology, all reflected the commitment to a non-experimental, data-recording practice.[47] To a greater or lesser degree, the observatories were temples of geophysical erudition, and sometimes of intellectual production, and nearly all published series of observations with which to justify their labour.

With the indisputable bequest of Britain's terrestrial magnetism forever latent and the magnetic records augmenting by substantial amounts (systematic at Kew since 1857; at Eskdalemuir, since 1908; Lerwick, 1921; and Abinger, 1925), the benefits of the observational tradition seem apparent.[48] Gradually the series of reliable observations began to accumulate (those of potential gradient at Kew became the largest in the world);[49] and happily, the observatories recruited some of the best young brains of Cambridge as assistants (Chapman at Greenwich and Dobson at Kew, before initiating their professional careers).[50] But if the leverage of observatories seems clear, the mode in which this is exerted is far from it. Or, in other words, and in the guise of questions, how far did their existence enable, say, Chapman to establish the causes of lunar and solar variations of terrestrial magnetism—the bases of his dynamo theory—in 1919? And to what extent did the systematization of magnetic and electric field observations permit investigators in fields as diverse as atmospherics, radio propagation, fading, and magnetic storms to formulate in the 1920s theories which, in rigour and profundity, call the estimates hitherto effected into question?

[45]For Chree in relation to geomagnetic theories and his aversion to speculations, see Sydney Chapman (1941, p. 633).

[46]W.G. Harper, 'Lerwick Observatory', *Met. Mag.*, 1950, 79, 309–14, p. 310; M.J. Blackwell, 'Eskdalemuir Observatory: The First Fifty Years', *Met. Mag.*, 1958, 87, 129–32; J. Crichton, 'Eskdalemuir Observatory', *Met. Mag.*, 1950, 79, 337–40.

[47]Harrison (2003, p. 13).

[48]The *British Meteorological and Magnetic Year Book* began to publish annual geomagnetic data in 1911, and it was followed by the *Observatories' Year Book* in 1922. See Jacobs (1969, p. 169).

[49]Chree (1915).

[50]Cowling (1971, p. 54); Houghton & Walshaw (1977, p. 43).

In response, one fact is unquestionable: the hypothetical influential value did not reside in its function of incentive for radio research, but in the accumulation (often without previous discussion) of geomagnetic data. This is not to say that some observatories did not appreciate the radio as a tool for observation; in fact, Lerwick was one of the three stations possessing equipment for direction-finding (since 1923) and atmospherics (from 1924 to 1929).[51] But in this case, the core of the programme was devised and implanted by the DSIR instead of the Meteorological Office, on which the observatories depended. The willingness of the meteorologists and geomagneticians with administrative responsibilities to invest time and energy in radio research was by no means manifest. When in 1920 George Clark Simpson, a leading authority on atmospheric electricity, took over the Meteorological Office, he adopted the apparently paradoxical decision (paradoxical for being an expert in electrical physics) of separating national weather service from radio research, and of accentuating the geophysical nature of meteorology to the detriment of physics.[52] At this point, radio research and meteorology in all respects parted company in Britain. Only from 1937 onwards when the Meteorological Office initiated the radiosonde programme (15 years later than in several European countries) were the truncated links formalized.[53] Ironically, in the interwar years, when relations between geomagnetism and radio physics became appreciably closer than ever, the theoretical geomagneticians who not only followed their discipline but also pursued a global view of atmospheric physics were to find part of their source of inspiration in the observatories managed by physicists in favour of compartmentalization who were apostles of routine observation and of data amassment and whose aspirations to research never transcended their disciplinary realm.[54]

To illustrate this fact, let us recall the beginnings of the most prominent geomagnetician—Sydney Chapman, who held the post of senior assistant at the Greenwich Observatory from 1910 to 1914, a position attractive both

[51]Together with RRB's at Slough and Aboukir in Egypt. See W.G. Harper, 'Lerwick Observatory', *Met. Mag.*, 1950, 79, 309–14, p. 313.

[52]The sense of disregard for radio research is vividly conveyed by R.A. Watson-Watt, 1957. *Three Steps to Victory: a Personal Account by Radar's Greatest Pioneer*. London: Odhams Press, p. 54.

[53]The delay in developing radiosonde for meteorological purposes in Britain has not yet been analysed, but it profoundly contrasts with its capacity for radio communications and with the situation in countries like France, Germany, and Russia. See Dubois, Multhauf, & Ziegler (2002, pp. 26–48) and D.N. Harrison (1969).

[54]For an impression of the overemphasis on observational facets at Kew, see F.J. Scrase (1969, pp. 181–2). See also Gold (1965, pp. 157–75).

financially and as regards status.[55] At the time of his appointment, at the age of only 22, Chapman harboured many doubts about his future dedication to astronomy or to pure mathematics. But his role in the reconstruction of the magnetic section and his valuable, if sporadic, contacts with professor Schuster (as a member of the Greenwich Board of Visitors) clearly proved decisive in determining his future interests. Significantly, at this time Chapman published his first three papers on solar and lunar magnetic variations (in terms of a dynamo theory) extending Schuster's work to the lunar case, using data from about 21 observatories.[56] He systematized Balfour Stewart and Schuster's earlier theories, while collecting and interpreting data against which the theory must be tested—a factual and tedious task implying the classification of observations. The high degree of Chapman's attachment to investigation powerfully contrasts with the detachment of most his colleagues at the Observatory: 'it seems an unfortunate fact', stated Chapman, 'that the efforts of magneticians are unduly devoted to the accumulation of data, the time and labor spent in their discussion being proportionally inconsiderable.'[57] Even if Chapman was subsequently interested in topics unrelated, in the first instance, to magnetic variations, as in fact he was, his most varied theories rested very firmly on the interpretation of data (either magnetic or of ozone and meteors); and his sojourn at Greenwich first, as his reiterated use of worldwide data later, only reinforce his confidence in the observational-type tradition.[58]

Despite initial appearances, the field of atmospheric ozone also constitutes a good illustration of the especially stimulating nature of the observatory-based tradition. Its study, which had been a main focus of attention in Oxford in the first half of the nineteenth century (especially for its influence on health), flagged at the end of the century, and remained in a lethargic state until the 1920s.[59] While in France physicists like J. Chappuis, Charles Fabry, and Henri Buisson won international recognition for their spectroscopic measurements of ozone, the contributions of British meteorologists were virtually non-existent.[60] It was only in the astronomical realm, and here almost fortuitously, as some

[55]Cowling (1971, pp. 54, 62–3).

[56]S. Chapman (1913, 1914, 1915).

[57]S. Chapman (1919, p. 2).

[58]E.H. Vestine, 'Geomagnetism and Solar Physics', in S.I. Akasofu et al., eds., 1967. *Sydney Chapman, Eighty: From His Friends*. Boulder: University of Colorado, 19–23, p. 22.

[59]Walshaw, 'The Early History of Atmospheric Ozone', in Roche (1990, pp. 316–19).

[60]Between 1875 and 1920, only two works pertaining to ozone were published in the *Quarterly Journal of the Meteorological Society*. The history of atmospheric ozone research has been little studied. See D.H. DeVorkin, 'Ozone', in Good (1998, pp. 641–46) and Schmidt (1988).

photographs of the spectrum of Sirius showed mysterious bands of absorption, that Alfred Fowler and R.J. Strutt—later Lord Rayleigh—were interested in ozone, demonstrating the gaseous origin of problem.[61] 'The first absolutely definite proof of its presence in the atmosphere', as euphorically asseverated Rayleigh, was not the result of firmly established practice,[62] as neither were ensuing studies confining ozone principally to the upper atmosphere, rather than to lower regions.[63]

However, the burgeoning interest in ozone research, promoted by Dobson in Oxford in the 1920s, is hardly understandable without the participation of the observational tradition. In this case, frequently unrecognized jobs served as a school for gaining expertise and dexterity. Supported partially by the Meteorological Office, the posts of assistant at Kew Observatory, acting director at Eskdalemuir, and meteorological advisor at the Military Flying School were able to provide Dobson with a geophysical medium in which to learn C.T.R. Wilson's method of measuring atmospheric electricity and the use of pilot balloons, to mention nothing of techniques in pursuit of magnetic precision.[64] These experiences were followed by a sojourn at the Royal Aircraft Establishment, Farnborough, where, at the age of only 27, he was appointed director of the experimental department. When F.A. Lindemann (later Lord Cherwell) recruited him to the Clarendon Laboratory in 1920, he was clearly not lured by the repute that Dobson had won at Cambridge or by the intellectual faculties of a physicist devotee of meteorology; he took him on for his sophisticated expertise in instrumentation and for the guarantee of observational mastery.[65] Certainly these qualities permitted Dobson to photograph meteors in 1921 with a precision exceptional by techniques of the time, and, at the prompting of Lindemann, to relate the variation of the density of air (inferred from pictures) with atmospheric ozone.[66] For Lindemann, the existence of a warm region at a height of 50 km could only be explained by the absorption of solar ultraviolet radiation, by what he called an 'ozone layer', his term for the vast region hitherto only suspected.[67] Far from being merely

[61]Fowler & Strutt (1917).

[62]Egerton (1949, p. 524); Strutt (1964, p. 1114); Dingle (1941, p. 486).

[63]Strutt (1918).

[64]Houghton & Walshaw (1977, p. 43).

[65]Russell (1992, p. 275).

[66]Lindemann & Dobson (1923a, 1923b).

[67]On Lindemann's instrumental role in early ozone theory, see G.M.B. Dobson, 1966. *Forty Years' Research on Atmospheric Ozone at Oxford—a History*. Oxford: Clarendon Lab. (reprinted in *Applied Optics*, 1968, 7, 387–405, pp. 387–8); and Thon (1958, p. 48).

anecdotic, observations of meteors were the commencement of a productive research line in ozone that set Oxford on a par with the team of Fabry and Buisson in Marseilles. Its chief promoters are readily identifiable as Dobson (masterminding a technical revolution for ozone measurements) and Lindemann (in his role of a Socratic oracle).[68] In all respects, one of the most active fields of physics in Britain—that was radiantly exhibited in the second international conference on atmospheric ozone held in Oxford in 1936—can be regarded as a by-product of the study of meteors, the desired object of a solid and inveterate observational tradition.[69]

Radio research, government, and imperial Communications

The years following World War I represent one of the most convenient and critical periods in which to study the approach of state policy towards scientific research, and in particular towards radio research. Between 1919 and 1923, the State, which had already assumed a new relationship with science from at least 1914, concretized its new manner and methods of dealing with the exigent imperatives of investigation.[70] The establishment of 23 research associations by 1924 clearly illustrates the movement of the British government towards a regulatory and supervisory role in scientific fields which it had eluded before the war. In this period, one may also observe significant changes in the military-imperial context within which radio research came to germinate, and in the engagement of particular individuals to the problems of defining and procuring national security over areas of communications requiring coordination and financial support instead of state guidance. As *Nature* editorialized in 1917, the new circumstances and public opinion compelled the State 'to accept responsibilities and exercise initiative to an extent hitherto undreamt of'.[71]

By the middle of the war, the State was impinging upon radio research in several ways. The fighting services concentrated part of this interest. But military research, unquestionably the most important at that time and often

[68] Lindemann's experimental research—and so his inspirational virtues—flagged by the mid-1920s. Works such as 'Meteors and the Constitution of the Upper Air', *Nature*, 1926, 118, 195–8 and 'Note on the Physical Theory of Meteors', *Astrophysical Journal*, 1927, 65, 117–23 were but two exceptions.

[69] On relations between meteors observations and ozone, see David W. Hughes, 'Meteors and Meteor Showers: an Historical Perspective, 1869–1950', in Roche (1990, p. 293).

[70] Although the DSIR was created in 1914, the state efforts to support fundamental research began some years before. See Alter (1987, especially chapter 4); MacLeod & Andrews, (1970); Hutchinson (1970); A. Hull (1999).

[71] *Nature*, 1917–18, 100, p. 266.

closely interlinked with civil research, was, in many ways, incomplete and organizationally inadequate. For example, the Admiralty had in its charge a network of 12 short-wave Bellini-Tosi direction-finding stations, and operators skilled in measuring the direction of arrival of radio signals (from 1916 to 1920, at Watson Watt's request they recorded over 12,000 observations of atmospherics).[72] But activities were made on the basis of cooperation, improvisation and the overlap with other bodies, rather than of a strategically calculated and orchestrated planning.[73]

The flagrant duplication of endeavour was a result of a decentralized if often effectual policy. With a work programme according nearly fully with the agenda of the Air Ministry, the Meteorological Office was striving to provide security (by means of warnings of thunderstorms) for pilots of war aircraft.[74] Even if the idea of using wireless direction-finders to locate the lighting flashes causing atmospherics initially appeared alluring to meteorologists, there could be no talk of an unconditional reception; the precipitate absorption of the Office by the Air Ministry (and G.C. Simpson's arrival) diluted the initial engagement.[75] Finally, the State was involved in the encouragement of radio research through the support of the National Physical Laboratory (NPL). This revolved around the application of direction-finders to a wide range of sectors ranging from meteorology to aircraft and navigation.[76] Reasons for the support of radio research were varied, but mainly pertained to notions of military security and to a sense of defensive preoccupation, combining technological reliability with the unpredictability of war an inquietude urged with greater vigour as the necessity for determining the position of ships and aeroplanes increased in scale and size.[77]

[72]The first systematic observations with a view to storm location by radiotelegraphic direction-finding were performed by Captain C.J.P. Cave in 1915. See R.A. Watson Watt (1923, pp. 1010–26). For a comprehensive account of the application of the direction-finder during the war, H.J. Round, 'Direction and Position Finding', *Jour. IEE*, 1920, 58, 224–47.

[73]On Britain's military research during World War I, see Alter (1987, pp. 201–11), MacLeod and Andrews (1971), and Varcoe (1970).

[74]For the commitment of the Meteorological Office to radio research, see R.A. Watson-Watt (1957, pp. 45–56) and Ratcliffe (1975, p. 553).

[75]For a valuable précis on the early development of direction-finders for atmospherics in Aldershot and the parallel school of thought headed by Bureau in France, see NAC [MG 30 B157, vol. 1/37] *John Tasker Henderson's Papers*, 'Direction Finding of Atmospherics'.

[76]DPA. 'Wireless Direction Finding. Report on the Work Carried Out at the National Physical Laboratory During the Period 1914–1918', by G.W.O. Howe, 8 March 1921.

[77]On the activities of the NPL during the war, see Pyatt (1983, pp. 62–73).

When the sound of bombs ultimately fell silent in 1918, the government found itself with a cable network for imperial communications, after four years of neglect and destruction, unduly undermined by delays and by ineffectiveness; at the same time, the cable system presented obstacles which precluded direct, instantaneous communication. While cables needed to redefine their postwar space, wireless schemes for imperial communications which had been deemed as temerarious in the early 1910s appeared now not only desirable but necessary. Plainly, Britain was endeavouring to solve the double dilemma between politics and commercial dominance by reconciling the apparently opposed interests of both technologies: amending, not procrastinating, its position as laggard in radiotelegraphy; retaining, not jeopardizing, its hegemony in cables.[78]

The Committee of Imperial Defence, its Imperial Communications Committee (ICC)—created in February 1919—, and the Wireless Sub-committee of the ICC, all dealt with reconciling these questions. However, in dealing with the Empire and communications, they were made up of representatives of several departments (the War Office, the Admiralty, the Post Office, the Treasury, the Air Ministry, the Board of Trade, the Colonial Office, and the India Office), each with its own idiosyncrasies and self-interests.[79] Their task, *proprement dit*, was not to fix the criteria according to which certain schemes would be accepted and others disesteemed, but to give recommendations, weigh policies, and articulate a range of suggestions congruent with the Dominion's interests.[80] These patterns were made explicit in regard to the issue of radio research in October 1919, when the ICC urged the Government to create forthwith 'a central research establishment', regarded as 'an urgent

[78]Contemporary accounts are: Frank James Brown, 1927. *The Cable and Wireless Communications of the World; a Survey of Present-Day Means of International Communication by Cable and Wireless, Containing Chapters on Cable and Wireless Finance.* London, pp. 1–8; J. Saxon Mills, 1924. *The Press and Communications of the Empire.* London, pp. 68–102; Charles Bright, 1923. 'The Empire's Telegraph and Trade'. *Fortnight Review*, 113, 457–74.

[79]The ICC was formed by Viscount Milner as the Chair, Capt. R.L. Nicholson (Direction of Communication Division Admiralty), Lt-Col. L. Evans (War Office), Col. L.F. Blandy (Air Ministry), R.C. Barker (India Office), F.J. Brown (General Post Office), H.C.M. Lambert (Colonial Office) and R.F. Wilkins (Treasury).

[80]Issues such as the establishment of high power stations throughout the Empire, licences for experimental stations, and the future conduct of radio communications were examined by the ICC between 1919 and 1923. See NAUK, [CAB 35/1], 'Committee of Imperial Defence: Imperial Communications Committee. Minutes of the 4th Meeting held on June 11th 1919, of the 5th Meeting held on June 19th 1919'.

imperial question'.[81] In the context of bipolarity cables–wireless, three factors were of paramount importance in influencing the government (through the DSIR) towards the formation of a Radio Research Board (RRB) in 1920: first, the desirability (indeed, the urgent necessity) of securing imperial long-distance communications;[82] second, a tenacious resolve to foster the research in valves (whose manufacture had emerged with great forcefulness during the war);[83] and third, a profound conviction (of influential personalities such as Admirals Henry Jackson and William Nicholson) in the potentiality of radio technology.[84] These factors contributed overall to a British assertiveness in radio communication (diminished with the war), to the reification of 'its hegemony' as a value to be preserved at all costs, and to an unreserved confidence that, with funds deposited in the hands of radio experimentalists, both industry and the Empire would benefit in recompense.[85]

From its inception, the RRB strove, in many ways successfully, to rise to these new challenges. In the first meeting held in February 1920, it was structured into four sub-committees, each with jurisdiction over specific research fields. This division was intended to foment any research of a fundamental nature having a 'civilian as well as a military interest', and this it achieved by means of the participation of universities and government laboratories. Moreover, each sub-committee had the assistance and material help of the fighting services and the Post Office.[86] However, the management organization of the Board and the sub-committees presented a military complexion, not so much in number as in posts and functions. Thus, for the period from 1920 to

[81]NAUK, [CAB 35/1], 'Committee of Imperial Defence: Imperial Communications Committee. Minutes of the 10th Meeting held on October 28th 1919, p. 3'.

[82]NAUK, [AVIA 8/14], 'Radio Research Board: Appointment', 1920; NAUK, [T1/12505], 'D.S.I.R.: Arrangements for the establishment of the Radio Research Board, 1920'.

[83]See, for example, NAUK, [CAB 35/1], 'Committee of Imperial Defence: Imperial Communications Committee. Minutes of the 11th Meeting held on November 17th 1919', p. 12; 'Minutes of the 14th Meeting held on March 5th 1929, section 12'.

[84]Admiral W. Nicholson proposed Henry Jackson to be Chairman of the RRB—in fact, he also suggested the name 'Radio Research Board' for being short and clear. See NAUK, [T1/12505], 'Report to the advisory council by the members delegated to confer with representatives of the departments concerned on the proposals for the establishment of a wireless research board', 3 Dec. 1919.

[85]See NAUK, [DSIR 11/19], 'Constitution of the Radio Research Board. Appointment of members. Functions and Allocation'; 'DSIR Advisory Council: Memorandum on the recommendation of the Imperial Communications Committee that a Wireless Telegraph Board of the DSIR should be established without delay', 23 July 1919.

[86]*Report of the Committee of the Privy Council for Scientific and Industrial Research for the year 1919–20*. London, 1920, pp. 88–90.

1928, Admiral of the Fleet Henry Jackson held not only the chairmanship but also gave personal attention to over one hundred papers published during his term of office.[87] Since the idea of forming the Board originated in a committee of defence, it is not surprising that the four members of the fighting services (the Admiralty, the War Office, and the Air Ministry), in fact, represented half the board, and took the largest share of responsibility. Not surprisingly, direction-finding and valves—two essential elements for the navy—were a *forte* in the RRB; and some of the achievements gained in propagation of waves would have been impracticable without the presence of these ingredients.[88] But also there were clear disadvantages if few of these particular investigations were performed along the lines they deemed as appropriate. In this respect, there were few cases in which theoretical research was a predilection.

In very general terms, and in the interests of brevity, radio research may be divided into several defined, yet sometimes overlapping areas: (A) *Field on the propagation of waves*; this included aspects connected with the radiation, absorption and attenuation of waves, as well as methods of accurate measurement of wireless standards. (B) *Field on atmospherics*; this covered the causes of atmospheric interference and the best means of reducing its detrimental effects. (C) *Directional wireless*; this embraced not only the theoretical and practical details of directional transmission and reception but also the elimination of errors in direction-finding being necessary in the assistance of navigation and aircraft. (D) *Thermionic valves*, which dealt with the valves themselves and the physical problems presented by their manufacture and design.[89] Three centres were the scene of the main spheres of research: Ditton Park in Slough (direction-finding, field strength, and, since 1924, atmospherics), Aldershot Station (atmospherics), and the NPL at Teddington (valves and radio standards).[90]

These fields, for the first financial years, produced a pattern of expenditure which was characterized by an important initial investment in the purchase of apparatus and by an annual constant allocation for the maintenance of services.[91] By 1922, the total public money invested in the implementation of radio activities amounted to about £6,650—a figure which included £2,500 expenditure on investigation into atmospherics, and other £2,500 on the

[87]R. Naismith, 'Early Days at Ditton Park', *R.R.O. Newsletter*, 15th Sept. 1961, 5, p. 1.

[88]Gardiner (1962, p. 9).

[89]See 'Summary of Information: British Radio Research Board, 1926', in W.F. Evans, 1973. *History of Radio Research* Board, *1926–1945*. Melbourne, Appendix 1.

[90]On the history of radio research at the British RRB, see Gardiner, Lane, & Rishbeth (1982), Naismith (1961), Gardiner (1962a, 1962b, 1962c, 1962d, 1962e) and Pyatt (1983, pp. 91–9).

[91]NAUK, [T 161/43], 'Radio Research Board: Expenditure'.

variation of bearings of fixed stations. This amount also includes an allocation of £800 granted to specific university staff for investigation on valves, such as Professor O.W. Richardson of King's College, London (£500), Professor Owen of the University College of Wales (£100), and F. Horton of the Royal Holloway College (£200).[92]

Radio Research Board: characteristics and constraints

The abovementioned four fields were developed in a context not too distinct from that of the industrial or military in which innovation and progress were nurtured from an imperial milieu particularly bereft of direct, safe and immediate communications, and, increasingly, of a system alternative to cables.[93] The similitude between the two contexts became especially marked in the thirties as academic, civil-type research activities gave way to military security-based research—such as the radar programme initiated by Watson Watt at Bawdsey Research Station in Felixstowe—which was crucial in the course of World War II.[94]

The effect of this progressive militarization was highly significant in the fields of direction-finding, atmospherics, and thermionic valves, upon which security in navigation and aviation depended more crucially, than it was in the realm of the propagation of waves, which held an honourable place as academic research under Appleton's guidance at King's College, and subsequently with Ratcliffe at Cambridge.[95] In the former areas, assignments were more commonly determined by specific necessities than by co-operative alliances between institutions (the usual pattern in the military world). For this object, RRB's facilities at the Compass Observatory Ditton Park, Slough (lent

[92]NAUK, [T 161/43], 'Report of the Advisory Council to the Committee of Council on the proposal of the R.R.B. for certain fundamental investigation on the physics of the valve, 1st March 1921'.

[93]On military interest in short-wave imperial communication, see NAUK, [AIR 5/318], 'Experiments by Navy and R.A.F. regarding short wave point-to-point communication'; and 'Shortwave development. Action concerted with R.A.F. Farnborough, 9th February 1925', by Robinson; and NAUK, [AIR 5/455], 'History of private short wave W/T experimental work by R.A.F. Personnel'.

[94]On the early history of the British radar programme, see R.A. Watson-Watt (1957). *The History of Radar Development to 1945.* Proceedings of a Seminar held at the Institute of Electrical Engineers, London, June 1985; Gough (1993); and M. Bragg (2002).

[95]An excellent review of British radio investigations in the early 1930s is R.L. Smith-Rose, 1934. 'Report of the British National Committee to Commission II on Investigations of the Propagation of Waves carried out in Great Britain from April 1931 to June 1934', in *Papers Ass. URSI, London, September 1934.* Brussels, 4, 153–65.

by the Admiralty), the NPL at Teddington, and the Aldershot Wireless Station (belonging to the Meteorological Office) could provide a suitable environment, whereas the fighting services generally supported and financed investigations at the highest level. The rotating radio beacon is an example of a method of direction-finding that was devised by the Air Ministry as an aid to coastal navigation, and which was experimented with at the RRB with the collaboration of the Admiralty and Trinity House.[96] Other systems, such as the Adcock direction-finder, were developed, for use both on long and short wave, with a view to providing exceedingly valuable assistance for the safety of life at sea.[97]

It was a common feature of British radio researchers to encourage and foment initiatives consonant with the contemporary internationalism in radio communication. They made the globalization of observations a duty and a virtue, though globalization had to be effected on terms that would permit a sufficiently high degree of reliability to extend such radio techniques as those employed at the RRB. Remarkably successful among the campaigns which were devised to achieve the implantation of systematic observations and the requisite international guarantee were the measurement of the strength of signals (coordinated under the auspices of the URSI), the observations of bearings of transmitting stations (proposed at the Assembly of the URSI of 1922), the network of recorders for atmospherics (contrived by Watson Watt), and the network of radio sounding measurements (*inter alia*, by Appleton in 1931). All present a recognizable and common pattern. Each had a technical recording device developed at the RRB which would be 'exported' throughout the Empire and to other countries (direction-finding sets, automatic recorders, and cathode ray directional recorders for atmospherics) with British radiophysicists playing the role of instructors and mentors. A great number of the most influential personalities in Britain's radio science held posts of responsibility in international agencies or organs.[98] In a majority of cases, the initiatives were promoted for a variety of reasons that ranged from the quest for worldwide observational data for the sake of their own investigations, to the more altruist

[96] *Report of the Committee of the Privy Council for Scientific and Industrial Research for the year 1925–26.* London, 1926, p. 93; ibid., *for the year 1926–27*, p. 75; ibid., *for the year 1927–28*, p. 97; Smith-Rose & Chapman (1928); Smith-Rose (1928, 1931).

[97] *Report of the Committee of the Privy Council for Scientific and Industrial Research for the year 1925–26.* London, 1926, p. 92; ibid., *for the year 1928–29*, p. 100.

[98] Appleton as President of the URSI since 1934; Eccles as Vice-president; Watson Watt, Chapman, and Ratcliffe as chairmen of numerous commissions.

Fig. 1: Radio Division: Cathode Ray Direction-Finder for the location of thunderstorms. Courtesy of the Science & Technology Facilities Council, Rutherford Appleton Laboratory.

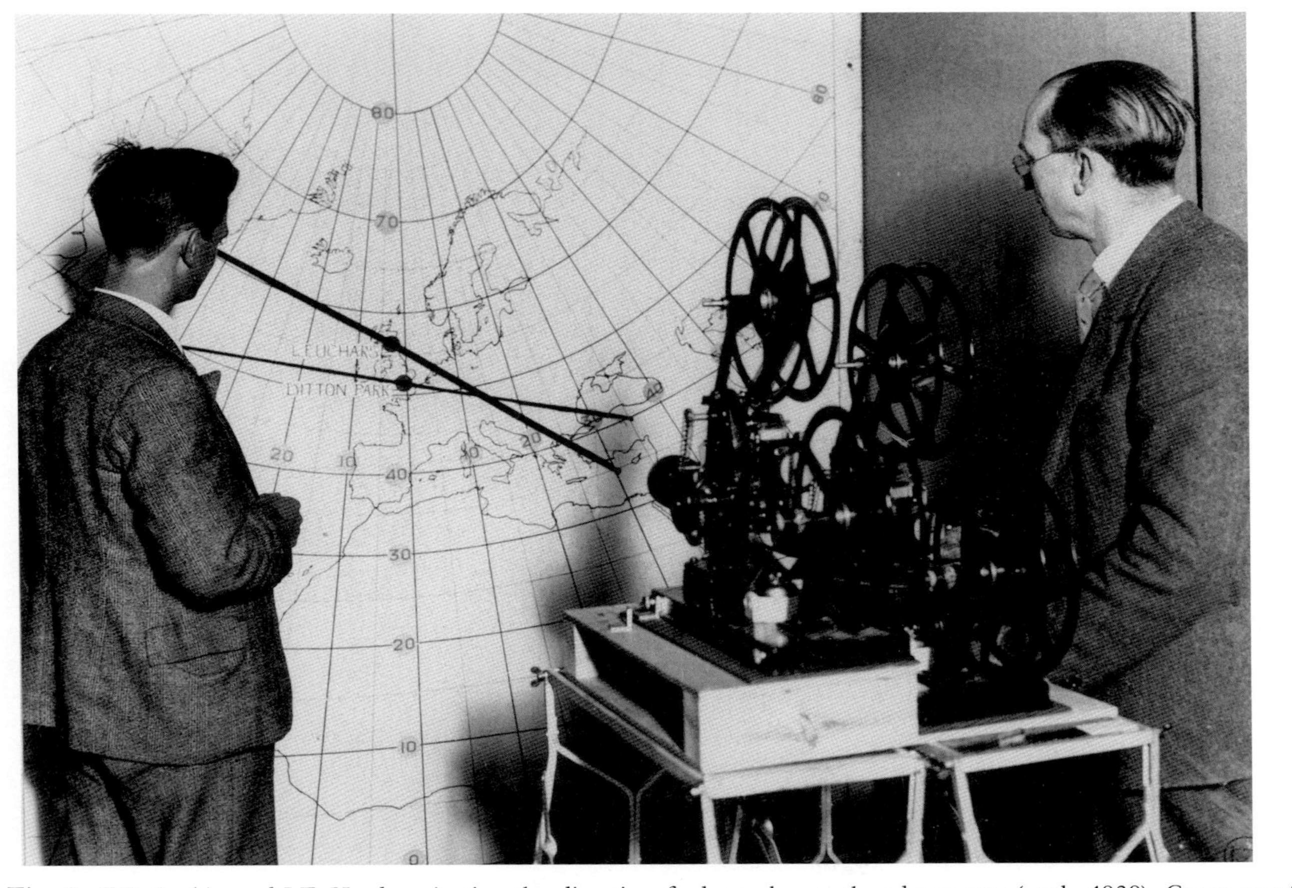

Fig. 2: F.E. Lutkin and J.F. Herd projecting the direction-finder to locate thunderstorms (prob. 1939). Courtesy of the Science & Technology Facilities Council, Rutherford Appleton Laboratory.

satisfaction of seeing how knowledge and technical advance were diffused over more isolated and peripheral regions.[99]

Watson Watt's recorder network for atmospherics illustrates well the potentialities and constraints of radio research at the RRB. The network extended its mesh not only to the colonies and Dominions, by means of university professors maintaining close contact with Watson Watt, but also to the meteorological services in numerous countries through the URSI.[100] Atmospherics—also called X's, strays, and statics, especially in the US—were signals produced by atmospheric electrical discharges, such as lightning flashes, negatively affecting wireless reception.[101] Cathode ray direction-finders were designed and constructed at Slough to classify the forms, features, and strength of these signals, and great emphasis was placed on observation as a way of locating cold fronts and thunderstorms—their alleged sources of origin.[102] The upshot of it all was a lively combination of controversial debate and ambiguous results with respect to their hypothetical value for weather forecasting.[103] Watson Watt and the members of his team (J.F. Herd, J.E. Airey, J. Hay, F. Lutkin) installed automatic recorders in Slough, Lerwick, Aboukir (Egypt), and Bangalore (India), and, from 1928 onwards, once technical constraints proved manifest, the network was supplanted by new devices equipped with cathode ray oscillographs which were disseminated through the Dominions, among which those of Watheroo Observatory (Australia, 1928) and the National Research Council of Canada (1931) were pioneers in their countries.[104] Simultaneously, Watson Watt effected arrangements in the URSI to extend measurements on the energy spectrum of atmospherics with the aim of mitigating their effects, in accordance

[99]On the early history of the International Union for Scientific Radio (URSI) in connection with ionospheric activities, see Beynon (1975).

[100]Some of the professors who participated in Watson Watt's network include J.K. Catterson-Smith (Institute of Science, Bangalore), Thomas Howell Laby (Melbourne University), and John Tasker Henderson (National Research Council, Canada). Recorders of the RRB pattern were also installed at the Observatories of Potsdam and Lindenberg (Germany), and at Calcutta University, and Cape Town University.

[101]See R.A. Watson Watt (1925).

[102]A valuable review of methods, instruments, theories and of the meteorological importance of atmospherics is summarized in Bureau (1926).

[103]On the meteorological relations of atmospherics, see Cave & Watson (1923). See also, for the Australian case, White & Huxley (1974, pp. 23–8).

[104]*Report of the Committee of the Privy Council for Scientific and Industrial Research for the year 1922–23.* London, 1923, p. 79; ibid., *for the year 1924–25*, pp. 75–6; ibid., *for the year 1927–28*, pp. 99–100; ibid., *for the year 1928–29*, p. 97; ibid., *for the year 1929–30*, pp. 98–9; ibid., *for the year 1930–31*, p. 82; ibid., *for the year 1931–32*, p. 83.

Fig. 3: Sir Robert Watson-Watt's Experimental Hut, Aldershot Source: 1919. From Royal Air Force.

with methods designed and tested at the RRB.[105] The 36 sets manufactured by the General Electric Company of America were warmly welcomed by the chief meteorologists in France (R. Bureau), the US (Reichelderfer), and by the Japanese, Italian, and Danish authorities, among others.[106] Paradoxically, and unlike these, the observation of atmospherics at the British Meteorological Office did not win immediate and widespread acceptance as an adjunct to the traditional methods of forecast and of localization of thunderstorms.[107] Despite the early, if somewhat desultory support of the Royal Meteorological Society, Watson Watt was not sufficiently persuasive to attract British meteorologists to his particular crusade, but the experience gained in Slough without doubt permitted him to confront the more promising field of radio-detection.

The plan of observations of bearings of transmitting stations—one of the stumbling blocks in direction-finding—forms a scheme of cooperation with characteristics and potentialities very different from those of Watson Watt's network.[108] Unlike this, it was the result of a bilateral agreement with the Norwegian radio authorities to analyse the effect of high Northern latitudes on the variation of bearings.[109] The elimination of errors in bearings was essential for transmission and reception, but also for the location of the position of objects on land, sea, and air.[110] Hence Smith-Rose and his team (R.H. Barfield,

[105]For comments on this proposal, and the general agreement in connection with radio research as an aid for meteorology, see 'Minutes of the Commission of Atmospherics', in *Papers Ass. URSI, Brussels, September 1928*. Brussels, 1929, 2(3), 17–21. See also 'Some Observations on International Research on Atmospherics', *Papers Ass. URSI, Copenhagen, 1931*, Annexe II, 60–2; and 'Commission III: Atmospherics', ibid., 56–60.

[106]For a critical report on the state of investigations on atmospherics in Europe, see H.C. Webster, 'Report on Investigations of the State of Radio Research in Europe at the Present Time, July–August 1933', in Evans (1973, Appendix 8, pp. 2–3).

[107]In 1929 Watson Watt delivered the Symons Memorial Lecture to the Royal Meteorological Society, praising RRB's work on the nature and origin of atmospherics and its relationship to meteorology, though the degree of active engagement on the part of meteorologists did not match his effusive speech. See R.A. Watson Watt (1929).

[108]Simultaneous observations were made with an identical receiving apparatus at the University College of North Wales (Bangor), Armstrong College (Newcastle), Leeds University, Cambridge University, Teddington, Bristol, and at the Post Office Experimental Station (Peterborough). See DPA. [Sub-Committee C], 'Reports of the Investigation on Variation of Bearings of Fixed Stations, 1921', by R.L. Smith-Rose.

[109]The plan agreed at the Brussels meeting of the URSI in 1922 included the extension of observations at Lerwick Observatory, together with the university network. See *Report of the Committee of the Privy Council for Scientific and Industrial Research for the Year 1924–25*. London, 1925, p. 76.

[110]For an excellent critical résumé with a historical summary and abundant bibliography on this topic: R.L. Smith-Rose, 1928. 'Radio Direction-Finding by Transmission and Reception,

J.F. Herd, F.M. Colebrook, R.M. Wilmotte, S.R. Chapman, H.P. Hopkins, R.A. Fereday, W. Ross, J.S. McPetrie, H.A. Thomas) benefited far more than Watson Watt had done from the military nature of investigation on direction-finding and from the profusion of projects and interests—both on the part of the Admiralty and the Air Ministry—resulting from World War I.[111]

There is ample evidence that an unprecedented emphasis on short-wave research, coincident with the first broadcasting in that waveband, made the study of it almost indispensable in all the centres of the RRB.[112] By the mid-1920s, J. Hollingworth and his team (at first R. Naismith, and afterwards J.S. McPetrie, F.M. Colebrook, and B.G. Pressey), who had specialized in field-strength measurements,[113] progressively abandoned the traditional long band and became involved in a variety of projects pertaining to the very short wave that had a bearing on innovation.[114] The burgeoning engagement with high frequency can be interpreted as a quest for innovation in a virgin and unexplored segment, but also as a mode of challenging the ventures initiated in the early 1920s by G. Marconi and his company and by military radiotechnicians at H.M. Signal School.[115]

with Particular Reference to its Application to Marine Navigation'. In *Papers of the Assembly of the URSI. held in Washington, October 1927.* Washington, 54–84.

[111]An indication of the importance of direction-finding at the RRB was the special reports devoted to this subject in the 1920s. See Radio Research Board, *A Discussion of the Practical Systems of Direction-Finding by Reception. Special Report No 1.* London: H.M. Stationery Office, 1923; *Variation of Apparent Bearings of Radio Transmitting Stations. Part I. Rad. Res. Spec. Rep. 2.* London, 1924; *Part II: Observations on Fixed Stations, Rad. Res. Spec. Rep. 3.* London, 1925; *Part III: Observations on Ship and Short Transmitting Stations, Rad. Res. Spec. Rep. 4.* London, 1926; R.L. Smith-Rose, *A Study of Radio Direction Finding. Rad. Res. Spec. Rep. 5.* London, 1927.

[112]Short wave affected all research fields. See for instance Smith-Rose & Hopkins (1938), Barfield & Ross (1937), and Appleton & Green (1930). See also NAUK [DSIR 36/4015], 'Ultra-short waves research, 1930–1937'; and NAUK [DSIR 36/2188], 'Propagation of waves: various reports on short waves, 1926–1933'.

[113]Effectively Hollingworth and Naismith gained their reputation by organizing field-strength measurements from Manchester, Glasgow, Aberdeen, and Slough in the early 1920s. See *Report of the Committee of the Privy Council for Scientific and Industrial Research for the Year 1922–23.* London, 1923, p. 78; ibid., *for the year 1923–24*, p. 75; ibid., *for the year 1924–25*, pp. 71–2.

[114]See for example: Hollingworth & Naismith (1929).

[115]On the production and utilization of short wave at the RRB, and the magnetron valve, see *Report of the Committee...for the year 1934–35*, pp. 75–6; ibid., *for the year 1936–7*, pp. 84–5. On short wave research by military bodies: NAUK [AIR 5/318], 'Experiments by Navy and R.A.F. regarding short wave point-to-point communications, 1925'; NAUK [AIR 5/516], 'Reports from units on short wave A/T organisation, 1926–28'; NAUK [AIR 5/321], 'Development of D.F. apparatus for very short wave working, 1925–1927'; NAUK [AVIA 23/375, 387], 'Short wave artillery sets: tests, valves, 1928'.

At the same time as this unparalleled emphasis on direction-finding and short wave, there was particular stress on propagation phenomena, thermionic valves, and circuitry. Here the renewed interest in the theory and mass production of valves following the war and Appleton and Barnett's unprecedented success of 1924 concerning the existence of the Heaviside-Kennedy layer contributed to promote the research endeavour that took place from the mid-1920s. This helped equally to increase the staff of radiophysicists in Slough and Teddington, but also in universities, especially at Cambridge and at King's College, London. Between 1925, in which Appleton and Barnett's classic paper was issued, and 1938, 66 papers were published at the RRB on propagation of waves alone,[116] and 16 on valves and radio apparatus.[117] Likewise, a great number of assistants and radio technicians were contracted, and joint research with numerous universities was developed.[118] By the end of 1933, the Radio Research Station at Ditton Park and the Wireless Division of the NPL at Teddington were amalgamated to form the Radio Department of the NPL under Watson Watt's superintendence.[119] But thereafter, pressed by the necessity for a defensive system of radiolocation, radio research at the RRB acquired a predominantly military note. By 1935, at the request of H.E. Wimperis of the Air Ministry, Watson Watt commanded a venture without precedent to detect aircraft by radio methods.[120] Now, he could boast, there were dozens of radiospecialists with experience in the use of pulse techniques, circuitry, and direction-finding, compared with the handful of ten years ago.[121] The case of the most outstanding officers, such as Smith-Rose, Barfield, and Naismith, who served with the fighting services during the war (e.g. in the

[116]In the preceding years only one had been published.

[117]The number of publications had risen from 32 up to 1925 to 229 by the end of 1938 (67 on propagation of waves, 21 on atmospherics, 35 on direction-finding, 31 on aerials and precision measurements, and 31 on valves and radio apparatus). For a list of papers published up to 31 March, 1929, see Evans (1973, Appendix 7). For a summary of the work in the first ten years, see *The Report of the Radio Research Board for the Period ended 31st March, 1929.* London: H.M. Stationery Office, 1930.

[118]C.L. Fortescue of the City and Guilds Engineering College, South Kensington (high-frequency measurements), E. Pearson of University College, London (direction-finding statistics), and A.C. Davies and F. Horton of the Royal Holloway College (dull emitting valves) in collaboration with the RRB.

[119]Pyatt (1983, p. 97).

[120]The memorandum 'Detection and Location of Aircraft by Radio Methods' is reprinted in R.A. Watson Watt (1957, pp. 470–4). See also Ratcliffe (1975, pp. 554–8).

[121]For a comparative study on radar development in Europe, see Allan Beyerchen, 'From radio to radar: Interwar military adaptation to technological change in Germany, the United Kingdom, and the United States', in Murray & Millet (1996, pp. 265–98).

Ionospheric Forecasting Service, supplying predictions of the ionosphere) is an indicator of the direction adopted by radio research.[122]

A first consequence of this burgeoning interest in radio was the growth in the number of postgraduate students, a majority from the Dominions, who joined Appleton and Ratcliffe's research groups. On this occasion, the new avenue for their training was comparable with that of a generation earlier as atomic physics and physical mathematics of Cambridge attracted the minds of overseas academics. A good illustration is the core of Britain's ionospheric research par excellence, Appleton and the Halley Stewart Laboratory, at King's College, London.[123] Here, between 1930 and 1936, the year in which Appleton moved to Cambridge, the number of research students exceeded 20.[124] As he strengthened the contacts with the academic elite and the radio industry, the number of young physicists pursuing radio research-based careers under his supervision multiplied. Plainly, Appleton and his group became a worthy aspiration for a new graduate with special interest in radio. It could, at best, fulfil his prospects and serve as a guarantee in the return to his native country. The sojourn at King's College (or, failing that, at Cambridge) delivered experience and knowledge, in the same way the Cavendish of Rutherford, J.J. Thomson, and W.L. Bragg had served others in analogous conditions for successive generations.[125]

In sum, the manifold changes following World War I—the expansion in international communications, the burgeoning radio industry, and the sharply declining reliability of traditional cable system—created a necessity for the coordination and furtherance of radio research. In all probability, without

[122]Barfield was seconded to the War Office to organize a direction-finding service, Smith-Rose was responsible for the Radio Department at the NPL and for the ionospheric service, and Naismith managed the RRS at Ditton Park. See Gardiner (1962e).

[123]White (1975, pp. 41–3) and Clark (1971, pp. 88–92).

[124]Appleton's team included, among others, G. Builder (Australia; an able expert in radio techniques); A.L. Green (London, England; ionospheric layers); F.W.G. White (New Zealand; attenuation of radio waves); E.G. Bowen (Swansea, Wales; atmospherics and radiation); O.O. Pulley (Sydney; the first manual P'f equipment), F.W. Wood (Western Australia; ionospheric prediction); Poole (Western Australia); E.C. Halliday (South Africa; tidal movements in E-region); W.R. Piggott (ionospheric absorption); L.W. Brown (collisional frequency of electrons); D.B. Boohariwalla (Bangalore, India); E.C. Childs; Biggs; Hutchinson; Williams; and P. Roberts and Gander (technicians).

[125]Ratcliffe's group was smaller: F.W.G. White; E.C.L. White; J.L. Pawsey (Australia); and W.B. Lewis; and from 1934 onwards, J.E. Best; F.T. Farmer; S. Falloon; H.G. Booker, and M.V. Wilkes. This latter little group joined Appleton when he was elected Jacksonian Professor at Cambridge in 1936. Other members of Appleton's team at that time were K.G. Budden, J.W. Findlay, and K. Weekes. See Wilkes (1975, pp. 54–5).

these ingredients, one might hardly envisage how the picture of the upper atmosphere, at least regarding its layered structure, could have been formed to the degree that it did in the interwar years. In the early 1920s, for the first time from the state realm, there began to prosper the strategy of coordinated endeavours of a multinstitutional kind which military authorities demanded. In particular, this tactic pursued both a formal response to defensive and imperial exigencies and a policy of immediate industrial application.[126]

The shrewdness of a leader

For the interwar years, Edward Victor Appleton was the essence and very definition itself of ionospheric physics in the whole world. In December 1924 his indefatigable search for the elucidation of a long-running debate culminated in a 'discovery'—as it was labelled at that time—that he tenaciously pursued and yearned for: he validated the Heaviside-Kennedy layer and helped to shape the picture of the upper atmosphere, which turned him into one of the founders of ionospheric physics.[127]

Appleton was always the first, and that was of his liking. He always remained at the top of his class: one hundred per cent in his first examination in physics at Hanson Secondary School, a First Class in the London Matriculation Examination (at the minimum age), First Classes in Natural Sciences Tripos and Physics Tripos at Cambridge, and the youngest professor in England (the Wheatstone Chair of Physics at King's College, London, at the age of 32). He did well in everything—physics, mathematics, but also Latin, Greek, German—and was incredibly energetic and industrious. His supervisors immediately perceived his great patience and extraordinary retentive memory, as well as his brilliance, qualities appropriate for investigation. They appreciated the strength of his leadership, his deep sense of justice, and his contagious enthusiasm; and Rutherford in particular judged him to be 'the most suitable candidate'[128] to occupy the next vacant chair at Cambridge, and perhaps to succeed him in the near future.

Appleton's success in ionospheric physics, to which he devoted his entire life (95 papers) and for which he was awarded the Nobel Prize in 1947, came

[126]The industrial application was demonstrated by the practice of securing authorship by means of patents—up to 37 appear in the name of members of the RRB between 1920 and 1936. For the list of patentees, see DPA. 'Patents by members of the Radio Research Board'.

[127]The best biographical works on Appleton are Clark (1971), Ratcliffe (1966), and Piggott (1994). See also Wilkes (1997), Ratcliffe (1966), and obituaries in *Science and Culture*, 1965, 31, 348–50, and *The Times*, 23 April 1965.

[128]Clark (1971, p. 49).

Fig. 4: Apparatus for measuring the angle of incidence of HF signals, 1933. A.F. Wilkins and E.C. Slow operating. Courtesy of the Science & Technology Facilities Council, Rutherford Appleton Laboratory.

Fig. 5: The van shown here was used in 1935 by R. Watson-Watt, A.F. Wilkins, and J.F. Herd for the first British radar experiment. Courtesy of the Science & Technology Facilities Council, Rutherford Appleton Laboratory.

Fig. 6: Pulse Ionosonde receiver designed by L.H. Bainbridge-Bell. RRS Ditton Park, 1933. Courtesy of the Science & Technology Facilities Council, Rutherford Appleton Laboratory.

Fig. 7: Frequency-change transmitter used at National Physical Laboratory (Teddington) then moved to Slough for first routine soundings in 1931. Courtesy of the Science & Technology Facilities Council, Rutherford Appleton Laboratory.

in part from a profound appreciation of team-based work and from a dignified regard for his pupils. He knew how to elicit the best from his research students ('all respond to stimulus and incentive'[129]) and rewarded them with joint publications (a total of 53, except in Wilhelm Altar's controversial case on the magneto-ionic theory). By the same token, they held him in high repute ('a curious mixture of affection and esteem'[130]) and valued his solicitousness and

[129]EUA [D34], *Papers*, Appleton Room. 'Science and the Public', manuscript, 1948.

[130]*Student*, 29 April 1965 (The students' newspaper of Edinburgh University—quoted in Ratcliffe (1966a, p. 7).

affability ('his kind understanding of the student's problems'[131]). Unlike others such as Watson Watt who deemed science as an intimate and confidential activity, he cultivated it as a homely one. It was this homeliness, with which he pervaded his laboratory and classroom, which infused his pupils with character and aplomb. Disorder and lassitude were not to his liking. Quite the contrary, Appleton endeavoured to secure the pleasantest conditions with that solicitude and deference that, according to Faraday—one of his masters—, was the best guarantee for 'the development of the true student-spirit in investigators'.[132]

A wise civility was the recipe Appleton applied. He once described the mode of proceeding successfully in life in general: 'In dealing with people I have always found two golden rules to apply. It is always best, at the start, to impute to the qualities you would like them to possess. You then almost invariably find that they display these qualities. In any case, it is better to trust and be deceived than to distrust and be mistaken… . One can achieve almost anything if one is prepared not to want the credit of it.'[133]

At the apogee of his career, Appleton's universe centred around the main institutions of ionospheric research in one of the main centres of radio research in the world: King's College in London and the Halley Stewart Laboratory. To these he added what was probably the first and most select worldwide specialist society for atmospheric physics, the Maxwell Society, which between 1932 and 1936 attracted the *números uno* in the subject (Chapman from Imperial College, Ratcliffe from Cambridge, Naismith from Slough, E.B. Moullin from Oxford, S.K. Mitra from Calcuta, C. Stormer from Oslo).[134] The College housed a research group and the laboratory the latest instrumentation and devices; the Society regularly organized lectures, and Appleton was in charge of it all. It is likely that then, at the zenith of his fame, he congratulated himself on having renounced nuclear research under Rutherford's direction in 1920, on having renounced remaining for years in the shadow of a renowned figure.

[131]F. Oldham, a student at King's College, 1922–25—quoted in Ratcliffe (1966a, p. 4).

[132]Quoted in EUA [D34], *Papers*, Appleton Room. Draft of 'The Scientist in Industry', Feb. 1948.

[133]EUA [D39], *Papers*, Appleton Room. 'The Organization and Work of the D.S.I.R.', 39th May Lecture to the Institute of Metals, 1949, p. 989.

[134]The Society was founded by Appleton and managed by F.W.G. White, its secretary, although no minutes were kept before 1939. For subsequent records, see KCLA [GB 0100 KCLCA KSM]. See also White (1975, p. 42); and AAAS [111/4/4], 'Biographical notes', by F.W.G. White, 1989, 1: 23–4.

Radio, more than any other field, provided him with uncontested leadership and social distinction.[135]

His skill in simple physical interpretations, his ability to analyse a subject as a whole, his resoluteness—despite his doubts and constraints—to lay the foundations of the knowledge of wave propagation by means of the magneto-ionic theory, all this distinguished Appleton according to his contemporaries.[136] Quite apart from the debate of primacy, the 'discovery experiments' of 1924 and of 1927 came with an air of novelty and disclosure cohering with his ever-increasing inclination to allude to the ionosphere as a 'new continent' to explore.[137] But whence this impetuosity for discovery, this eagerness to be the first, this 'continuing anxiety of a man', in his biographer's words, R. Clark, fearing that competitors might surpass him?[138] According to Ratcliffe, radio wave propagation, the last of his preferences before the crucial experiments of 1924, after valves and atmospherics, held out incomparably greater possibilities for investigating the fundamental structure of the upper atmosphere.[139] And that, to Racliffe, was the result of his 'real shrewdness'; here lay 'his genius'.[140]

Equally, the combination of sagacity and adroitness was, in Tuve's words, his main virtues: 'It is correct that Appleton was the first to announce in print that the ionosphere comprised two layers. My own impression would be that Appleton should be credited with having hazarded this proposal in print, with a 50–50 chance of being right. . . . The fact is that until observations were available on different wave lengths, no one could say which interpretation would be correct [stratifications of the ionosphere or multiple reflections in one only layer]. . . . Appleton was lucky, perhaps, but more than anything else he was astute. Radio has been his life work, and as such it was entirely correct for him to hazard such guesses as he wished, in order to be first in the field.'[141]

The recognition of the importance of the ionospheric layers to radio communication probably helped to convince the Nobel Prize committee to bestow

[135]Clark (1971, p. 23) states that from the start of his career Appleton looked for a field in which he would be able to gain, and hold, the position of unchallenged leader.

[136]H. Rishbeth, 'What became of Appleton's ionosphere?', *Jour. Atm. Terr. Phys.*, 1994, 56(6): 713–26, p. 713.

[137]P.S. Excell, 'Sir Edward Appleton and Joseph Priestley: two giants of electrical science', *Jour. Atm. Terr. Phys.*, 1994, 56(6): 693–704, p. 698.

[138]Clark (1971, p. 25). See also on the same page Appleton's letters to Van der Pol, 1920.

[139]Ratcliffe (1996a, pp. 10–11).

[140]Ibid., p. 6.

[141]Tuve Papers [container number 4], Manuscript Division of the Library of Congress, Washington, USA—quoted in Gillmor (1994, p. 140). See also EUA [E20], *Papers*, Appleton Room. G. Breit to E.O. Hulburt, 2nd February 1935.

an award in a field that had proved fruitful in many disparate branches of physics and that had not been decorated yet. In 1947 Appleton was honoured 'for his investigations of the physics of the upper atmosphere, especially for the discovery of the so-called Appleton layer'.[142] As E. Hulthén, a member of the committee, emphasized in the presentation speech, 'no conclusive proof of the conducting layer was forthcoming' before 1924, and Appleton found determinant evidence ('a height of 100 km' for the *Heaviside layer* and 'of about 230 km' for the *Appleton layer*).[143] Evidence that, in the eyes of his contemporaries, proved convincing.

Perhaps essential for men like Appleton was the profound conviction that science discloses the fundamental aspects of nature. The pursuit of unity within the variety of experience, the overemphasis on discovery and on the 'first direct evidence', to which Appleton returned again and again,[144] evinced both a hidden earning and a desire for transcendence. The 'direct proof of the existence' of the Heaviside-Kennedy layer, the finding of E and F layers, epitomized that ambition; and here the British had acted as physicists, the Americans as engineers. Appleton asseverated in 1948 that, despite the collective nature of science, 'the start of any new development was the result of a single person', a personal adventure, an act of revelation; and that it was knowing 'how or where to start that mattered.'[145] Later, if it succeeded, 'an army of people would follow up and consolidate his position'; it was in the later stages that teams of scientists were required. Scientific leaders, whose personal repute was consecrated by international recognition, would not succeed unless they had the qualities of curiosity, imagination, intuition, and the love of orderliness. Paraphrasing Bacon, he distinguished between *experimenta fructifera* (useful for knowledge) and *experimenta lucifera* (illuminative in man's life). And, doubtless, Appleton regarded this latter or the pursuit of it as a transcendental act.

There are those who think that Appleton, more than anybody, established the fundamentals of ionospheric physics. When, in 1965, Ratcliffe wrote his obituary for the Royal Society, he confessed that what they knew about the ionosphere up to 1939 was 'almost entirely due to him, or at any rate to

[142]Clark (1971, pp. 145–7).

[143]E. Hulthén, 1964. Physics 1947. In *Nobel Lectures Physics: 1942–62*. London: Elsevier Publ. Co., 75–8, pp. 75–6.

[144]The allusions to the discovery of layers, as an intrinsic feature of the atmosphere, are interminable. See, for instance, Appleton to Van der Pol, 12 October 1924—transcribed in Clark (1971, p. 40)—and E.V. Appleton, 1963. 'Radio and the Ionosphere'. In C. Domb ed. *Clerk Maxwell and Modern Science*. London: Athlone Press, 70–88, p. 72.

[145]EUA [D34], *Papers*, Appleton Room. 'Science and the Public', manuscript, 1948.

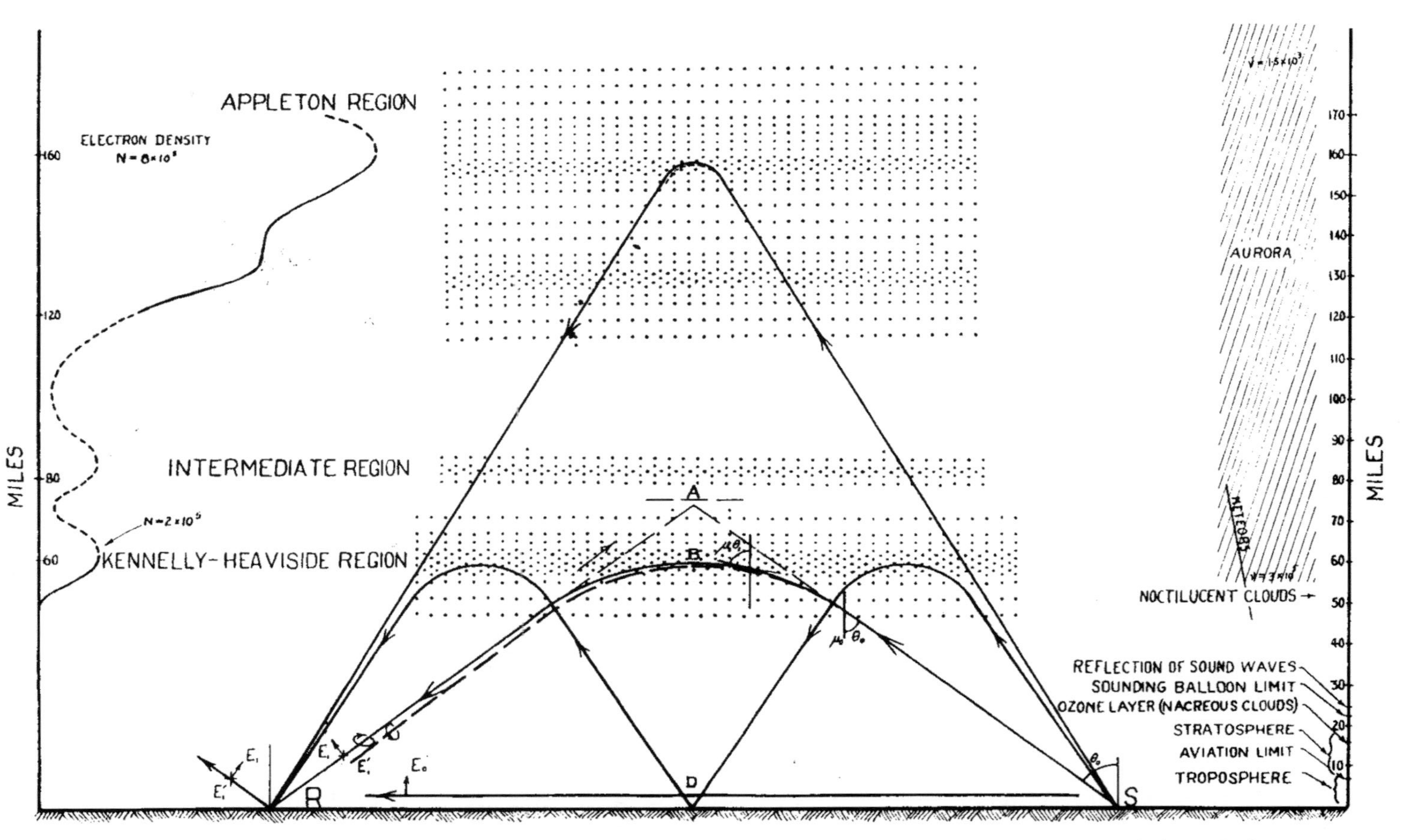

Fig. 8: The layered structure of the upper atmosphere, circa 1930. See the sky waves as being transmitted through free space and reflected from horizontal and stratified mirrors.

him and to research schools which he started and inspired.'[146] On this point, the acknowledgement of his institutional and organizational authority, even more than his theoretical contributions, is de rigueur.[147] Appleton had helped to strengthen and consolidate the international spirit of the URSI from the pinnacle of the presidency since 1934. Previously, he had arranged an expedition to perform ionosphere soundings at Tromsö in Norway on the occasion of the Second International Polar Year in 1932.[148] He decisively contributed to the planning of ionospheric investigations all over the world as chairman of the Mixed Commission on the Ionosphere after World War II. Through these activities, he could reinforce the solidity of his leadership, but also he would guide the endeavours of the investigators and alert the rest of the radiophysicists' community to the necessity of international teamwork.

Modelling the ionosphere: the Chapman layer and electron density profiles*

From the beginning of the twentieth century, identifying the mechanism behind the formation of the ionized layers of the atmosphere was a priority. Model-building was to have a predominant influence on ionospheric physics, with early theories being infused with a strong emphasis on the simplification and idealization that had inspired the atomic models.[149] Its origin was rooted in the confluence of various geophysical branches, and in the need for a multidisciplinary approach to solve existing problems.

The concept 'model', however, was particularly ambivalent, even polyvalent in meaning, when adopted by mathematical physicists or radiophysicists. The distinction is not absolute or evident but indicates a tendency. For the former, science seemed to involve idealized, counterfactual schemes for data

[146]Ratcliffe (1966a, pp. 15).

[147]For further details on Appleton's theoretical and technical contribution, see Strangeways (1994), and Bradley, Excell, & Rowlands (2000–01, p. 121).

[148]Letters, manuscripts, and reports on British expeditions to Tromsö are deposited at the DPA. See for example: 'Proposed International Polar Year Programme of Observations on the Ionization of the Upper Atmosphere'; and 'Report of U.R.S.I. Sub-Commission on Radio-Work during the Polar Year 1932/3'; and Minutes of the National Polar Year Committee, 1932–35.

*This section is a partial version of the article published in Annals of Science by Aitor Anduaga, 'Sydney Chapman on the Layering of the Atmosphere: Conceptual Unity and the Modelling of the Ionosphere', Copyright Clearance Center (2009). The author would like to thank T&F Informa UK Ltd for permission to reproduce the work here.

[149]In physics, the terms 'model' and 'theory' often have overlapping meanings. For some authors, models are frequently used in the development of fundamental theories, and function as a step towards a ready-made theory. See Redhead (1980).

manipulation, sometimes serving only to save appearances, and in most cases 'to represent fields of observation of the natural world'.[150] For radiophysicists, however, the representational model (and by extension, science) had a clearly ontological sense, in which the primary objective was not to keep up appearances but rather to mirror reality (hence their insistence on discovery), reproducing it as closely as possible, and approving (or rejecting) the model in accordance with its degree of approximation.[151]

Since 1924, issues of realism and reductionism had been eclipsed by a discovery frenzy, when Appleton initiated research which would lead to the finding of the layers (E, F_1 and F_2) of the ionosphere. However, the first mathematical models soon appeared, and with them a variety of different approaches in science. From the epistemological perspective, ionospheric physics was preliminary; radiophysicists scarcely comprehended the nature of atmospheric knowledge, its presuppositions and foundations, and its extent and validity.[152] Mathematical physicists, such as Chapman, E.O. Hulburt, and P.O. Pedersen, had set themselves the task of quantitatively demonstrating how the layers were formed, based on the vague information available[153] Little was known then about the composition of the upper atmosphere (oxygen, nitrogen or hydrogen?), or the solar radiation that might ionize it (ultraviolet rays or corpuscular streams?); but they nevertheless wanted to find a model for layer formation, even though it was not the *final answer* to the problem in question.[154]

There were several reasons for wanting to know the way in which the electrons responsible for wave reflection were produced in the upper atmosphere.[155] Plausible arguments and acute observations from radio sounders showed that their concentrations at the peaks of the layers altered from time

[150]Akasofu, Fogle, & Haurwitz (1968, p. 69).

[151]There is abundant literature on models in science, logic, and philosophy. For our purposes, see the standard papers: Bunge (1968, pp. 165–80) and Hesse (1963). See also Czarnocka (1995).

[152]For the concept 'preliminary' and the use of models in physics, see Stephan Hartmann, 1995. 'Models as a Tool for Theory Construction: Some Strategies of Preliminary Physics'. *Poznan Stud. Phil. Sc. Hum.*, 44, 49–67, p. 50.

[153]Reviews of Chapman's theoretical works around 1930 include Massey (1974); Bates (1973); J.A. Ratcliffe, 'Ionospheric Physics and Aeronomy', in Akasofu et al. (1968, pp. 27–30); D.R. Bates, 'Ionospheric Physics and Aeronomy', in Akasofu et al. (1968, pp. 31–4).

[154]On the hypothetical, speculative, and extrapolative nature of the knowledge on the ionosphere in the 1920s, see Hulburt (1974).

[155]A clear and comprehensive account on the production of electrons and on the Chapman layer is Ratcliffe (1970, pp. 93–102).

to time. A formula that might predict the form of a simple layer and its temporal evolution would explain these variations. The production model would consequently contain the four variables surmised from observations, namely height, time of day, season of the year, and latitude. Another reason for interest was entirely pragmatic. A production function should illustrate the rate of production of electrons (and thereby its density) for a given height in a wide variety of conditions. Hence Chapman in particular sought to describe the most possible idealized and oversimplified atmosphere, so that it could serve as a theoretical standard for forthcoming ionospheric models.[156] For Chapman, finding a production function meant founding the basics of ionospheric formation, something that would serve as 'a starting point for further investigation into the influence of factors here neglected'.[157]

Chapman drew the inspiration for his formation theory from other models produced by European and American physicists in the 1920s, which did not gain widespread acceptance for various reasons. In 1926, H. Lassen of the University of Cologne described how a layer was formed, eliciting figures for maximum ion density.[158] He identified hydrogen ions as responsible for refraction, and pinpointed a single layer at a height of 100 km. But like most of Germany's early ionospheric research, it went practically unnoticed abroad.[159] At the Naval Research Laboratory in Washington, one of Chapman's biggest contenders (and then a friend), E.O. Hulburt, set forth a model for the diffusion, recombination, and attachment processes.[160] He made a large number of estimates of the ultraviolet energy and its effects, which more or less fit the facts. Interestingly, they were the 'first estimates of all factors involved in the production and loss processes'[161] in layer formation. Yet, for a time the theoretical formulas proposed by P.O. Pedersen in Denmark in 1927 seemed the best: he derived expressions for the peak electron-density and the total conductivity of the layer, and demonstrated that its concentration depended on the zenith angle of the solar radiation.[162]

[156]Ratcliffe, 'Ionospheric Physics and Aeronomy', op. cit., p. 28.

[157]S. Chapman (1931, p. 27).

[158]Lassen (1926).

[159]For a review of Lassen's work and German ostracism in ionospheric physics, see: Dieminger (1975, pp. 27–8).

[160]Hulburt (1928).

[161]Waynick (1975, p. 17). For a review of Hulburt's model, see: Beynon (1975b, p. 49).

[162]Pedersen's well-documented and excellent monograph was without doubt at the forefront of ionosphere theory at that time. See P.O. Pedersen, 1927. *The Propagation of Radio Waves along the Surface of the Earth and in the Atmosphere.* Copenhagen: Danmarks Naturvioenskabeliage Samfund, especially Chapter V.

Four years later, Chapman set out to seek a more general model based on Pedersen's somewhat overlooked work.[163]

Because knowledge regarding gas composition, temperature, and density was scarce in 1931, Chapman was free to introduce assumptions in the most convenient manner.[164] He considered the upper atmosphere as an isothermal plane-stratified region, and each stratum as being composed of a single kind of molecule. Furthermore, he assumed that density varied exponentially with height. Sunlight—supposedly monochromatic—ionizes the molecules, and as a result produces electrons and positive ions. For a time, the electrons remain free and then recombine with the ions; an equilibrium is eventually established, in which the rate of production is equal to that of loss. Since radiation gradually loses strength as it descends, and gas density increases the nearer it is to earth, Chapman thought there must be a maximum of free electrons at a certain height. He expressed it in a mathematical equation;[165] however, as it was complicated, he illustrated it by means of graphical curves. The *Chapman layer* pertains to the ideal stratum formed in the aforementioned conditions; the *Chapman profile* to the curve of electron density inferred.[166]

In an attempt to temper the realist impulses of radiophysicists, Chapman warned about the constraints of his model and endorsed the desirability of more comprehensive investigations in future. Only a pragmatic attitude, strengthened by a multidisciplinary view of the problem, seemed to offer the basis for progress: 'The problem is an ideal one which scarcely represents adequately all the factors of importance in any actual case; it is thought likely, however, to be of value as an approximation.'[167] In this vein, Chapman preferred to consider model-building as a useful process, appropriate not for

[163]Chapman knew about Pedersen's work as soon as it was published. In 1928 he alluded to Pedersen's 'recent valuable and comprehensive discussion of radio propagation' in a footnote. See S. Chapman (1928, p. 370).

[164]On the composition and pressure of the upper atmosphere in the 1920s, see Pedersen (1927, Chapter IV). At that time most estimates for the pressure and density in the upper atmosphere were taken from H.B. Maris, 1928. 'The Upper Atmosphere'. *Terr. Mag. & Atm. Elect.,* 33, 233–55.

[165]Indeed, the so-called *Chapman function* was introduced in a second paper in which he extended his calculations to a spherical earth and atmosphere. Chapman (1931, pp. 483–501).

[166]Some of the most useful textbooks covering the Chapman layer are K.G. Budden, 1961. *Radio Waves in the Ionosphere.* Cambridge: CUP, 3–9; J.K. Hargreaves & R.D. Hunsucker, 2000. *The High-Latitude Ionosphere and its Effects on Radio Propagation.* Cambridge: CUP, 13–17; and S.A. Bowhill & E.R. Schmerling, 1961. 'The Distribution of Electrons in the Ionosphere'. In L. Marton, ed. *Advances in Electronics and Electron Physics.* New York: Academic Press, 265–326.

[167]S. Chapman (1931a, p. 27).

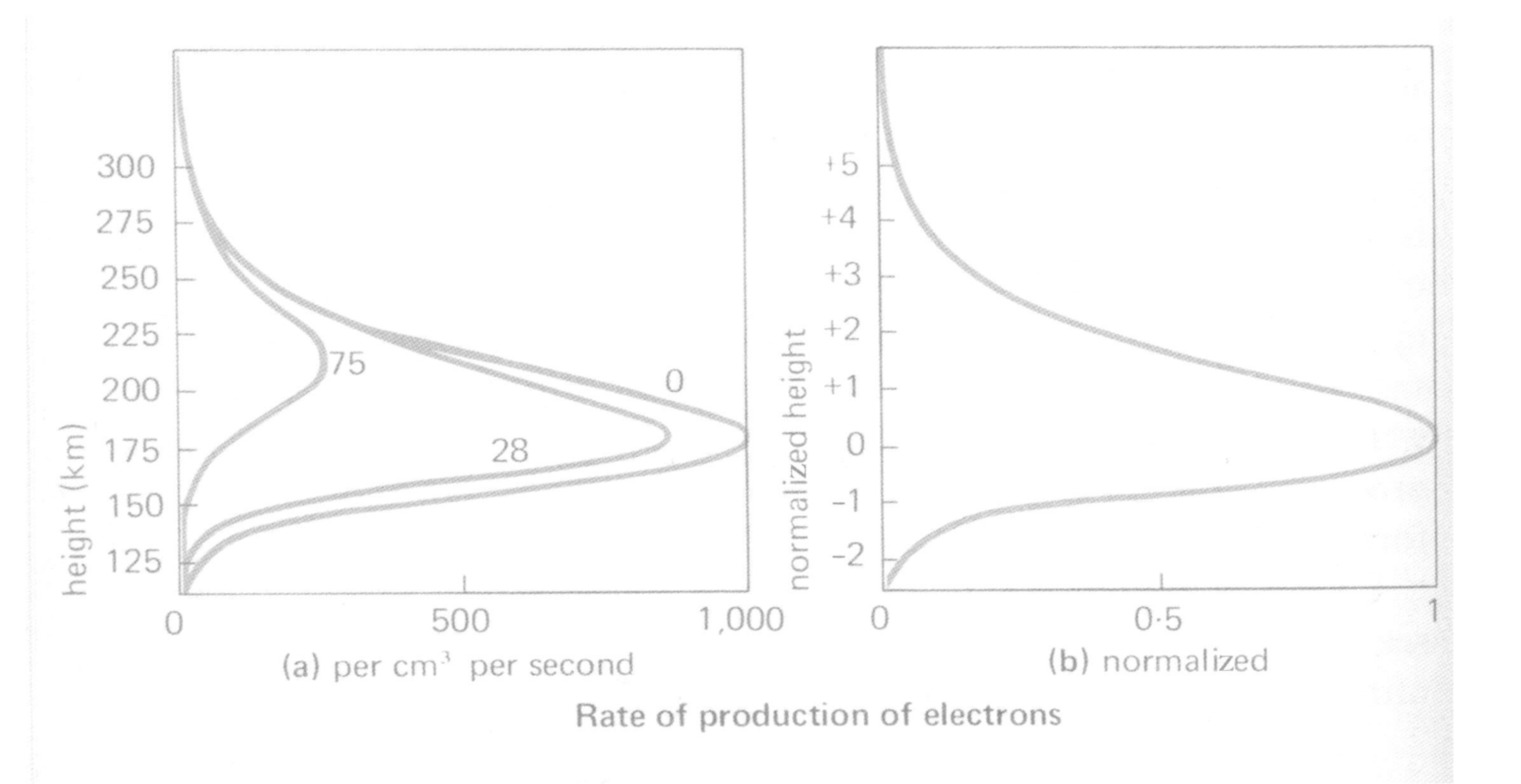

Fig. 9: A Chapman Layer. The rate of production of electrons calculated for a hypothetical layer in which the scale height is 25 km and the peak rate of production is 1,000 per cm3 per second at a height of 175 km when the sun's rays arrive vertically. The three curves correspond to different solar zenith angles. Source: Ratcliffe (1970), p. 94.

faithfully reproducing a phenomenon but rather for analysing it, explaining it (at least partially), and predicting its behaviour. Unlike many radiophysicists, for Chapman a model was a simple instrument of intelligibility without any ontological pretension, replaceable if one finds a better theory.

Chapman's equations explained everything about the production, recombination, and dynamic equilibrium of electrons, but they did not disclose the sources generating these processes. By 1931, the development of quantum theory had enabled many atomic physicists, such as E.A. Milne, H. Kramers and J.A. Gaunt, to calculate the absorption and photoionization coefficients for electrons and ions in ionized gases. Photochemical research gained strength; when a gas was ionized with ultraviolet light in the laboratory, the concentrations of the charged particles decayed as the light was removed. Physicists contrived simplified atomic models to account for the ion chemistry involved. The next step was to translate this into the upper atmosphere. Just as radio became an instrumental tool for experimentalists, so the ionosphere became (in Cambridge radiophysicist John Ashworth Ratcliffe's words) a 'low-pressure laboratory without walls' for theorists, a natural laboratory in which to confront ideas with data.[168]

Aeronomists, like atomic physicists, adored mathematical models: having regarded photoionization approaches as valid, Chapman applied them to the ionosphere—for reasons of simplicity, clarity, and economy, and doubtless influenced by atomic theoreticians. Thus, he propounded that solar photoionization (of ultraviolet origin) was the source of the F-region, while neutral corpuscular streams from the Sun (indispensable for his dynamo theory) were responsible for the E layer. The new model partially agreed with the measurements: the E and F_1 layers seemed 'to be a Chapman layer' (an expression by which he was amused), whereas the F_2 layer (and at times all layers) definitely was not. Model-building, as in atomic physics, led inevitably to an oversimplified, idealized image of nature that was not necessarily realist.

An upright man to erect the structure of the atmosphere

Sydney Chapman, chief mastermind behind geophysics in the twentieth century, was descended from a nonconformist family of strict principles. From his parents he imbibed the ideal of tolerance and a firm sense of rectitude, later re-expressed as pacifism (exempting him from military service during World War I), and finally, a code which put human beings, rather than God, at the

[168]For the concept of a very low pressure laboratory, see Ratcliffe (1970, p. 97) and Appleton (1937, p. 451).

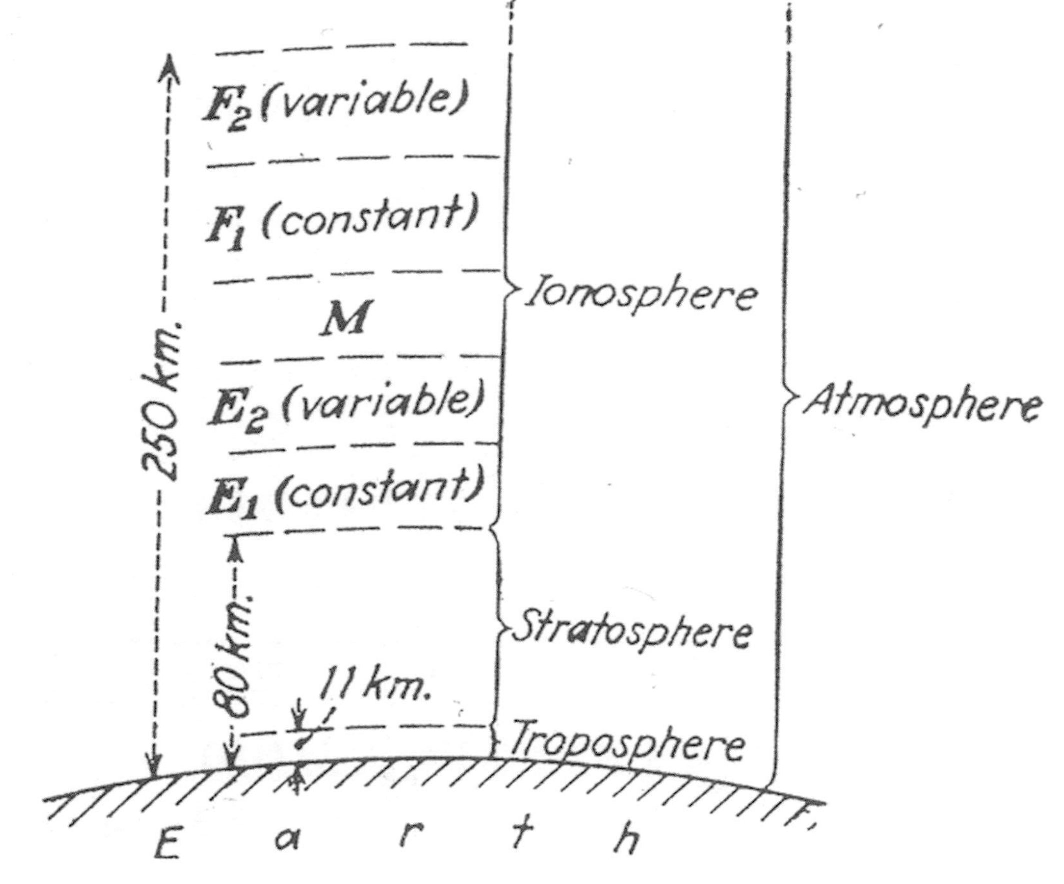

Fig. 10: Location of various ionized layers in the upper atmosphere. Source: Glasgow (1936), p. 492.

centre of faith and belief. During World War II, Chapman would forswear pacifism in order to combat fascism, and throughout his long life relentlessly pursued social justice and the ethical implications of science.[169]

Chapman's tolerant and open-minded attitude towards life extended also to science, in which he cultivated a wide and varied range of interests.[170] A few glimpses of this amplitude can be caught from the contributions he made between 1912 and 1919, probably his most creative period. During this time, he substantially improved the kinetic theory of gases, initiated his research into geomagnetism and lunar atmospheric tides, and developed the dynamo theory of diurnal variations (both S and L), to mention only the most significant.[171] Here, Chapman emerges as a brilliant, avid young man who focused on theory in order to counterbalance the dominance of observation in geophysics, and who confessed to feeling tremendously lonely in such a stance—unlike in his 'maturing' age around 1930.[172] Nevertheless, this solitude was a symptom of his pioneering spirit, which prompted him to open up new avenues in geophysics. As S.I. Akasofu, one of his foremost pupils, wrote in retrospect, 'as early as 1918, Chapman comprehended geomagnetic storms in almost the same way as we do today'.[173] Chapman regarded himself as more of a theoretician than most of his predecessors (such as Arthur Schuster, his professor of physics at Manchester, and, of course, Chree and Maunder). However, he was just as much an applied mathematician as a geophysicist. His particular virtue lay in his ability to formulate and express problems in mathematical terms while visualizing and analysing them from a physicist's perspective.[174]

Chapman's achievements in geophysics, which was always for him an extension of applied mathematics, came from his talent for simplifying a problem down to its bare essentials. As a rule, he would recognize a physical problem of key importance (for he had 'a physicist's sense of when to assert his intuition'[175]), idealize it (for 'mathematics has to be subservient to the needs of application'[176]), treat it fully, and express it clearly and precisely. His simplicity

[169]The most comprehensive biographical sources on Sydney Chapman are Akasofu et al. (1968), Cowling (1971), and Ferraro (1971). See also Ferraro (1968, pp. 1–5) and Good (1999, pp. 715–17).

[170]Bush & Gillmor (1995, pp. 1992–93).

[171]BLOA [CSAC, 11/5/74], Letter from T.G. Cowling to S.I. Akasofu, 7 October 1970.

[172]BLOA [CSAC, 11/5/74], Letter from S.I. Akasofu to T.G. Cowling, 13 October 1970.

[173]Akasofu (1970, p. 600).

[174]Kendall (1970, pp. 1871–2).

[175]S.G. Brush, Cambridge, Massachusetts, in S.I. Akasofu et al. (1968, p. 69).

[176]K.E. Bullen, Sydney, Australia, in , in S.I. Akasofu et al. (1968, p. 69).

in relation to concepts and theories was in harmony with his directness and spontaneity in his professional life—'pedestrian, though workmanlike' when lecturing[177]—and in his human relations—without artifice or pretence—; these aspects were often inseparable. His lectures were a combination of personal traits and scientific exposition. In a talk with the beguiling title of 'Dressing the bird, or adapting the problem', Chapman imparted a majestic lesson of candour and straightforwardness as he solved a highly complex problem 'by skilfully whittling it down to a manageable simplicity until the happy convergence of intuition, knowledge, and good luck that characterizes creative work.'[178]

The tidy and methodical nature of his mind was to appear to the eyes of his contemporaries as a great attribute for methodizing and structuring the upper atmosphere. One of his best pupils—and later colleague—, T.G. Cowling, stated that Chapman had creative, imaginative gifts, not only for mathematics but also for 'new and pungent words'.[179] His orderly mind was the only quality he needed to marshal facts, data, and concepts. Chapman once proposed abandoning the term 'meteorology', which seemed to him 'excessively polysyllabic', 'cumbrous to write, inconvenient in speech', in favour of 'aeronomy',[180] and replacing 'geophysics', too sibilant a word, with 'geonomy', which 'flows smoothly from the tongue'.[181] He was an inventor of words, a purist of language, a virtuoso in etymology. Words designated concepts, and these were fundamental in his elucidating task; he renamed *geomagnetism* 'terrestrial magnetism' and coined the terms *scale-height, stratopause, turbosphere, turbopause, thermosphere, mesosphere, mesopause, neutrosphere, thermosphere, protosphere,* and *heterosphere*, among many others.[182] Not all, of course, were accepted, but he was responsible for a large part of the common nomenclature.

Modesty, humility in his search for knowledge, absolute confidence in young researchers, respect for people and foreign cultures, and an exemplar of simplicity—with little regard for titles and honours—, these were Chapman's inherent characteristics. His austere and disciplined life, his continuing habits: walking, swimming (often in icy water) and his passion for travel (he was always ready to sleep in a sleeping-bag under the stars) and cycling (he rode

[177]Cowling (1971, p. 57).

[178]R.C. Tanner, Willington, Surrey, England, in S.I. Akasofu et al. (1968, p. 91).

[179]Cowling (1971, p. 73).

[180]S. Chapman, 1946. 'A Plea for the Abolition of "Meteorology"'. *Weather*, 1, 146–7.

[181]S. Chapman, 1946. 'Some Thoughts on Nomenclature'. *Nature*, 157, 405.

[182]S. Chapman, 1950. 'Upper Atmosphere Nomenclature'. *Jour. Geo. Res.*, 55, 395–9; *Jour. Atm. Terr. Phys.*, 1950, 1, 121–4 and 201; *Bull. Am. Met. Soc.*, 1950, 31, 288–90. See also H. Flohn & R. Penndorf, 1950. 'The Stratification of the Atmosphere'. *Bull. Am. Met. Soc.*, 31, 71–8, 126–30.

from Montreal to Washington in 1939 to attend the URSI meeting!) were legendary among academic circles. His numerous friends respected his integrity and his human dimension as much as they revered his intellectual capacity. To mention but one example, when his colleagues at the Universities of Alaska and Colorado paid tribute to him for his eightieth birthday in 1968, they eulogized not only his colossal oeuvre (seven books and over 400 articles) but also, and most specifically, his steadfast belief in humanity.[183] Chapman valued affability and directness as the requisite qualities to be cultivated by men, and above all, scientists.

His profound sense of humanity and his clarity of conscience rendered him, according to his geophysicist colleagues all over the world, an inimitable, unique personality. One of his many admirers, E.K. Smith, always had the impression that Chapman was a man who 'had experienced the joys of the world and its sorrows'.[184] For most of his geophysicist colleagues, the rare combination of scholar and humanitarian, the 'kindness and scientific probity combined with his eminent position in science', were exceptional traits. 'This man is made of modesty, honesty and integrity', stated V.V. Beloussov from Moscow;[185] 'he has the most sensitive social conscience of anyone I know', averred W.H. McCrea from Sussex.[186] Chapman excited admiration invariably linked with affection. 'A man of three-dimensional stature', 'an incorruptible man', among other epithets, a man—and this is less common among academicians—'who inspires those who have the good fortune to know him'.[187]

Nevertheless, it was not his human dimension or his feats of physical endurance that rendered Chapman the foremost authority in geophysics during the interwar period. His worldwide eminence derived from the rare condition of being a versatile mathematician who had solved major geophysical problems in several fields. For our purposes, however, Chapman's uniqueness lies in his unified picture of the atmosphere.

Synthesis and conciliation: the unified picture of a layered atmosphere

Chapman's concern with painting a coherent, cohesive picture of the atmosphere, which already manifests itself in 1920 in a joint work with

[183]Akasofu et al. (1968).
[184]Ibid., p. 89.
[185]Ibid., p. 67.
[186]Ibid., p. 83.
[187]P.B. Hays, Ann Arbor University, Michigan, Ibid., p. 76.

Cambridge mathematician E.H. Milne,[188] resulted from his pursuance of, and steadfast adherence to, a strict ethical code. It ensued from the incessant pursuit of the ideal of integrity, which in the personal realm denoted coherence and rectitude in his practices and beliefs, and in the scientific realm meant a conviction of the interconnectedness of all atmospheric phenomena and processes. The venture would inevitably lead him to pay particular attention to causes rather than to consequences. As E.N. Parker, an eminent astrophysicist, states, 'his interest in geomagnetic, ionospheric and auroral observations was only part of his interest in understanding their origin';[189] and the elucidation of their origin permitted the elucidation of their interrelationship. Chapman had previously endeavoured to ascertain the sources of ionization in ionospheric layers, and that was precisely the prime task in his theory of ozone formation.[190]

At first sight, one might be surprised at Chapman's concern about a then mysterious, unknown molecule concentrating below the upper atmosphere. Why was ozone so important? What rendered it essential? However, there was *one* fundamental reason for desiring to know the mechanism, or the chemical process, responsible for the production of atmospheric ozone. Advances in atomic physics, made largely with a view to astrophysical applications, showed that the amount of energy required to ionize, for example, helium, nitrogen, or hydrogen is much larger than that required for oxygen. Hence the former could scarcely contribute to ionization, while atomic oxygen, if present, would be the *only* constituent capable of yielding the enormous supply of ions necessary. A mechanism that explained the production, and therefore the existence, of atomic oxygen would be of vital importance in connection with the formation of ionized layers. For Chapman, the ozone layer seemed to constitute the missing link for the unifying aspirations of the upper atmosphere.[191]

Chapman traced the missing link to academic discussions of the upper atmosphere held at the Royal Society and Physical Society of London, altars

[188]An attempt to review an integrated picture of the outer layers of the atmosphere was made by Chapman & Milne (1920, pp. 357–9).

[189]E.N. Parker, 'Geomagnetism and Solar Physics', in S.I. Akasofu et al. (1968, p. 25).

[190]There are few texts including introductory chapters on the history of ozone. A very agreeable exception is A.Kh. Khrgian, 1973. *Fizika atmosfernogo ozona.* Leningrad, translated from Russian into English by D. Lederman, 1975. *The Physics of Atmospheric Ozone.* Jerusalem, 6–20, and in particular Chapman's theory, pp. 18–19.

[191]For a critical judgement on Chapman's theory, see Christie (2000, pp. 12–16).

and prime foci of science.[192] There, Dobson and Lindemann reawakened his interest, which later became a passion, in exploring the harmony reigning between the variability of ozone and the versatility of geomagnetic conditions.[193] The topic of ozone was by no means unheard of in British academic and institutional forums; since 1926, the Royal Society, with the help of the DSIR, had financed Dobson's network of photoelectric spectrophotometers over Europe (Switzerland, Germany, Sweden), then extended to the US and to the Empire (Australia, New Zealand, South Africa, India, and Egypt).[194] By 1930, Dobson computed from these daily measurements the distribution of ozone according to latitude and seasonal variations;[195] and at the invitation of F.W.P. Götz at Arosa, Switzerland, embarked on the task of establishing its vertical distribution. Chapman was therefore familiar with, and knew first-hand, the state of the question.

His aim for a unified picture and his idea of interconnectedness were no doubt what Chapman had in mind when he expounded at the Paris Conference on Ozone in 1929 his *theory of upper atmospheric ozone.* Over the next two years, in three brilliant papers, Chapman revised and improved his original scheme.[196] If a concentration of ozone gas (O_3) exists with a maximum level at a height of 40 to 50 km (as data showed), subject to annual and daily variations (as indeed it did), the gas will be in a chemical equilibrium sustained by five reactions, justly in accordance with the laws of chemistry: one reaction that would explain its formation during triple collisions ($O + O_2 + M = O_3 + M$), and another four for removal processes by reaction and by photodissociation. Moreover, the absorption of ultraviolet radiation will be responsible for the dissociation of molecular oxygen ($O_2 + h_v = O + O$). Since the photodissociation reaction ($O_3 + h_v = O_2 + O$) resulted in atomic oxygen, Chapman's successful treatment of the chemical equilibrium additionally provided crucial information about the ingredient sought for the ionosphere.[197]

[192]The formation of atmospheric ozone was debated by meteorologists (Dobson, Simpson) and Chapman. See 'Discussion on the Electrical State of the Upper Atmosphere', *Proc. Roy. Soc.*, 1926, 111, 1–13, esp. pp. 11–13. See also 'Discussion on Ultra-Penetrating Rays', *Proc. Roy. Soc.*, 1931, 132, 331–52.

[193]A failed attempt to connect magnetic activity and the amount of ozone was Chree (1926). See also, Dobson, Harrison, & Lawrence (1926, 1927, 1929).

[194]On Dobson and his international network on ozone research, see: Dobson (1966, pp. 391–97).

[195]Dobson (1930).

[196]S. Chapman (1930a, 1930b, 1930c).

[197]For schematic account of Chapman reactions, see Wayne (1985, pp. 114–18).

And it did more. From observations of oxygen absorption lines in the upper atmosphere, J.C. McLennan at the University of Toronto had proved the presence of atomic oxygen in the auroral regions above 100 km and in the airglow at an unknown height in 1927.[198] Chapman therefore could now assign its origin to photodissociated oxygen; and this, in accordance with his calculations on ozone formation and destruction by ultraviolet radiation, should 'exist mainly mixed with O_2 and a far smaller proportion of O_3' above 100 km. Thus, Chapman predicted that in the ionosphere, oxygen was largely dissociated ('a conception then entirely novel'[199]), and introduced the problem of the ratio of atomic and molecular oxygen ('a classic aeronomic topic').[200] Atomic oxygen, opportunely generated in Chapman's photochemical theory, provided a clue as to the connections between the ozone layer and the ionospheric layers or, to use the terms of that time, between the lower and upper atmosphere. It was an unparalleled achievement.[201]

In June 1931, in a featured address to the Royal Society of London, Chapman resumed the problem of unity and the need for synthesis. The setting was the illustrious and respected Bakerian Lecture, which, despite having hitherto witnessed numerous dissertations on physics, had never dealt with geophysics.[202] Chapman's objective was not to persuade the auditorium of the realist nature of the ionized layers but rather to convey to physicists that the recent advances in the field encompassing geomagnetism, radio propagation, spectroscopy, and atomic physics, argued for an endeavour to attain the unified atmosphere. The theories of the formation of ozone and of the ionized layers (culminating in the standard Chapman layer) emerged as the building blocks of his physical model; and the dynamo theory essentially characterized the geo-magnetizing component that, according to him, is required for any sound knowledge of the atmosphere.

Chapman's physical model was not entirely original but rather the synthesis of elements of knowledge eventually coalescing into a single unity. 'There are'—thus began his abstract, in his usual clear and precise style—'three layers in the upper atmosphere in which dissociation is produced by the absorption of solar radiation. These are the layer of ozone, with its maximum concentration

[198]McLennan (1928).

[199]B. Haurwitz & B. Fogle, 'Meteorology', in Akasofu et al. (1968, p. 12).

[200]M. Nicolet, 'Ionospheric Physics and Aeronomy', in Akasofu et al. (1968, p. 36).

[201]On ionospheric photochemistry, see: Wayne (1970, pp. 216–24).

[202]S. Chapman, 1931b. 'Some Phenomena of the Upper Atmosphere'. *Proc. Roy. Soc. Lon.,* 132, 353–74.

at about 50 km, and two ionized layers at about 100 km and 220 km'[203] There is also—and this is new—'a region, of unknown height, where oxygen exists in a metastable atomic state', in which aurorae are displayed. However, the mode in which the pieces were articulated and interconnected was original: 'the dissociation of molecular oxygen which results in the formation of the ozone layer has an important influence on the whole of the overlying atmosphere.'[204] The basic ideas were simple, so too the results, which even seemed obvious, but, as Cowling stated, 'it was obvious only because Chapman had set it out so clearly'.[205]

Chapman resumed, and extended, his mathematical theories of layer formation. Here, in contrast to the physical models of the early twentieth century, which were mainly explanatory and largely speculative, his model provided qualitative and quantitative predictions. Moreover, it fitted the radio data, especially those pertaining to the lower layer, but also some from the upper layer too: according to our estimates, he concluded, 'the atomic oxygen will absorb most of the ionising ultra-violet radiation, at heights not much above 200 km.', which 'is in fair agreement' with Appleton's measurements.[206] In light of the necessity and exigency—and both words are apt—of adjusting to radio measurements, Chapman became a forthright opponent of those few—mainly geomagneticians—who still doubted that radio was vital for the advance of atmospheric understanding.

Hitherto the synthesis. Next came the conciliation. In trying to determine the ionizing agent of the lower layer, Chapman based his studies on the evidence of magnetic variations. He interpreted that the two fields responsible for the solar and lunar daily periods (S and L) were produced in different layers, and ionized by different agents. The dynamo theory of L requires a certain electrical conductivity of the layer in which the currents flow, he said.[207] Whereas conductivity estimates from radio measurements were ambiguous, magnetic restrictions circumscribed the region associated with L to a height of 100 km. Chapman identified it with the lower layer (or E layer) in an attempt to conciliate both sets of data. Furthermore, he propounded that this zone must be ionized by solar corpuscles, in conformity with Milne's theory, and not

[203]Chapman (1931c).

[204]Ibid., p. 464.

[205]Cowling (1971, p. 68).

[206]Chapman (1931c, p. 365).

[207]Chapman (1931c, p. 366).

by ultraviolet radiation, as in the upper layer.[208] Here, he encountered strong opposition from Appleton, who proposed that photons, not corpuscles, were the causative radiation (a fact demonstrated during the solar eclipse of 1932).[209] For a time Chapman remained reluctant to accept Appleton's proposal. It threatened his conviction regarding the geomagnetic influence. Nevertheless, his unified conception remained intact.

[208]For indications favouring the ionization by neutral atoms from the Sun, see E.A. Milne (1926). See also Massey (1974, p. 2,147).

[209]Appleton & Naismith (1932); Massey (1974, p. 2,147); Ratcliffe (1970, pp. 143–6). For the definitive test carried out in Canada, Henderson (1933).

2

RADIO COMMUNICATIONS, GEOPOLITICS, EDUCATION, AND MANUFACTURING IN THE BRITISH RADIO INDUSTRY

Introduction

In the peroration of his article on research and the British firms of the interwar years, Michael Sanderson—the author of the unparalleled study *The Universities and British Industry, 1850–1970*—bemoans that the initiative and resilience showed by the latter have not been appreciated by historians and contemporaries.[1] In attempting to understand this inconsideration, Sanderson alludes to a vision among historians 'understandably overshadowed' by the darker chimerae of depression, unemployment, and the ineptitudes of public policy, but also to a distorted picture which 'virtually ignores, belittles or even slanders the vast bulk of industrial research' carried out within the firms. To tell the truth, these distortions and their variants, e.g. entrepreneurs' lack of enthusiasm for science and research, derive from a set of clichés that characterizes post-1870 Britain and her often mentioned 'economic decline'.[2] This chapter will try to redress that injustice.[3]

From the perspective of the foregoing chapter, the experience of state-funded radio research and development would seem to suggest that in its solidity and cohesion might be discerned the structural foundations of Britain's advanced position in upper atmospheric sciences. In fact, a hypothetical 'declinist' might assent to this impression from his orthodox view on the subject: he would argue that this considerable rise of research emanated almost exclusively from the DSIR and its attendant Research Associations, and that even the supposed research promoted from within the firm in great part had a state origin, in clear consonance with the plethora of information extant about

[1]Sanderson (1972b, p. 151).

[2]For a rigorous critical review of the misconceptions and misinterpretations of the 'declinist' positions, with a selective and commented bibliography, see David Edgerton, 1996. *Science, Technology and the British Industrial 'Decline', 1870–1970*. Cambridge: CUP.

[3]There is a vast literature on Britain's industry in the interwar years. See, for example, Buxton & Aldcroft (1979); Aldcroft (1966); Hobsbawm (1987, pp. 174–211, as well as the chapter 4 'Society since 1914,' pp. 233–51).

the public sector. *Per contra*, in a field as innovatory as radio's, a heterodox critic would tend to overemphasize the role of the individual inventor attached to a company but likewise detached from any business strategy, underestimating moreover state participation. Both judgements, we believe, are erroneous. Next we shall attempt to show that the engagement of the private industry towards not only research and innovation but also technological education was more than satisfactory, and consequently it is necessary to dispense with putative declining entrepreneurship, its tactlessness with respect to for exports and patents, or insensitivity to innovate in order to explain plainly present advancements.[4]

Even more, we are convinced that such advancements would be clearly inexplicable without, and this must be underlined, the existence of an entrepreneurial, rather than government, commitment to the issue of technical education and instruction. Strikingly, this anticipated perception dissents from what has come to be seen as one of the essential causes of the failure of highly technical teaching in Britain, that is, the lack of industrialists' demand for a technically trained labour force (besides the government heedlessness).[5] The accusations of apathy, indifference and insensitivity to the education issue have come, overtly at least, not only from historians but also from contemporaries, especially from those engaged in the government. Thus, in 1919 the Board of Education had not doubts in pointing to the culpable: 'The root cause of the unsatisfactory condition of the system of technical instruction is to be found in the small and feeble response which the Education Authorities have received from the industries they have tried to serve. There has been no organized demand from the industries for the provision or betterment of technical instruction. Employers as a whole have not only taken no trouble to consider and formulate their needs, but have not even realised the necessity of technical instruction.'[6] Here, we shall try to demolish those arguments.

In the preceding chapter we have analysed from several directions the traditions, origins and motivations which distinguished state-funded upper atmospheric radio research in civil and military centres. But a quest for structural

[4]On technological education and industry, the aforementioned M. Sanderson, 1972a, *The Universities and British Industry, 1850–1970*. London: Routledge & Kegan Paul, 243–313. See also Divall (1990). For comparative studies, see Lundgreen (1990) and Ahlstrom (1982).

[5]For a study against this standard view, Marian Bartlett, 1995. *Education for Industry: Attitudes and Policies Affecting the Provision of Technical Education in Britain, 1916–1929*. Oxford University, Ph. Diss., esp. chapters 3 and 4. See also Nae-Joo Lee (1992).

[6]PRO/ED [24/1863], 'Proposals for Developing the National System of Technical and Commercial Instruction,' 1919, p. 2—quoted in M. Bartlett, op. cit., p. 89.

reasons which may foment and enable that research is not completely defined by a specification of such substantive components, not even when the repertory of characteristic features includes the meticulous depiction of the personalities who most contributed to that enterprise. To fully ascertain the structural foundations one must take into consideration the commercial, industrial, and innovative aspects of radio, the accepted view of the contemporary socio-economic situation, and the business strategies to which that situation gave rise. One must, moreover, specify the nodal and focal points, the intersections and interstices, the connective tissues and confluent lines between radio industry, technical education, and academia.[7]

The imperial chain

There is a certain tendency, especially among the historians of radio industry, to analyse the progress of the sector as a lineal sequence of inventors, devices, and firms that takes place decorously within an unbiased and unproblematic framework. Habitually, there are reasons for this practice. It is used to show the high degree of innovativeness and the revolutionary spirit of wireless technology, usually with special reference to continuum wave and vacuum valve. The approach, without doubt, well serves the objective designated, but unfortunately its upshot is fragmentary and equivocal. For while it is true that the telegraph was mostly welcomed as a legacy peaceful by nature, radio, in contrast, emerged as a deterrent weapon in the aggressive ambiance of the great military and economical powers.[8]

This disparity between telegraphy and radiotelegraphy became even more pronounced in the war and following the new balance of power in the post-war years. In both its commercial and military guises, wireless was perceived as a potentially powerful instrument to consummate imperial dreams and thwart colonies and nations. It was dissension rather than peacefulness which

[7]Essential readings on research and development in British industry in the interwar years include R.S. Sayer, 1950. 'The Springs of Technical Progress in Britain, 1919–1939'. *The Economic Journal*, 60, 275–91; S.B. Saul, 1979. 'Research and Development in British Industry from the End of the Nineteenth Century to the 1960s'. In T.C. Smout ed., *The Search for Wealth and Stability*. London: Macmillan, 114–38; D.E.H. Edgerton & S.M. Horrocks, 1994. 'British Industrial Research and Development before 1945'. *Economic Historical Review*, 47, 213–38; and David Mowery, 1986. 'Industrial Research in Britain, 1900–1950'. In R. Elbaum & W. Lazonick eds. *The Decline of the British Economy*. Oxford: Clarendon Press.

[8]For an excellent study of the growth of cable and radio technologies within a global political context, see D.R. Headrick, 1991. *The Invisible Weapon. Telecommunications and International Politics, 1851–1945*. New York: OUP. See also Hugill (1999).

emerged as the dominant feature not only in radio communication but in geopolitical strategies generally.[9] It would be rather odd if it had not been so. On the eve of World War I, the great world powers jealously preserved their telecommunication networks.[10] Through their excellent, if conservative telegraphic services, France and Germany laboured vigorously endeavoured vehemently for a national scheme of radio communication (a substitute for the cables they lacked). Against them lay their weak position in the field of long-distance radiotelegraphy; both countries had several regional networks scattered around their colonies (usually linked by British cables), but suffered from a lack of real global network.[11] In this race, the USA was the most latent menace. The Navy's resolute commitment to continuous wave operation had secured radio communication with its colonies and all its warships. The US Navy, Federal Telegraph, GE, and AT&T were only four of several rival corporations which, already by 1914, began to threaten Marconi's efforts in demonstrating the oneness of the British Empire.[12]

Marconi's was one of the wonders and bulwarks of the Empire's industry.[13] Its assets had grown spectacularly from the handful of firms acquired at the time it was created in 1900 to its preponderance in the international market in 1910. Besides its privileged position in the Empire, it operated shore stations throughout the US and Latin America; it dominated Italian radio and the wireless traffic through the English Channel following its agreement with Telefunken in 1911; and owned an important share of most European companies, among them the French Compagnie Générale de Télégraphie Sans Fil (CSF).[14] Its leadership was materially strengthened even further by the bankruptcy of its principal American rival, United Wireless, in 1912, its capital being

[9]S. G. Sturmey, 1958. *The Economic Development of Radio.* London, esp. chapters 3 and 4. See also Barty-King (1979, pp. 141–256) and Wedlake (1973).

[10]Headrick (1988, pp. 130–7) and Girardeau (1951, pp. 1–3).

[11]Pickworth (1993, pp. 427–32) and Griset (1989, pp. 24–30).

[12]Indispensable readings on American radio are Hugh G.J. Aitken, 1985. *The Continuous Wave: Technology and American Radio, 1900–1932.* Princeton: Princeton U.P.; and Susan J. Douglas, 1987. *Inventing American Broadcasting, 1899–1922.* Baltimore: Johns Hopkins U.P.. See also Howeth (1963, chapters 12 and 13) and Brock (1981, pp. 165–76).

[13]The standard history of the Marconi Wireless Company is W.J. Baker, 1970. A *History of the Marconi Company, 1874–1965.* London: Methuen. See also R.N. Vyvyan, 1974. *Marconi and Wireless.* Yorkshire: E.P. Publishing; and Harry Edgar Hancock, 1950. *Wireless at Sea: The First Fifty Years.* Chelmsford: Marconi International Marine Communication Co. For the early years, see Pocock (1988), Donaldson, (1962), and Guagnini (1994).

[14]Jome, H.L., 1925. *Economics of the Radio Industry.* New York: A.W. Shaw Co., pp. 20–30.

acquired by the American Marconi's. At such a juncture, the Company reached the pinnacle of its power.[15]

Yet, in these first years the British Post Office, one of Marconi's increasingly numerous enemies, protested against its limitless ambitions and aggressive tactics in the only way circumstances permitted: restriction of licences for radio stations and control of wireless traffic. The argument was familiar: the private firm was a potential competitor of the state telegraph industry; its monopolistic position inconvenienced; and perhaps most important, its reputation was synonymous with substitute for cables. The argument raised vacillating reactions on the part of the British Government, however. For, although the Company endangered the monopoly of submarine cable system on which Britain's security apparently depended, neither deserved to be abandoned to fate. After all, both Marconi's and the Post Office aimed for global communication and Empire's modernization, and were sensitive to commerce, research, and imperial defence.[16]

The changing nature of the relations between Marconi's and the British government in the field of long-distance communications is reflected by what was known as the imperial chain. By 1910, Marconi had presented to the government several proposals to build a series of stations throughout the Empire, which were rejected as being too novel.[17] The question of radio communication and the advisability of gaining independence from submarine cables were resumed at the Imperial Conference of London in June 1911. Here, disaccord was the common denominator.[18] Whereas the Post Office defended a government-controlled service, other voices, such as that of the premier of New Zealand, Joseph Ward, preferred a joint solution that would permit Marconi's to built stations and the Post Office to operate them.[19]

With the imperial chain, which marked the beginning of an alternative strategy for cables, emerged a new correlation of forces originated

[15]James G. Harbord, 'Radio in World Communications', in Codel (1930, p. 98); and Harvard University, 1928. *The Radio Industry*. New York: A.W. Shaw Co., 75–86.

[16]On British government support to the submarine cables, see Coates & Finn (1979) and Garratt (1950).

[17]NAUK, [WO 32/7044], 'Report of cables (Landing Rights) Committee on Marconi Company's proposal to establish a network of wireless stations throughout the Empire', Code 76(A), 1910.

[18] 'Minutes of Proceedings of the Imperial Conference, 1911', in Great Britain, *Parliamentary Papers,* 1911, 54: 307–15.

[19]NAUK, [CAB 37/107/63], General Post Office, 'Scheme for Imperial Wireless Stations', 26 May 1911.

by the rising sector of the radio industry.[20] By 1913, Marconi's submitted a new joint scheme that was to be approved by the Committee of Imperial Defence, even though it did not meet completely the *senatore*'s ambitions. The new contract, subscribed with the Post Office in July 1913, foresaw the construction of six stations in England, South Africa, East Africa, Egypt, India, and Malaya or Singapore.[21] Subsequent war events mostly aborted these attempts.

By the beginning of the war, the government's defence policy assumed a character for which the desire to encourage wireless and at the same time to preserve an existing cable network all had their share of importance in an integrated system of communications.[22] So long as Britain could have the privilege of choosing (opting for submarine cables, for example, if secrecy was required, or taking multiple routes according to circumstances, but never at the expense of weather or foreign companies), agreement seemed worthwhile.[23] In certain respects, Britain regained, if not a supremacy, at least a leading position, for the radiotelegraphy was more effective and of greater scope, and, as a result, the Admiralty and the British Merchant Navy could communicate with their ships everywhere.[24] But it is also true that compared with the American wireless high presence in the Pacific and the Caribbean, the British response seemed to be tardy and, because of scandals of corruption and stock market rigging in parliament, rather controversial.[25]

Despite the agreement of July 1913, it was only after the war, with the new alignment of powers, that the shortages of the imperial chain began to be realized. Then, with Germany defeated, deprived of colonies and relegated to the running of purely commercial radio, but with the US clearly ahead and France following in its footsteps, Britain's hegemony survived with difficulty.[26] French radio stations were highly efficient, and a balanced policy based on agreements

[20]Vyvyan (1933, p. 71); Hugill (1999, pp. 97–101); *Electrician*, 1912, 68, 905, 69, 414–15 ; 'The Imperial Wireless Scheme', *The Wireless World*, 1913, 1, 256–8, 308–12, 359–60; 1914, 2, 60.

[21]NAUK, [T1/11582], 'Post Office. Agreement between the Marconi Wireless Telegraphy Co., Ltd., and the department for the erection of a chain of wireless stations throughout the Empire', 1913. See also [WO 32/7045], 'Establishment of chain of wireless stations throughout the Empire: Copies of correspondence relating to contracts. Reports and Agreements with Marconi Company. Selection of sites, with plans', 1911–1912.

[22]One of the best summaries of this story is in G.S. Shoup, 1928. 'Wireless Communication in the British Empire'. *Trade Information Bulletin*, 1928, 551, 1–28.

[23]Geddes (1974, pp. 29–30).

[24]Wedlake (1973, p. 84).

[25]Headrick (1991, p. 1319).

[26]Ristow (Ebering, 1927).

with the CSF permitted a rising flow of direct communications with their colonies on a scale, and with a quality, that refutes any notion of French radio backwardness after the war.[27] Part of the efforts were directed to building high-powered transmitters (the Saint-Assise station near Paris, and another three in Beirut, Bamako, and Tananarive), which allowed French radio to cover one third of the world. The US, for its part, had no competitor in the field of high-powered radio or in financial potential, although its radio industry seemed to be always embroiled in political strife.[28] Important corporations like the RCA and the Navy, with the support of their own State, invested in advanced radio technology, with the aim of expanding trade and American influence around the world, and thereby countering the British cable hegemony.[29]

After the war, the British government tried to create a *rapprochement* between cable and wireless interests. This was made urgent by the fact that, with the growth of the war-time volume of traffic, the cable system had been unable to operate efficiently. In July 1919, the Postmaster General obtained the authorization from the House of Commons to initiate an imperial radio service for commercial use equipped with a technology—the Poulsen arc—that had been removed because of its obsolescence in France, for example.[30] Under the expertise of Post Office engineers, the arc, now somewhat modified, accommodated itself to the requirements of the stations at Leafield and Cairo, while in fact it did not allay critical voices. In the meantime, American companies introduced huge wave continuous generators and oscillating triode vacuum tubes in their stations.[31] The combination of technical innovation and political wisdom was vital; for the years 1918–23 were a period of great dynamism in the radio industry.[32]

What had begun as a measure against the inadequacy of cable system had grown into a dispute about the most appropriate radio technology generally.

[27]René Duval, 1980. *Histoire de la Radio en France.* Paris: Alain Moreau, 28–9.

[28]According to David Sarnoff, President of the RCA, the country went from being 'a mere communications tributary' to the British cable systems to occupy 'an undisputed position of leadership in radio communication'. See D. Sarnoff, 'Art and Industry', in Codel (1930, p. 186).

[29]David Sarnoff, 1928. 'The Development of the Radio Art and Radio Industry since 1920'. In E.E. Bucher et al., *The Radio Industry. The Story of its Development.* Chicago: A.W. Shaw Co., 97–113, p. 101–2; Jome (1925, pp. 44–65); Aitken (1985, p. 25).

[30]Baker (1970, p. 207).

[31]W.H. Eccles, 1926. 'Wireless Development since the War'. *EW & WE*, 3, 740–2; Aitken (1985, p. 25).

[32]The antagonism between Marconi's and the Post Office characterizes the contemporaneous accounts of the empire chain. For a pro-Post Office approach, see Iles (1927). For a more personal and passionate view, see Marconi (1928, pp. 51–6).

The question was not only a simple gloss on the proceedings of Marconi's versus the Post Office. The problem, paraphrasing historian Paul Kennedy, lay also in the 'imperial overstretch'.[33] In fact, the increasingly manifest disproportion between Britain's existing communication facilities and its imperial responsibilities had already been a matter of preoccupation for the Empire Press Union in 1919. Under this pressure, the government designated to Sir Henry Norman the preparation of a scheme for imperial communications in November 1919.[34] The attitude of the allotted Norman Committee proved prejudiced, especially when associated with Marconi's. The Committee served the Post Office's interests in three ways. First, it recommended a mixed imperial chain equipped with valves and arcs to be constructed and operated only by the Post Office.[35] The plan seemed unrealistic, for the supply of equipment could not be concretized without Marconi's participation.[36] Second, it rejected a new scheme devised by Marconi's without taking into consideration the technical advances—such as the valve transmitters and Franklin aerials—that the Company had hitherto accomplished.[37] As Marconi's engineer R.N. Vyvyan recalled in 1933, 'the report was a very bitter attack on private enterprise…, contrary to the advice of previous commissions'.[38] The decision deprived Marconi's of the monopoly that such a scheme might confer.[39] Finally, Henry Norman, widely praised by the circles of the cabinet, deflected in theory away from trading purposes, while providing the Post Office with a fruitful succulent market in which to consummate its commercial ambitions. What the cabinet called 'a detached scientist', in the eyes of Godfrey Isaacs, Marconi's manager director, was 'a keen commercial competitor'.[40]

In the early 1920s, one might think that the objective of Marconi's was to utterly control the Empire's communications by abusing patents and monopoly concessions, while that of the Post Office was to obstruct it even against the Empire's interests. This seems a coherent argument but is not the whole truth: 'both strategical and commercial requirements' coloured the grey landscape of

[33]Paul Kennedy, 1987. *The Rise and Fall of the Great Powers: Economic Change and Military Conflict from 1500 to 2000.* New York—quoted in Headrick (1985, p. 185).

[34]Shoup (1928, pp. 2–3).

[35]Sturmey (1958, pp. 104–7).

[36]'The proposed Imperial Wireless System', *The Wireless World*, 1920, 8, 41–3.

[37]NAUK, [POST 88/38], 'Marconi Wireless Telegraph Company's proposal for wireless network to serve British Empire', 1920.

[38]Vyvyan (1933, p. 74).

[39]'Imperial Wireless Communication', *Electrical Review*, 1920, 86, 323.

[40]'Mr Isaacs and Imperial Wireless', *The Times*, 4 March 1920. See also Isaacs (1923).

the imperial communication network. Of all requirements, Eccles stated, those of the Admiralty were 'the most stringent, for they desired quick communication with every naval base in the Empire'.[41] 'Wireless was not ordinary wire telegraphy', argued on the other hand Lieut. Colonel W.A.J. O'Meare; it dealt with a strategic affair requiring the fighting forces' involvement, rather than the Post Office's.[42] Far from diminishing, pressure on 'the Empire Wireless Chain' was accentuated.

Whereas Britain assessed geopolitical and monopolistic aspects, the Dominions and colonies were exasperated with the procrastination in the establishment of wireless services.[43] 'The impatience if not irritation' was ostensible and generalized.[44] Communication was an instrument against isolation, but also the principle of dignity versus servitude. In the words of André Touzet, professor at the University of Hanoi, 'no matter how profound is our friendship with the English, the subjection to which we were reduced in having to turn to foreign intermediates for our communications with the metropole was somewhat humiliating for our national self-esteem'.[45] While bestowing command on Britain, the existing chain system was inimical to the periphery's interests. That is why the Norman scheme, endorsed at the Imperial Conference of London in June 1921, was vigorously rejected by the Dominions and then disesteemed by the Imperial Communications Committee itself.[46]

Between 1921 and 1924, the fuse lit by complaints in the Dominions was kept alight. Gradually, both the conservative Unionists and the first Labour government had moved towards more flexible, resolute positions through the appointment of advisory boards, such as the Wireless Telegraphy Commission in 1922 and the Imperial Wireless Telegraphy Committee in 1924,[47] whose reports were debated in parliament. All recommended high-powered state-run stations, but refused a pact with Marconi's, despite the fact that this held almost all of Britain's patent rights. These negotiations were slow and tortuous. Furthermore, they must be conciliated with the newly demanding

[41]W.H. Eccles, 1922. Imperial Wireless Communication. *Jour. Roy. Soc. Arts*, 70, 509–22, p. 512.

[42]Ibid., p. 521.

[43]Shoup (1928, pp. 3–4); Sturmey (1958, pp. 108–11); Barty-King (1979, pp. 183–5).

[44]NAUK, [POST 88/41], 'Imperial Wireless Telegraphy Committee', Robert Donald, Chairman, 1924, 2–3.

[45]André Touzet, 1918. 'Le réseau radiotélégraphique indochinois'. *Revue Indochinoise*, 245, 7–21, p. 21—quoted in Headrick (1985, p. 187).

[46]NAUK, [CAB 32], 'Records and Minutes of the Imperial Conference, June 1921'.

[47]NAUK, [POST 88/40], 'Wireless Telegraph Commission, First Report (Cmd. 1572), Viscount Milner, Chairman', 1922; NAUK, [CAB 27/240], 'Cabinet Committee on the Report of Imperial Wireless Service', 1924.

cabinets of Australia, New Zealand, Canada, India, and South Africa individually.[48] As a result, the imperial chain reached stalemate.

Meanwhile, the gap between its most direct rivals grew wider: the US had 21 high-powered stations and a total power of 3,400 kilowatts, France had 12 and 3,150 kilowatts of power, whereas Britain only had three and the low figure of 700 kilowatts.[49] In this context, the Post Office planned the erection of a super-powered radio at Rugby that would be the only valve-transmitting station in the world, regarded as the quintessence of the imperial scheme by the government, but as an expensive white elephant by detractors.[50] In the light of these data, what would be the future scheme for communications with the Empire—direct or chain-linked, imposed or consensus, state or private; and as *The Times* editorial questioned, predicated on cables or wireless?[51]

Shortwave and Empire's unity

At this point it may be useful to recapitulate. Britain remained in a leading and influential position because it had the hegemony of the world cable network, a strategic and commercial capacity for radio communication, and an Empire through which it could grow. But its wireless power developed in parallel with, albeit falling behind, its main rival, the US, met neither exigencies of the Dominions, nor needs of the metropolis.

At the beginning of 1924, the procrastinations in long-distance communications appeared more likely to confirm the pessimism in the Dominions than to vindicate the Post Office's confidence in the power of long waves. It was so until the irruption of a new interloper, the short wave. In May 1924, Marconi's announced a revelation to the world: the completion of a system for direct communication by means of short wave.[52] The beam system—so called because of its directional antennas—embodied the purest expression of the commercial side of scientific research.[53] For Marconi's, the critical distinction was not

[48]Shoup (1928, pp. 4–10).

[49]Charles Bright, 1923. 'The Empire's Telegraph and Trade'. *Fortnight Review*, 113, 457–74—quoted in Headrick (1985, p. 184).

[50]NAUK, [WORK 13/1281], 'Rugby Post Office Wireless Station. Erection, Chessums Ltd., Dec. 1923'; Sturmey, (1958, p. 115); Vyvyan (1933, pp. 172–3); Johnston (2002).

[51]*The Times*, December 1922, quoted in Barty-King (1979, p. 185).

[52]Short-wave messages transmitted from Cornwall were received by Marconi's engineers in Australia, India, South Africa and America. Marconi publicized these successes at the Royal Society of Arts in London and in Rome. See Marconi (1924a, 1924b).

[53]For Marconi's beam system, see Marconi's Wireless Telegraph Co., n.d. *The Marconi Beam System for Long-Distance Communications*. London: Marconi House; Morse (1925); Headrick (1994,

between science and business but between *short wave* and *long wave*. Short wave involved placing the radio 'on a par with cables for the first time';[54] to challenge the long-held concept of 'long wavelengths plus high power equals long ranges'[55] with its corollary of stations of vast size and expense; but it also involved adopting 'enormous risks and [giving] guarantees of performance'[56] for a system that had never been used on any commercial circuit. Long wave, by contrast, was equated with 'a blind alley',[57] with enormous costs and high power requirements. It symbolized the past and obsolescence.[58] The Labour government accepted the superiority of the beam system at the expense of long waves—in the face of Post Office opposition—as the safest measure for Empire's defence and unity. And in the Dominions, with the erection of short- wave stations by Marconi's in Australia, Canada, South Africa, and India, there was a clear preference for direct communication, an air of contentedness, a sentiment of complicity.

Before 1920, short-wave range had been a waveband of marginal rather than prime interest. This was largely due to three reasons: first, physical reasons—its high attenuation at short distances suggested that the same might happen at long distances—;[59] second, technical reasons—the impossibility of generating high frequencies with the current arcs and alternators and the low sensitivity of receivers—;[60] and, third, because of the nature of its rival, the long wave, as an instrument for aid to navigation—an ordinary aerial was more satisfactory than a directed beam, since it emitted signals in all directions.[61] This was perhaps the most important of all, for the exploration of low frequencies and the design of equipment for communications between ships and between ship and shore was to become Marconi's most imperative task. In time this commercial strategy proved counter-productive. For short-wave tests demonstrated that receptors installed in ships were able to pick up signals

pp. 22–4); Vyvyan (1933, chapter 8); Baker (1970, chapter 26); Shoup (1928, pp. 10–21); Iles (1927, pp. 275–80).

[54]British Information Services, 1963. *Britain and Commonwealth Telecommunications*. London, in Barty-King (1979, p. 193).

[55]Baker (1970, p. 212).

[56]Vyvyan (1933, p. 83).

[57]Digna Marconi, 1962. *My Father Marconi*. London: F. Muller, 215.

[58]For the advantages of short wave, see Kintner (1925), Hallborg, Briggs & Hansell (1927, pp. 467–69), and Hallborg (1927, pp. 501–2).

[59]J.A. Fleming, 1925. 'Propagation of Wireless Waves of Short Wavelength Round the World'. *Nature*, 115, 123.

[60]Morse (1925, pp. 88–90).

[61]Marconi's Wireless Telegraph Co., op. cit., p. 4.

reappearing at distances of 2,300 kilometres during the day and over 4,000 at night. Short wave united audibility with commercial viability.[62]

Yet this apparently sudden appearance of short wave has to be taken with caution. Three qualifications are necessary. First, it was a micro-invention, or, in other words, a new development of an established technology, rather than an invention *per se*.[63] Marconi himself had demonstrated that short wave could be directed in a beam and concentrated in a parabolic reflector as early as 1896.[64] Secondly, the Company's commitment to short wave research derived from necessities of war. In 1916, Marconi had encouraged one of his most qualified engineers, C.S. Franklin, to re-explore the use of short wave with the aim of eliciting appropriate instruments for transmission, incapable of being intercepted by enemy.[65] In the next year he succeeded in generating wavelengths of three metres with the use of thermionic valves.[66] Moreover, Marconi's engineers linked beam transmission with ionospheric propagation. In the early 1920s, the transmitter at Poldhu station and Marconi's yacht the *Elettra* became mobile laboratories in which Franklin, Round, and Eckersley demonstrated that their signals travelled long distances with little loss of strength, even when they were inaudible at shorter distances.[67] According to Appleton, these general conclusions, as valuable as they were for subsequent formulations, 'were due largely to' the perspicacity of these engineers 'investigating the wavelength range of 10 to 100 m. at all hours of day'.[68] Thirdly, it is clear that insofar as postwar vacuum tubes had become an affordable product for radio amateurs, overseas communication was no longer a rarity.[69] Paradoxically, the restriction of authorities for the use of wavelengths of less than 200 metres—known as 'useless bandwidth'—on both sides of the ocean only accelerated the process, since it permitted amateurs to expend all their time and effort on virgin wavebands without fear of penalty. Many of them who were war servicemen seized the opportunity to communicate with overseas relatives.[70]

[62]A.H. Taylor, 1933. 'High-Frequency Transmission and Reception'. In Henney (1933, p. 434).

[63]For a definition of technologic micro-invention, see Mokyr (1991, pp. 291–9).

[64]Morse (1925b).

[65]G. Marconi, 1922. Radio Telegraphy. *Proc. IRE*, 10, 215–38, p. 215.

[66]Franklin (1922).

[67]Round, Eckersley, Tremellen, & Lunnon (1925).

[68]EUA [D6], *Papers*, Appleton Room, 'Short Wave Wireless Transmission. The Discovery of the Low Attenuation of Short Waves'.

[69]Eccles (1923, 1924); Sturmey (1958, pp. 28–9).

[70]Clinton B. DeSoto, 1936. *Two Hundred Meters and Down: The Story of Amateur Radio.* West Hartford, Conn., 70–8; Aitken (1985, p. 512); Clarricoats (1967, pp. 62–72); 'Who Discovered the Long Range of Short Waves?,' *EW & WE*, 1926, 3, 715.

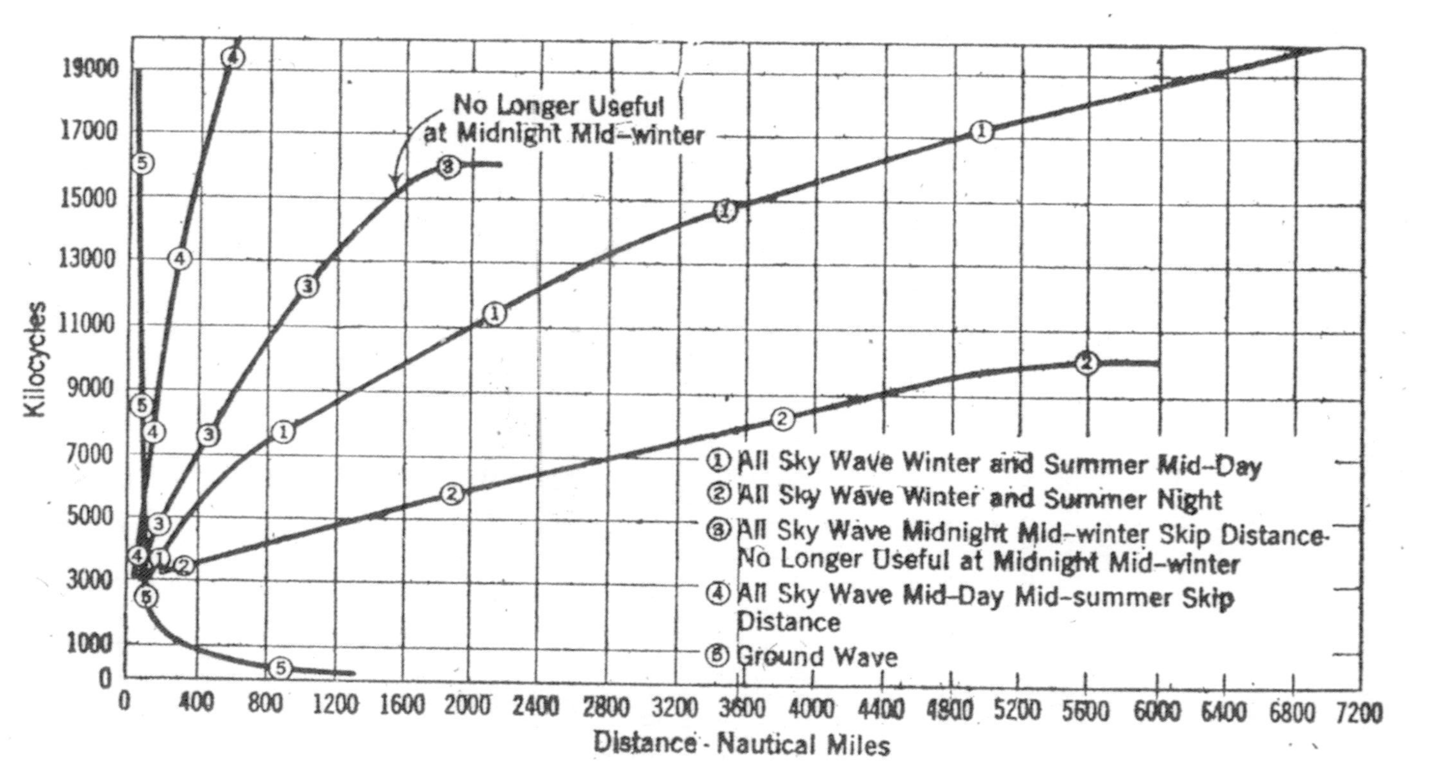

Fig. 11: These curves show the relative distances (communication range) that short waves of different frequencies may carry during the summer and winter months. From Duncan & Drew (1929), p. 757.

It is not necessary to describe in detail the accounts and disputes over the priority in the use of short wave to find the beam system as an innovation of far-reaching consequence in industry and geopolitics as in ionospheric physics. Just as in the radio trade it denoted the access to an 'unknown terrain'[71] (the progressive opening-up of new markets, new applications, new prospects), in physics the term had connotations of 'mystery',[72] 'enigma',[73] and 'revolution'[74] (the means towards an unexplored field).

The short-range tests performed by Round in 1920 and by Franklin and Marconi in 1923,[75] revealed puzzling circumstances in phenomena that had not been hitherto perceived: the 'skipped distance'[76] inexplicably varied with wavelength (being greater the shorter the wavelength), and signal strength was stronger during the day than during the night (whereas with long wave the reverse was the case).[77] But one fact was unquestionable: the signals bounced off the upper atmosphere to reappear far from their transmitting source—what was known as the 'skip' effect. The *sine qua non* of long-distance propagation was to concentrate the waves in place of scattering them in all directions—as a rifle with telescopic sight instead of a machine-gun—in order that the Heaviside-Kennelly layer reflected them on to the precise target. As short-wave tests proceeded, the new high frequencies became an operational tool for measurement, each fragment of waveband a high-calibre weapon to explore the structure of the atmosphere. The ionosphere became the object of exploration and investigation. In this respect, as had augured A.H. Taylor, president of the Institute of Radio Engineers, the advent of experimental short wave was 'to cause a general overhaul of wave propagation theory' in the next decade.[78] Underlying the booming of short wave, however, were commercial imperatives. And underlying these was, among other aims, an ideal that was always intimately associated with the wireless development: Empire's unity.

[71]G. Marconi, 1935. 'Fenomeni Accompagnanti le Radiotrasmissioni. Discorso Pronunziato l'11 Settembre 1930 a Trento'. In Jacot & Collier (1935, p. 220).

[72]Berg (1999, p. 47).

[73]'Success of First Empire Transmission. BBC Chairman Speech to the Royal Empire Society', *World-Radio*, 1933, 16, 109.

[74]Marconi to Solari, June 1923, in Jolly (1972, p. 247).

[75]'Captain Round's Lecture', *EW & WE*, 1923, 1, 105–6; Smith-Rose (1923, pp. 121–2); Fleming (1925).

[76]At first strength fell off rapidly as the distance from the transmitting station was increased, but later the signals suddenly began to increase at a critical distance—the so-called 'skipped distance'.

[77]EUA [D5], *Papers*, Appleton Room, 'Beam Wireless', 21 November 1924.

[78]A. Hoyt Taylor, 'Short Waves', in Codel (1930, p. 277).

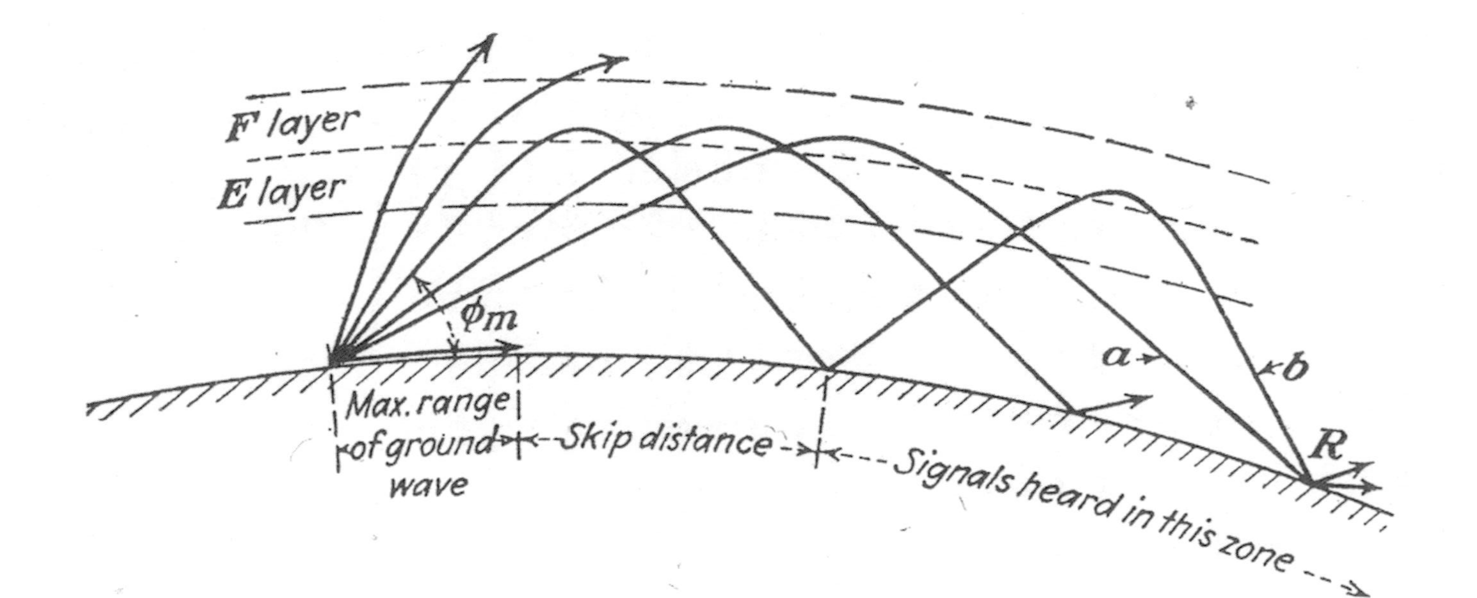

Fig. 12: Propagation of short waves, illustrating skip-distance effect. A ray having a vertical angle greater than ϕ_m will not return to earth. From Glasgow (1936), p. 495.

Plainly, with the implementation of this technological vanguard, Marconi's had the elements for an 'Imperial communications network'.[79] 'Wireless telegraphy', remarked the Donald Committee in their 1924 Report, 'is the victory of science over space, because it annihilates distance.'[80] The short-wave scheme, asserted *The Times* in its leader of 24 July, guaranteed the end of 'deplorable and destructive delays', 'the cessation of the paralysis caused by conflicting counsels and purposes'.[81] The two issues to which the Labour government gave immediate prominence, and to which Marconi's engaged when they signed the contract of 28 July 1924, were the improvement of communications and their reduction of cost for the benefit of imperial trade, and the normalization of relations with the Dominions. The first object it pursued through the establishment of four duplex beam circuits between England and the Dominions;[82] the second, by permitting Marconi's to handle communications with Canada, Australia, India, and South Africa, which delegated the control of their stations in the hands of private firms associated with the company.[83] Imperial unanimity, so remote with the chain scheme, was in 1926 a reality.[84] 'Just as wireless on the small scale became, in the shape of broadcasting, an important social influence in Great Britain', averred W.H. Eccles, Vice-Chairman of the Norman Committee and one of the artisans of the BBC policy, 'so wireless on the large scale... strengthened the unity of the British Empire.'[85]

The late 1920s witnessed a marked *idealization* of short wave as a vehicle of unification over large areas of the British Empire. The journals and magazines of the period show an increasing tendency towards a kind of devotional acceptance, an appeal to the fraternity through waves. On the one hand, the notion that, thanks to short wave, something happening in Britain could be transmitted—the mere possibility of hearing voice!—and experienced in the Dominions, and vice versa, was completely new and exciting.[86] Increasingly, the wireless scheme combined a highly sophisticated analysis of short-wave pre-eminences with a deep scepticism about the capacity of submarine cables

[79]'The Use of Short Waves for Long Distance Communication', *Marconi Review*, 1929, 1, 18–21.

[80]NAUK, [POST 88/41], 'Imperial Wireless Telegraphy Committee', 1924, Cmd. 2060, p. 5.

[81]*The Times*, 24 July 1924.

[82]For Marconi's short-wave imperial scheme, see 'A Chapter in the History of the Marconi Beam', *The Marconi Review*, 1928, 1(1), 3–12; 1(2), 1–10; 'The Marconi Short Wave Beam System', *The Marconi Review*, 1928, 1, 29–32; Sturmey (1958, pp. 111–13).

[83]Barty-King (1979, pp. 5–97); Headrick (1994, pp. 24–5); Jolly (1972, pp. 248–9).

[84]Jacot & Collier (1935, pp. 193–4).

[85]W.H. Eccles, 1926b. Radio Communication and Imperial Development. *Nature*, 117, 659–62, p. 662.

[86]Berg (1999, p. 54); Deloraine (1930).

for telephony.[87] The doctrine that came to prevail in the Empire was that established by the British Post Office, the greatest authority on the whole field of wireless (albeit often labelled as unprogressive): short-wave radio telephony, inaugurated in 1927 across the Atlantic, was essentially a public service for imperial intercommunication.[88] But simultaneously, the long-wave Rugby station was a supplementary platform.[89] The establishment of radio links in telephony, in the words of one of Post Office's managers, had made Great Britain 'the switching centre through which Europe and America communicate with each other and with the British Dominions'.[90]

On the other hand, commercial broadcasting was producing a remarkable intensification of the *collective identity*, a unique experience for sharing common information and entertainment, but also a risk of alienation through propaganda (what some politicians saw as a dangerous 'radiocracy').[91] Radio contributed to the formation of a new audience, one far more anonymous, intimate and intermixed than that of the traditional press, but at the same time more vulnerable and accessible.[92] Short-wave broadcasting accentuated those trends.[93] By the mid-1920s, shortwave stations were springing up everywhere except in Britain.[94] At the Colonial Conference of May 1927, the colonial representatives criticized the BBC's managers for the absence of an imperial service. The prevailing view in the BBC was, however, that of the *approfondissement* of the unitary spirit: an organic vision for the BBC's Director

[87]See Ryan (1923), Miles (1926), and Sacazes (1928).

[88]By 1932, any British subscriber could communicate with 90% of the telephones of the world. See E.B. Moullin, 'Development of Radio Communication', *The Times*, 19 November, 1932, p. 25.

[89]Lee (1930); E.H.S., 1926. 'The Post Office Wireless Services'. *Post Office Electrical Engineers' Journal*, 19, 58–66. By the mid-1930s, the British Post Office would establish radio-telephone circuits for intercommunication between the Dominions, the US and Egypt—a total of 135 cities had radio-telephone transmitters in the Empire. See Headrick (1994, p. 25).

[90]Post Office, 1939. 'History of the Engineering Department'. *Post Office Green Papers*, 46, 3–41, p. 40. See also Vyvyan (1933, pp. 176–84). For the international radio telephony, see Bown (1937), and Hugill (1999, pp. 134–8).

[91]G. Beer, 'Wireless: Popular Physics, Radio and Modernism', in Spufford & Uglow (1996, p. 151).

[92]'By means of the short-waves the outposts of our scattered Empire are linked'. Rigby (1944, p. 7).

[93]For short-wave broadcasting, see Phipps (1991, pp. 215–16) and Pawley (1972, pp. 126–36). See also Wedlake (1973).

[94]The American RCA, Germany, and Italy, for instance, took short-wave broadcasting more seriously than Britain did. See Isbell (1930) and 'Shortwave Broadcasting Stations of the World', *Electronics*, 1931, 2, 180–1.

of Publicity—the development of 'a more conscious sense of imperial unity' by means of short-wave broadcasting;[95] a personal bid for the BBC's Director General John Reith, whose greatest ambition had been 'to broadcast to the Empire from London';[96] and a pragmatic attitude from the Chief Engineer P.P. Eckersley—'a regular and reliable service rather than an affair of stunts and surprises'.[97]

Insofar as to listen to American radio programmes debilitated and disfigured British identity, it seemed to be imperative to have this rapprochement between these aspirations and the desire to fulfil the 'sentimental feeling abroad'.[98] In the terms expressed above, materialized in December 1932 through the erection of the short-wave transmitter at Daventry and the subsequent Empire Service.[99] The decision adopted at the Colonial Conference of June 1930, satisfied the colonies much more than the Dominions, although it received laudatory descriptions generally: 'a noble ideal',[100] 'wireless bonds' on the road to unity for the Empire,[101] and 'a connecting and co-ordinating link between the scattered parts of the Empire',[102] among other honours.

The beam system had its beginnings in a concept of profound innovativeness, a concept which so questioned the technological categories of the time—the superiority of long wave, the monopoly of cables—that the ensuing effect was the expansion of the world's communications network.[103] Most of

[95]G. Murray, 1926. 'Empire Wireless Possibilities'. *Unit. Emp. RES. J.*, 17, 167–9, p. 168.

[96]Pawley (1972, p. 50).

[97]Briggs (1961, pp. 312–14).

[98]The engineer in charge of the Durban Station, South Africa, vividly conveyed this feeling of melancholy: 'When one hears the sombre notes of Big Ben chiming away out in the wide open spaces of these far-flung Dominions, and the voice of London's announcer penetrating the dense, almost impenetrable forests of our tropical possessions it cannot do otherwise than create a feeling of infinite pride in this colossal heritage, and a deep appreciation of the fact that other worlds exist besides our own'. N.D. Cumming, 1931. 'Radio Communication: What it Means to the Empire and the World'. *Unit. Emp. RES. J.*, 22, 506–7.

[99]'Short Waves', *World-Radio*, 1933, 16, 141; 'Big Ben: A New Link of Empire', *World-Radio*, 1933, 16, 731; N. Ashbridge, 1933. 'Six Months of Empire Broadcasting'. *World-Radio*, 16, 677–80

[100]According to J.H. Whitley, the Chairman of the BBC governors. See J.H. Withley, 1933. Empire Broadcasting and the new Service. *Unit. Emp. RES. J.*, 24, 85–95, p. 91.

[101]The Countess Eileen de Armil, 1931. 'Wireless Bonds, Why Not an Empire Broadcasting?'. *Unit. Emp. RES. J.*, 22, 544–5. See also C.G. Graves, 1931. 'Empire Broadcasting'. *Unit. Emp. RES. J.*, 24, 479–80.

[102]In the words of J. Reith. See Walker (1992, p. 24).

[103]Harold H. Buttner, 1931. 'The Role of Radio in the Growth of International Communication'. *Electrical Communication*, 9, 249–54.

postal and military services adopted short wave as part of their systems of telephony and telegraphy. In Britain, the Post Office decided to operate the beam stations, albeit not without reservations, because of its high investment (£490,000) in the long-wave Rugby station.[104] At a certain level, Marconi's experience had its continuation in the establishment of 'Empiradio' services, a series of short-wave stations (Radio House, Carnavon, Ongar, Brentwood, Dorchester, and Somerton) interconnected commercially and technically with foreign stations so that they formed a global network.[105]

In short, short wave, by its origins and by the very nature of its art, was innately and intimately united to Empire. At a luncheon organized by the Royal Empire Society to celebrate the marvels of wireless, Appleton asked himself whether its cause had been vouchsafed to mankind for the benefit of the British Empire. If that was so, he continued, 'the Empire had made the fullest use of its opportunity'.[106] Imperial needs had stimulated the innovativeness and commercialization in the field of radio communications.[107] And the greatness of imperial communication rested, according to him, on the solidness of a laudable trinity: Marconi's (by means of its beam system), the Post Office (radio telephony), and the BBC (Empire Service), which incarnated the supreme ideal of the Empire's unity.[108]

The introspection of an ambitious man

There is a certain irony in considering Marconi's as one of the pillars of imperial communication, and therefore of imperial development. At least by its imperial attachment. Marconi's took on a different complexion, when viewed from London or from outside the Empire. From Washington, it embodied the imperial British policy translated to the communications sphere. It implied an extension and prolongation of the hegemony that Britain had accomplished in the age of cables.[109] From Berlin, the perception was similar, with the difference that it belonged to an enemy country.[110] In both places,

[104] *The Times*, Editorial, 30 January 1930, p. 15.

[105] 'Commercial Short Wave Wireless Communication', *The Electrician*, 1929, 103, 429; Dowsett (1929).

[106] 'The Marvels of Wireless', *Unit. Emp. RES. J.*, 1933, 24, 398–402, p. 399.

[107] E.V. Appleton, 1933. *Empire Communication. The Norman Lockyer Lecture, 1933*. London: British Science Guild, 7–8.

[108] E.V. Appleton, 1937. 'Empire Radio Communications'. *Unit. Emp. RES. J.*, 28, 585.

[109] Denny (1930, PP. 368, 378–85) and K. Clark (1931).

[110] The German governments had tried to compensate for Marconi's predominance through the creation of Telefunken.

the appropriate analogy was not that of one more competitor responding to a market situation but the *rival par excellence*, fusing business and geopolitics, and whose presence it was necessary to neutralize.[111] In the long run this simple identification between Marconi's and the Empire proved an impediment to the realization of its ambitions.[112]

From London, however, Marconi's represented a mixture of imperial emblem and capitalist ensign, figuratively conceivable as a huge entrepreneurial octopus whose functional cerebrum reposed in London but whose goals and tentacles transcended the Empire. Marconi's was without doubt wedded to Empire: it was a British firm; the strategic decisions on financial policy were taken at Strand House, London, and the firm closely cooperated with the Admiralty in matters of defence. But the nature of its ambitions, halfway between imperial supremacy and world expansion, generated an amalgam of empathy and animadversion within Britain. Indeed, the company had never regarded itself as an instrument of the government but as an innovative agent of communications systems, with aspirations throughout the Empire, but also worldwide. The characterization of 'Marconi equals Empire', so perceptible abroad, was ambiguous and imprecise behind closed doors.

This alleged ambiguity of the company resulted, among other reasons, from the pragmatism and unlimited ambition of its founder.[113] Guglielmo Marconi, exemplar of radio experimenter, was a son of a well-to-do Italian businessman and of an Irishwoman heir to the wealthy Jameson family of whisky distillers.[114] The patenting of apparatus and the exclusive dedication to experimentation did not present undue difficulties for the son of an opulent family. Easily recognizable by a defect of his speech—he dropped the 'g' of words ending in 'ing—[115]' and by his heavy addiction to tobacco, Marconi never concealed his attachment to Italy (where he was educated) nor to Britain (where he achieved fame).[116]

Marconi was a man of two nationalities, but of one *idée fixe*: 'from the time of my earliest experience, I was always convinced that radio signals would

[111]J. Harbord, 1926. 'America's Position in Radio Communication'. *Foreign Affairs*, 4, 465–74.

[112]Aitken (1985, pp. 251–2).

[113]Marconi's figure is reflected in a vast secondary literature. See Giovanni di Benedetto ed., 1974. *Bibliografia Marconiana*. Genova: Giunti; and Giorgio Tabarroni et al., 1974. *Marconi: Cento Anni dalla Nascita*. Torino: ERI, 219–41.

[114]Jolly (1972, pp. 1–13); Süsskind (1974, pp. 67–72).

[115]O.E. Dunlap, 1937. *Marconi: The Man and his Wireless*. New York: MacMillan Co., 298.

[116]G.T., 1937. Italianità di Marconi. *Bologna. Rivista Mensile del Comune*, 24(7), 28–9.

some day be sent across the great distances of the earth.'[117] To transcend the limits of distance appeared to him as an obstinacy; to monopolize the power of communication—in patents, in the Empire, in the world—, as a consummation of his ambitions. Marconi stamped his visionary character on the radio art. To be sure, in terms of commercial innovation, he perfected existing methods in place of reaching out for radically new discoveries.[118] He considered himself, however, as an inventor, rather than as an innovator or business manager.[119] Invention required creativity, and creativity required ambition. Through discovery, as he liked to say, one 'realizes the infinity of the Infinite'.[120]

It was not his enormous perseverance or his perspicacity that made Marconi the personification of wireless—his *raison d'être*—, although he needed both for the attainment of his successes. The glory surrounding Marconi derived from his reputation as an inventor who had twice changed the course of the art of technology: first, in the historical transatlantic experiment of 1901, and then, in the short-wave tests of 1920.[121] In both, however, he pursued the doctrine that practical consequences were the criteria of knowledge, meaning, and profit.[122] Marconi valued a clear utilitarian mentality as the greatest blessing an inventor with entrepreneurial aims could enjoy. This inner clarity of commercial pragmatism could appear to the outside world, especially to scientists and politicians, as a capricious and unpredictable element, far removed from the purest values required in the pursuit of physical principles and of Empire's unification. One of his most distinguished consultants, Professor Ambrose Fleming, opined that 'his predominant interest was not in purely scientific knowledge *per se*, but in its practical application for useful purposes'.[123] His 'instinctive intuition' was the only compass he needed. Marconi of course did possess multifarious interests, but wireless was omnipotent, and the practicality and commercialization of communication, omnipresent.[124]

[117]Dunlap (1937, p. 305).

[118]W.R. MacLaurin, 1949. *Invention and Innovation in the Radio Industry.* New York: Macmillan Co, 42–9.

[119]Jacot & Collier (1935, p. 226).

[120]D. Marconi, 1996. *My Father, Marconi.* New York: Guernica, 215; Pietro Caccialupi, 1939. *Il Dominatore dell'Infinito (Guglielmo Marconi).* Milano: La Prora.

[121]Giancarlo Masini, 'L'Uomo che Cambiò il Mondo', in Tabarroni (1974, pp. 153–94); Dowsett (1934).

[122]W.J.G. Beynon, 1975. 'Marconi, Radio Waves, and the Ionosphere'. *Radio Science*, 10, 657–64, pp. 659–60.

[123]Fleming (1937, p. 62).

[124]For a comparison of styles between Marconi (as a *practician* and engineer–entrepreneur) and Fleming (university professor and consulting engineer), see Hong (1996).

Empire and business, the latter understood as the commercial application of wireless, must in his view run along parallel paths. The beam system and its associated categories of simultaneity and fidelity could not and must not be reduced to the communication from London to the Dominions and colonies. Marconi's commitment to the imperial vision was unequivocal, but not so much by sharing common principles and beliefs as by the duty and responsibility before a unique commercial venture. Presaging certain opposition against his pragmatism, Marconi concluded in a lecture at the Royal Empire Society that the progress of wireless was equivalent to imperial development and that both were congenital. His beam system enabled direct communication—'an amazing achievement and fraught with the profoundest significance of each integral portion of the vast Empire'.[125]

Although Marconi had been successful in reconciling the entrepreneurial spirit with imperial imperatives, notably through the successive wireless chains, his adherence to Mussolini's fascism turned into a theme too onerous for easy digestion. In Britain, it was irreconcilable with imperial values and ideals. But, in Italy, if Marconi had any hesitation about it, it was labelled as 'pro-British'.[126] The maintenance of an apolitical position, by which he had always been characterized, could eventuate in, or degenerate into, passivity and unresponsiveness with socially catastrophic consequences.[127] 'In its early stages he looked askance at fascism as rowdy and opportunistic, consequently feeling a grave distrust of the man who led it. But he was aware that Italy, torn by strikes, riots, and civil dissension, was spiralling downward in a vortex that was destroying her prestige and causing other nations to regard her as a less than first-rate power. These dual circumstances offended his essential sense of order and his national pride.'[128]

Optimism, persuasiveness, affirmation of life constituted Marconi's message and the core, he believed, of every desideratum that might at first seem chimerical. By 1926, after the completion of the short-wave imperial scheme, he would disengage from the company's obligations and interests; the materialization of the beam system gave way to alleviation, to inner peace, to the fulfilment of duty.[129] Short-wave threatened to ruin the cable companies, and it

[125]'The Marvels of Wireless', *Unit. Emp. RES. J.*, 1933, 24, 398–402, p. 399.

[126]As Baron Alberto Fassini incriminated him on one occasion. See D. Marconi (1996, p. 234).

[127]For Marconi's relationship with fascism, see Gabriele Falciasecca & Barbara Valotti, 2003. *Guglielmo Marconi: Genio, Storia e Modernità.* Milano, 105–11; Leslie Reade, 1963. *Marconi and the Discovery of Wireless.* London: Faber & Faber, 114–28; W.J. Baker (1970, pp. 295–6).

[128]D. Marconi (1996, p. 234).

[129]Jolly (1972, p. 253).

fulfilled his expectations.[130] One might wonder, in the guise of an *ex post* reflection, whether Marconi would have investigated long-distance communication if he had earlier had his share of responsibility in controlling imperial communications. To what extent his ambition was nurtured by the Post Office's opposition to the company, to what extent by the government's complacency about the hegemony of the cable system, or by the pressure exerted by the cable companies, is not clear. What seems clear, however, is that the supremacy of the cable network accentuated, and magnified, his fixation. Paradoxically, Marconi coveted what he had always combated: cable's monopoly.

During 1927 and 1928, Marconi's plans took on a more sombre tone. His first reverse, the report of the Committee of Imperial Defence of 1927, caused him great anxiety. Owing to security and defence imperatives—the transmission by cable guaranteed the confidentiality and secrecy of messages—the Admiralty preferred cables to radio.[131] Shortly afterwards, as a result of the Imperial Wireless and Cable Conference of January, 1928, the British government announced the merger of all communications corporations, among them the cable companies, the Post Office short-wave network, and Marconi's.[132] To Marconi's surprise, the new enterprise, baptized Imperial and International Communications, or I&IC, was born with the approval of the company's board.[133] It implied the end of Marconi's activities in the transmission of messages as a source of revenue, and its limitation to manufacture and research. According to Sir Basil P. Blackett, Chairman of the Eastern Telegraph Company, the new giant conglomerate was 'not merely a symbol but an agent of Imperial unity'.[134] Imperial communications final picture was not the monopolistic, Marconi-controlled, short-wave-based, high-fidelity chain that had attracted him from his beginnings but a strange construct regulated by the government, an alliance *contranatura* with his eternal rival.[135] For his consolation, the Empire had 'the finest system of world communication ever to exist'.[136]

[130]Geddes (1974, p. 36). According to Baker, Marconi's aversion to the cable companies was akin to 'that attributed to the devil for holy water'. See W. J. Baker (1970, p. 223).

[131]NAUK, [CO 323/990/4], 'Imperial Communications Committee: Papers and minutes of sub-committee between beam wireless and cable service. Interim Report, 1927'.

[132]J. Gilmour (1928); Donald (1928); E.V. Appleton, 1932. 'Cable and Wireless'. *The Times Trade and Engineering Supplement*, 21 May 1932, p. 17.

[133]According to Baker, the reason for this surprising decision lay in the weakness of a board dominated by cable interests and in which Marconi himself was an ordinary director without decisive power. See W.J. Baker (1970, pp. 230–3), Barty-King (1979, pp. 203–10), Headrick (1994, pp. 26–7), Sturmey (1958, pp. 118–19).

[134]'Wireless and Empire Unity', *Unit. Emp. RES. J.*, 1930, 21, 639–40, p. 639. See also Belfort (1929).

[135]'Editorial: The Wireless and Cable Amalgamation', *The Wireless World*, 1928, 22, 329.

[136]W.J. Baker (1970, p. 232).

Ionosphere research in the commercial context

It is not adventurous to enunciate the commercial and industrial priorities that gave ionosphere research at the Marconi company its distinctive feature. The proliferation of radio setmakers and the poor communication service in navigation (precisely one of the few areas not covered by the Post Office's monopoly) caused Marconi's to concentrate on the conditions for waves propagation and technology rather than on the manufacture of radio sets.[137] The deep-rooted sense of innovativeness and originality of its engineers was another influential aspect for the emphasis on point-to-point communications. Marconi's crucial transatlantic experiments and Franklin and Round's pioneering short-wave work undoubtedly contributed to the high prestige that investigation enjoyed within the company. Likewise, commercial imperatives associated with the imperial wireless scheme marked the gradual *displacement* from long wave to high frequency, the scattering of waves, the direction-finding equipment, and the wave propagation through the ionosphere. As a result, Marconi's engaged in fundamental research and it did so in such a way that it would be dishonest and erroneous to visualize the picture of ionospheric physics without its presence.

At about the same time as Marconi's engineers returned from war service, the company pondered turning to research pursuits of a commercial kind that it had so successfully fostered before World War I. One of the first decisions was the creation of twin research departments, which initiated their operations with a budget of £40,000 and which were allotted to Franklin and Round. From 1922 to 1931, their staffs (six engineers and 12 technicians that of the former, and 28 that of Round) acquired a reputation by their contribution to the design of devices and to the performance of the BBC's broadcasting transmitters. Still, the company was gradually abandoning its research engagement during the 1930s. Profits of £796,234 in 1929 fell to £260,300 by 1937, a decline in revenue which was attributed, by sources linked to the company, to 'the stringent curtailment of research activities'.[138]

Both Franklin and Round were two of the most loyal *Marconian elite corps*, with similarities disclosing a particular modus operandi that distinguished the Company's research policy. Indeed, similitude was not predicated on appearance and temperament (in fact, both were diametrically opposed), but on allegiance and individualism. Franklin, small, frail, diffident, reserved, conceived invention as a form of inner communion; Round, robust, extrovert, rather

[137]Vyvyan (1933, p. 169).
[138]W.J. Baker (1970, pp. 271).

Churchillian even to the cigar, regarded it as a logical deductive reasoning.[139] As two prolific engineers, however, both remained steadfastly devoted to Marconi. While their nonconformism to administrative regimes made it difficult for them to integrate into investigation projects hierarchically established, both severally fixed their priorities and responded only to Marconi's indisputable authority.[140] Both followed his precept of applying for patents on everything that they did (65 stood to Franklin's credit, and 117 to Round's).[141] Their patents, in fact, were utilized by the company as an instrument of power for cementing its dominant position, which far exceeded that of any of its rivals.[142] Hence it is reasonable to think that the independent nature of their departments could contribute to inculcate within the company a style facilitating fundamental research; a style of *laissez-faire* far removed from the short-term requirements of the manufacturing firms.

The total correspondence between this style of investigation and the commercial context that gave rise to it is beyond dispute. But it is explicitly more congruent in the case of another department, the Propagation Section, concerned with experimental and theoretical investigations on the behaviour of the ionosphere. The commercial stimulus in this section is manifest. The unpredictable vagaries of the ionosphere, still unexplained by physicists, assumed a capital importance for the implementation of the beam system in the early 1920s, as the first short-wave stations furnished an abundant mass of data on anomalies (echoes and scattering) in signals transmission.[143] The most illustrative work on ionospheric predictability, by K.W. Tremellen, dates from 1927 and is known as the 'shadow charts'. Drawn on transparent paper, these charts showed a picture of ionization densities (through a grey scale), as being slid over a Mercator map, over any given radio path. By 1932, the readiness to recognize their commercial value to the company (as instrument to predict the best frequency for transmission) was beyond question.[144]

The most distinguished contributions, however, were without a doubt those which Thomas Lydwell Eckersley, a brilliant mathematician of Trinity

[139]Baker & Hance (1981, pp. 897–8).

[140]W.J. Baker (1970, pp. 270, 279–83).

[141]On Henry Round, see 'Electronics Pioneer Helped to Track Enemy at Jutland', *The Guardian*, 19 August 1966; 'Captain H. J. Round—An Electronics Pioneer', *The Times*, 19 August 1966; 'Captain Round—Pioneer of Broadcasting', *The Financial Times*, 19 August 1966; 'Capt. H. J. Round, Pioneer of Electronics, Dies', *The Barnet Press*, 26 August 1966.

[142]MacLaurin (1949, p. 44).

[143]Eckersley & Tremellen (1929); and Bureau (1927).

[144]W.J. Baker (1970, pp. 289–91).

College, Cambridge, and head of the section, undertook after joining Marconi's in 1918. A brother of Peter Eckersley, the BBC's chief engineer, his interest in the ionosphere derived from a scientific tradition ingrained at the company: the study of direction findings (D/F) as an aid to navigation.[145] By about 1920, the company had devised all the technology required—the aerial system of F. Adcock with Tremellen's assistance, and previously the Bellini-Tossi direction-finder—and, in this area at least, it led the world.[146] Its efforts were directed to maximize the sensitivity of antennas and to minimize the effect of interferences. For Marconi's, the D/F emblematized the commercial application of science to navigation. Hence the importance of solving the uncertainties which it usually generated.[147]

While the invention of the amplifier and the disposition of antennas increased the range of directionality, at the same time they introduced errors of great magnitude and variability (D/F night errors).[148] As a lucid theoretician imbued with the experimental work of D/F, Eckersley ascribed those errors, as early as 1921, to the reflection of waves from the irregularities in the then hypothetical Heaviside layer.[149] As a result of the reflection, argued he, the polarization of the wave was altered and thereby it gave rise to D/F errors. His conclusive formulation that 'the existence of a ray reflected at night time from some upper conducting layer of the atmosphere' was 'beyond doubt'[150] was immediately recognized as a major contribution, albeit not persuasive enough for demonstrating the presence of a reflective layer. Immersed in imperatives of transmission, Eckersley pursued the causes of errors, rather than spectacular existential proofs (as the height of layer might be).[151]

[145]The D/F was one of the most productive fields of research in the company. See H.J. Round, 'Direction and Position Finding', *Jour. IEE*, 1920, 58, 224–57; 'Wireless Direction Finding as an Aid to Aerial Navigation', *The Marconi Review*, 1930, 25, 17–29; 'Wireless Direction Finding Systems for Marine Navigation', *The Marconi Review*, 1931, 29, 1–11; W.J. Baker (1970, p. 50, 166).

[146]In 1933, about one quarter of the British merchant ships (more than 4,000 vessels) were equipped with direction-finders, mostly by Marconi's. See Vyvyan (1933, p. 151); Davis (1930).

[147]'The Radio Direction-Finder and its Application to Navigation', *The Wireless World*, 1922, 10, 825–7.

[148]Smith-Rose (1926, p. 831).

[149]Previously, during World War I, as a commissioner of Wireless Intelligence in Egypt, Eckersley had familiarized himself with the study of the direction of arrival of radio waves from enemy transmitters.

[150]T.L. Eckersley (1921, p. 248).

[151]It is certain that, owing to the absence of data on the dielectric constant and the conductivity of the ground, he was not able to estimate the height. But that was not likely his objective. See A.L. Green (1946, pp. 187–91).

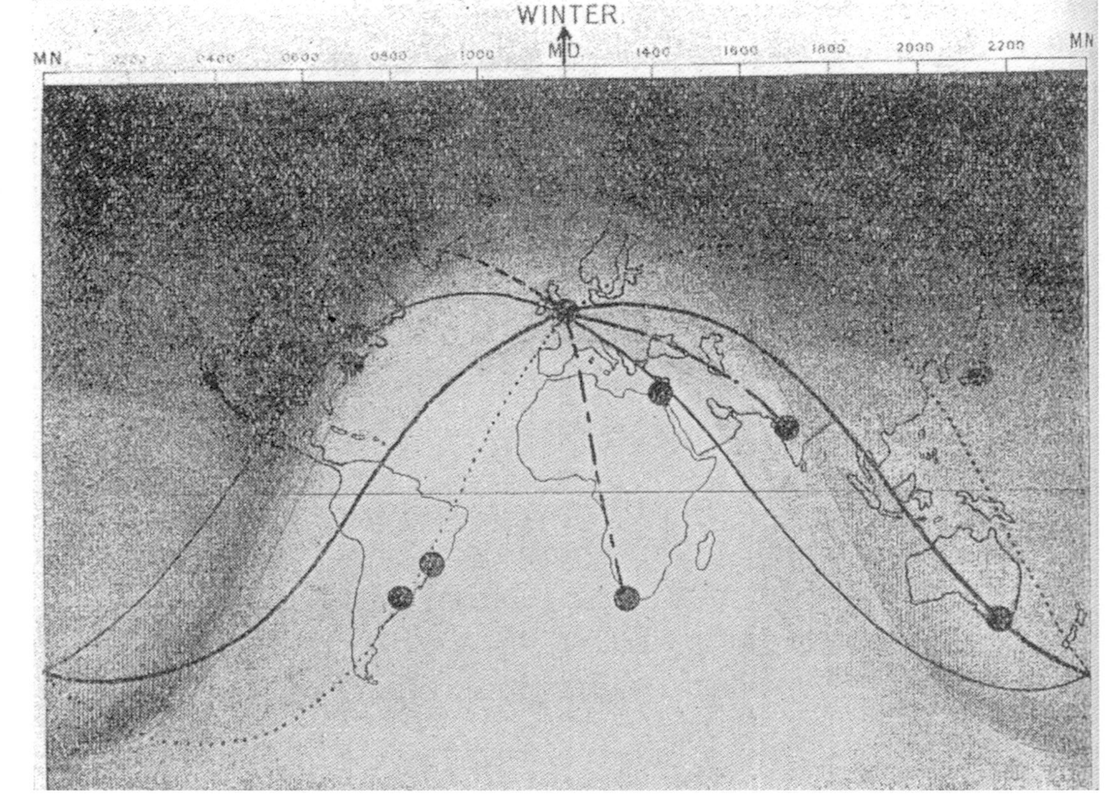

Fig. 13: A mercator projection with a shadow chart laid over it giving typical grand circles for routes from England to Cape Town, Poona, Melbourne, Montreal, and Rio de Janeiro (Tremellen's chart). From Ladner & Stoner (1932), p. 64.*

* There are instances where we have been unable to trace or contact the copyright holder. If notified the publisher will be pleased to rectify any errors or omissions at the earliest cpportunity.

Plainly, it was because of, and not despite, the company's imperial commercial policy that Eckersley's investigations prospered. And this could be extended to the remainder of the team, the engineers K.W. Tremellen, F.M. Wright, G. Millington, and R.F.Knight, which in two decades produced a prolific number of papers (over 50, mostly by Eckersley).[152] By manipulating the beam system, usually achieved in the form of observations made on commercial beamed transmissions of Morse code and of pictures, Eckersley inferred that the scattering from the irregularities of the ionosphere was the agent producing D/F errors.[153] To proceed with this research, his team devised and constructed a spaced-loop D/F being free of polarization errors. To the extent that these investigations were intended, as an integral part of commercial development, to remain prominently at the forefront of long-distance communication in the Empire and in the world, their objectives were completely congruent with those of the company's executives. And the same could be said of the case of the Somerton-New York facsimile transmitter, a technique of a purely commercial type that enabled eliciting the heights of the ionospheric layers by measuring the time delays of echoes observed on the facsimile pictures.[154] But this intellectual and commercial conformity implied neither a total submission nor a mere resignation to utilitarian ends. For while Eckersley and his team adapted well to the course of industrial needs, many of them of immediate practical benefit, at the same time they had the freedom of action and the financial means and autonomy to sound out, and delve into, fundamental research. In this respect, even as abstract a topic as the magneto-ionic theory, or several others such as whistlers, collision frequency of electrons, and the polarization of echoes, could flourish within the commercial framework set by the company.[155]

Although Eckersley's investigations anticipated and stimulated essential questions on the physics of the ionosphere, they did not lead to immediate recognition. Far from it. The best example was the scant attention that his

[152]For an incomplete list of Eckersley's works, see Ratcliffe (1959, pp. 73–4).

[153]These conclusions are included in Eckersley's two classic papers: 'An Investigation of Short-Waves', *Jour. IEE*, 1929, 67, 992–1029; and 'Studies in Radio Transmission', *Jour. IEE*, 1932a, 71, 405–54.

[154]T.L. Eckersley (1930).

[155]Significant theoretical contributions are T.L. Eckersley & G. Millington, 1939. 'The Application of the Phase Integral Method to the Analysis of Diffraction and Reflection of Wireless Waves Round the Earth'. *Philosophical Transactions*, A 237, 273–309; T.L. Eckersley, 1925. 'A Note on Musical Atmospheric Disturbances'. *Philosophical Magazine*, 49, 1250–60; and T.L. Eckersley, 1931. 'On the Connexion Between the Ray Theory of Electric Waves and Dynamics'. *Proc. Roy. Soc. Lond.*, A 132, 83–98.

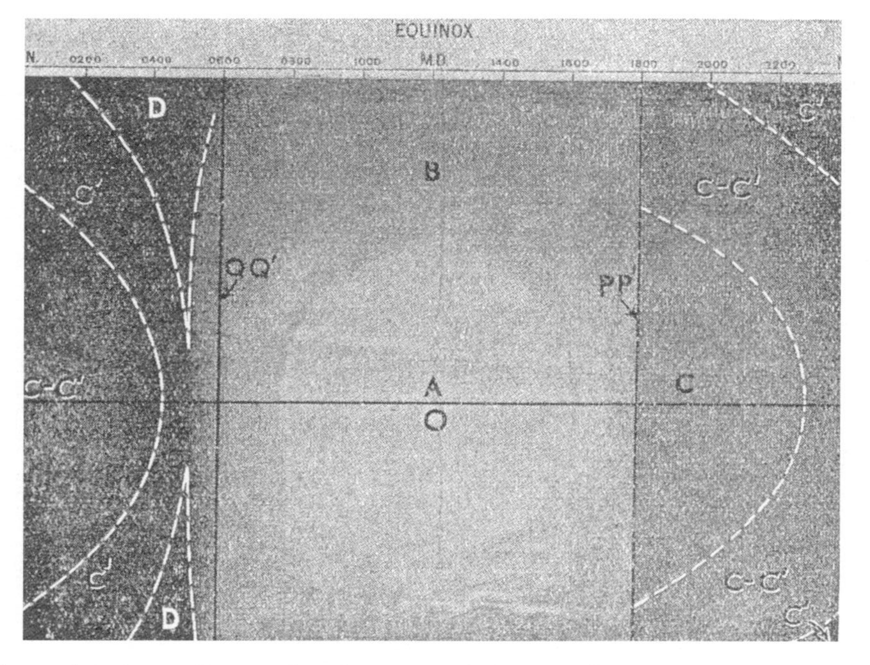

Fig. 14: T.L. Eckersley and K. Tremellen produced a series of charts to fit a mercator projection of the world, depicting the changes in the stages of the ionized layers. The three charts shown below are reproduced shaded in a conventional manner to indicate the various day and night layer changes. To simplify the charts, the infinite possible states of the layers were limited to four main grades: (A) that corresponding to intense daylight; (B) twilight; (C) darkness; (D) late darkness. From Ladner & Stoner (1932), pp. 53–4.*

* There are instances where we have been unable to trace or contact the copyright holder. If notified the publisher will be pleased to rectify any errors or omissions at the earliest opportunity.

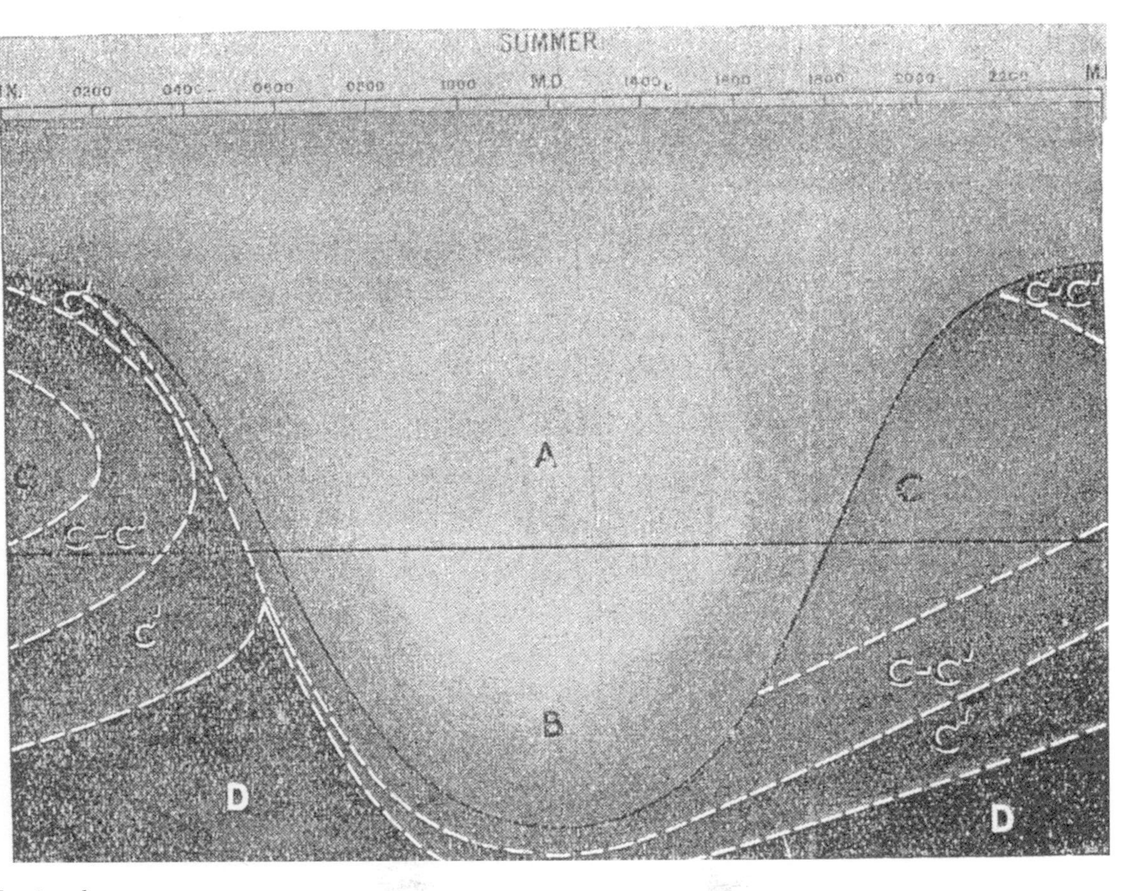

Fig. 14: *Continued*

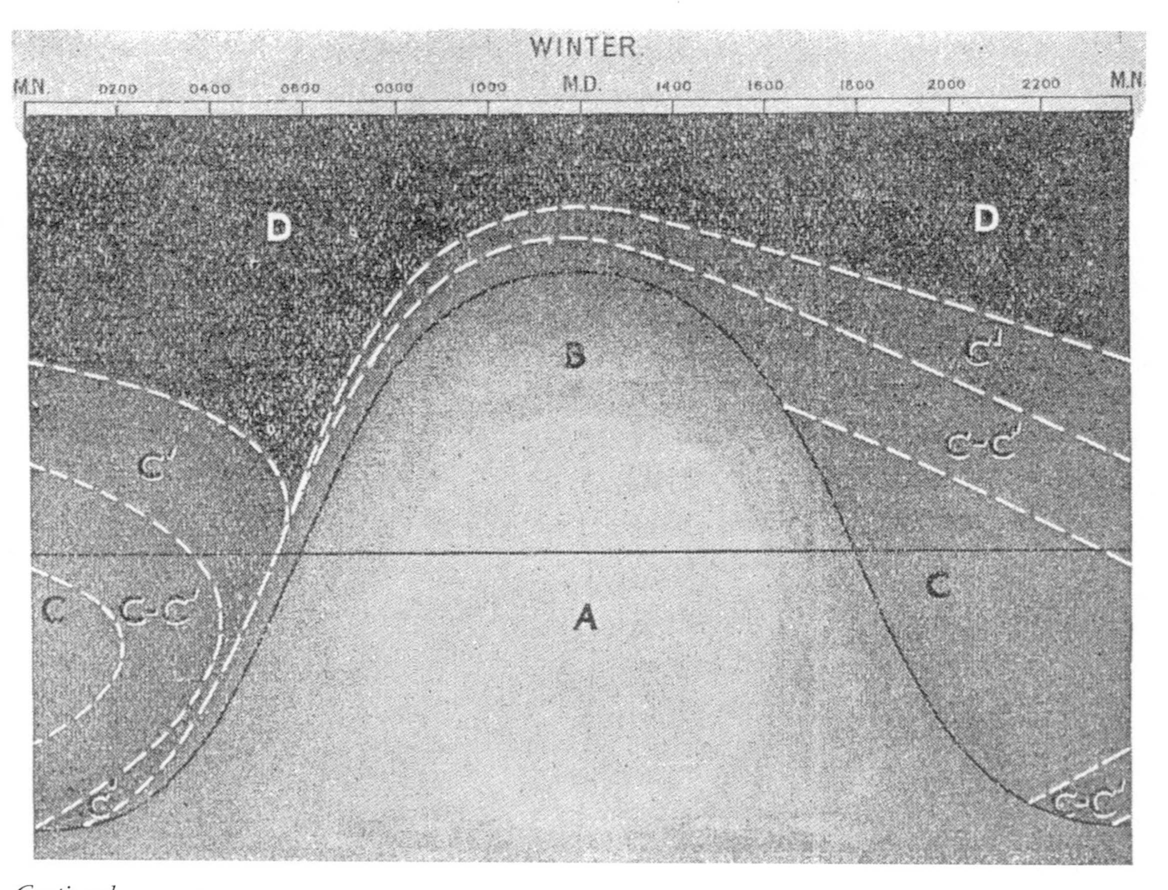

Fig. 14: *Continued*

works on scattering attracted when they were published. That research, which was to be reformulated with renewed vigour after the war, was regarded by experts as the pathfinder of subsequent systems of ionospheric scatter transmissions.[156] Precisely the same deferential tone was upheld by one of the most prominent figures on ionospheric physics, J.A. Ratcliffe, for whom Eckersley was the first to appreciate fully (and, as it turned out, perceptively) the implications of direction-finding errors in terms of waves reflected.[157] There can be no doubt that factors associated with the method of writing and with the propensity for including all his erudition in a single paper drastically and negatively affected the due appreciation of his oeuvre. In all he did, Eckersley tended to reflect on wave propagation in his own way and hence rarely endeavoured to facilitate comprehension and diffusion. This attitude is well illustrated by his papers on the magneto-ionic theory, in which he used a particular nomenclature that was quickly discarded as abstruse and convoluted. And that, clearly, was seen as an inimical element.

The favour that Eckersley and his team showed to short-wave investigation was wholly in keeping with the vehemence with which Marconi himself had pursued that line of research. In both cases, it reflected a determination to explore the commercial and scientific potentialities of a waveband that had experienced a complete marginalization under the dominance of long wave.[158] The Propagation Section analysed the effects of the eleven year sunspot cycle on the performance of Marconi's short-wave beam services.[159] In similar vein, in the early 1930s, Marconi insisted that ultra-short wave, no less than ordinary short wave, should be investigated because of its commercial value and its high degree of secrecy. In this way, Marconi would demonstrate, in a series of tests carried out in Italy, the usefulness of microwaves for short communication, while at the same time assigning Eckersley and his team to enquire into the behaviour of v.h.f. and u.h.f. waves from the fundamental prism.[160] Eckersley surveyed ultra-short-wave refraction and diffraction at a time in which the company's interests diversified from navigational aids to television and aeronautical radio communication.[161] The more diffuse the boundary between

[156]W.J. Baker (1970, p. 291).

[157]Ratcliffe (1959, p.72).

[158]Isted (1974, p. 315).

[159]Tremellen (1938).

[160]For Marconi's microwave tests, see G. Marconi, 1933. 'Sulla Propagazione di Micro-onde a Notevole Distanza'. *La Ricerca Scientifica*, 2(4), 71–2; 'Radio Communication Over 170 Miles With Ultra-Short Waves', *Nature*, 1932, 130, 269; and W.J. Baker (1970, pp. 291–5).

[161]Eckersley (1937, 1932b).

science and business, the more habitual such theoretical incursions would become. And 'this new economical means of reliable radio communication, free from electrical disturbances, eminently suitable for... moderate distances', as Marconi defined microwave, was seen as a commercial gold mine.[162]

Wireless training in Britain's military tradition

The authority of Britain's radio industry was such that by the beginning of World War I it dominated the industry of the entire world through the Marconi organization; much of this credit was due to the Empire's communications and defence requirements.[163] The Royal Navy became the first military institution in the world to adopt radio telegraphy extensively.[164] In order fully to appreciate the importance of these conquests, the commitment of the fighting services to technical training before the war has to be borne in mind. The military academies, whether the Royal Naval College at Greenwich or electrical engineering at the torpedo school, HMS Vernon, provided a solid education in applied science not only to those responsible of imperial communications and defence but also to all naval officers.[165] The Royal Navy, which in the late 1890s had already recognized the need for a wireless telegraph—even before Marconi—, and its technical education system did much to foster the development of radio telegraphy throughout the Empire.[166]

So the foundation of training centres for wireless in the years that followed the outbreak of war may be regarded, in every sense, as the continuation of a long-established practice. It is hardly surprising that British armed forces explored, even if with certain reservations, the potentialities of telephone and wireless technology and adapted them for Army use as the Navy had been doing successfully hitherto. When the trench warfare reached its halfway point in 1916, representatives of the War Office resolved to address the heavy demand for signal equipment through a Signal Service Committee uniting several organizations linked to wireless technology.[167] For the cause, emblematic institutions like Marconi's and the NPL gathered further strength with newly formed teaching centres, such as the Royal Engineers Wireless Training Centre

[162]G. Marconi, 1933b. 'Radio Communications by Means of Very Short Electric Waves'. *Nature*, 131, 292–4.

[163]Pocock (1988, pp. 162–75).

[164]R.F. Pocock, 1986. 'Radio Telegraphy in the Royal Navy, 1887–1900'. In *Papers Presented at the 13th IEE Weekend Meeting on the History of Electrical Engineering*. London, 12/1–12/6, p. 12/5.

[165]For the adoption of wireless in this tradition-bound body, see Pocock & Garratt (1972).

[166]Chrishop (1986, pp. 3/1–3/10).

[167]Nalder (1958, p. 87 and pp. 121–3).

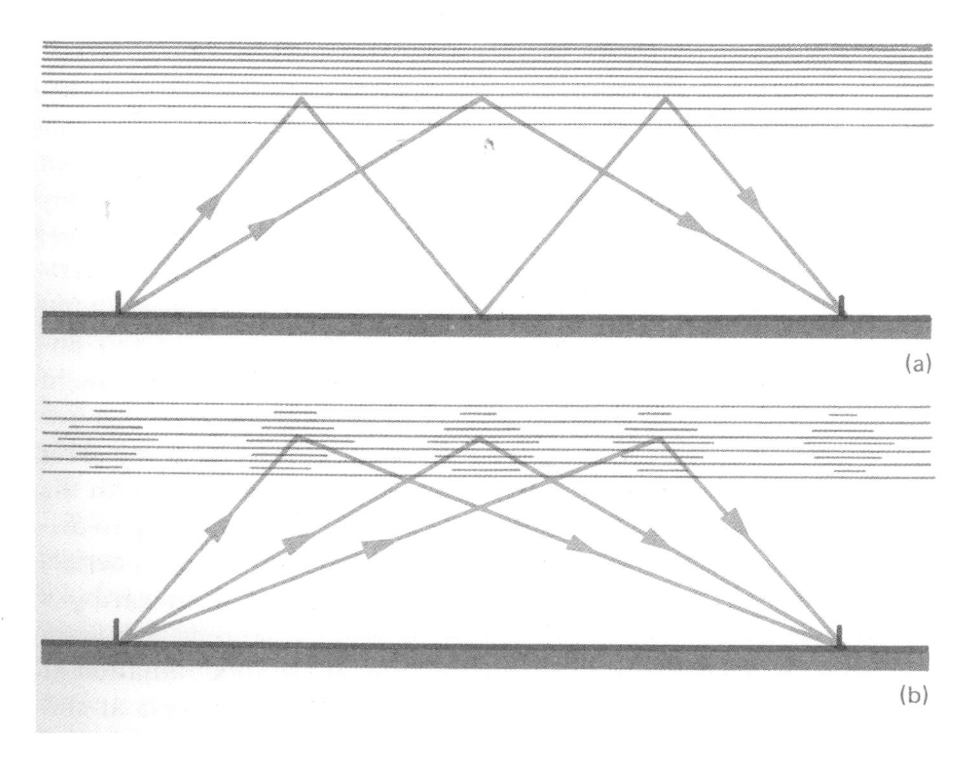

Fig. 15: Multipath transmission. A signal may travel by more than one path either because (a) it is reflected repeatedly between ground and ionosphere or (b) the ionosphere is irregular. From Ratcliffe (1970), p. 211.*

and the Royal Flying Corps Wireless Telegraphy School. Among the functions allotted at this time it is interesting to find assistance in the training of operators at the Wireless Depot and School and collaboration with the army wireless companies created in July 1916.

Despite their initial reluctance, the readiness of military authorities to participate in the training tasks that were derived from the expansion of wireless is praiseworthy. Apparently, a certain opposition about its use as a means of communication and artillery guidance prevailed among the senior officers, which on paper could have delayed imminent assimilation of its advantages.[168] The instruction of a great number of observers at the Wireless Officers and Mechanics Schools at Brooklands between 1916 and 1917, with practical demonstrations in tuning transmitters and receivers,[169] caused admiration in promoters as it did in other Army seniors, among them some of the most traditionalistic, regressive elements.[170] Over the next two years, there were operational requirements that the military commands could not underestimate. The short-term consequence of the irruption of the new technology was a demand for wireless operators on a grand scale.[171] The increasingly urgent need for communication led, in the last two months of 1916, to the recruitment of over 600 of all ranks, many of whom were sent to the theatre of operations before completing their preparation.[172] These needs and requirements were responsible for the amalgamation of all training centres and their concentration in Farnborough.[173] The syllabus of training expanded further by the addition of direction-finding and wireless telephony. In scarcely two years, the school had witnessed an increase from eight to 186 officers on the strength of their recommendations, while the number of learners multiplied by six (from 648 to 3,783).[174] These were the days when numerous officers and men would return from France to offer to the school the benefit of their experience.

[168]Moreau (1991).

[169]NAUK [AIR 1/502/16/3/4], 'Training of increased number of Observers and status of training at Wireless School, Brooklands, 1916 Apr.–1917 Jan.'; NAUK [AIR 1/754/204/4/77], 'Establishment of R.F.C. Wireless School at Brooklands, 1915 Dec.'.

[170]NAUK [AIR 1/131/15/40/221], 'Wireless—periodical reports from R.F.C. Wireless School, Brooklands and reports of the Wireless Telegraphy Committee, 1916 May 19–July 26'.

[171]Vincent-Smith (1919).

[172]Burch (1980, pp. 18–19).

[173]NAUK [AIR 1/120/15/40/89], 'Proposed establishment for the R.F.C. Wireless School at Farnborough'.

[174]NAUK [AIR 29/718], 'No. 1 Radio School, Cranwell formerly No. 1 Signals School, formerly No. 1 Electrical and Wireless School, 1915 Nov.–1945 Dec.'.

The foundation of a new centre at Flowerdown, Winchester (replaced in 1929 by the Electrical and Wireless School at Cranwell) coincided with, and in all probability was instrumental in, an unparalleled output of wireless operators. The project for its erection had been gestating since the limitations of Farnborough had been much in evidence, and the transfer of the old school had taken place in 1918.[175] But it was only following the war, when the wave of operators returning from overseas was recognized as highly qualified manpower, that the directors of the school began to facilitate their retraining in civil society. Military instructions paved the way for a fruitful field of fresh openings to which wireless operators, if it had not been for the war experience, would probably never have had access.[176] Both Flowerdown and Cranwell proved to be a fertile source of personnel well prepared to feed the needs of the Merchant Navy, as well as civil and military aviation.[177] Once again, the importance of training appeared unequivocally. In 1920, in what was without doubt the main wireless training camp in the country, its *post-bellum* syllabus covered a fundamental theoretical formation and a comprehensive instruction in the electrical workshops and Instrumental Laboratory.[178] Taking the high figures of wireless operators in the interwar years into consideration (see Table 1), the effort was commendable, although it was probably merely the mirror of a phenomenon of greater magnitude. It is hard to imagine that the ambitions of hundreds of aircraft apprentices, who would become the backbone of radio and radar communication throughout the Empire soon afterwards, had been consummated in one Air Force insensitive to technical training.[179]

[175]NAUK [AIR 2/77], 'Geographical: British Isles: Towns etc. Named 'F' (Code A, 57/3): Wireless School and Buildings at Flowerdown'; NAUK [AIR 2/298], 'Projected move of Electrical and Wireless School from Flowerdown to Cranwell. (1926–1929)'; NAUK [ADM 338/33], 'Electrical and Wireless School, RAF Flowerdown and RAF Worthy Down, Winchester, Hampshire'.

[176]On the qualities of wireless operators at that time (though in an American context), see Packman (1915, 1916).

[177]M.R. Orme, 1920. Radiotelegraphy and Aviation. In *The Yearbook of Wireless Telegraphy and Telephony*. London: Wireless Press, pp. 988–94. In October 1938, the Radio School at Yatesbury absorbed the No. 1 Wing (Airmen) of the Electrical Wireless School at Cranwell. On this new centre, see NAUK [AIR 29/719], 'No. 2 Radio School, formerly No. 2 Signals School, formerly Electrical and Wireless School Yatesbury, 1938 Oct.–1945 Dec.'.

[178]Michael B. Egan, 1920. 'The Training of a R.A.F. Wireless Operator'. *The Wireless World*, 7, 621–37. On the courses and syllabuses, see 'The R.A.F. Wireless School: A Digest of the Official Syllabu's, *The Wireless World*, 1920, 8, 285–6.

[179]The educational component has frequently passed unnoticed for radar historiography. See, for example, Alan Beyerchen, 1996. 'From Radio to Radar. Interwar Military Adaptation to

Table 1 Numbers of officers and students at the Wireless Operator's School at Farnborough, the Wireless Training Camp of Flowerdown, and the Electrical and Wireless School at Cranwell

Farnborough			Flowerdown			Cranwell		
	Officers	Learners		Officers	Learners		Airmen	Apprentices
1915	4	200	1918	186	3,783	1931	361	338
1916	8	648	1919	60	1,400	1932	295	410
			1921	15	500	1933	212	399
			1925	15	850	1934	175	351
						1935	101	297

Source: NAUK [AIR 29/718], 'No. 1 Radio School, Cranwell, formerly No. 1 Signals School, formerly No. 1 Electrical and Wireless School, 1915 Nov.–1945 Dec.'.

University and the radio industry

An even more eloquent form of commitment towards industrial needs was apparent in the universities and technical colleges. Here, the response from the educational institutions in developing specialities for the service of the local and national radio industry was considerable, albeit insufficient for the requirements of the sector.[180]

To a remarkable degree, Cambridge educational policy had been directed to pure science and research. Despite a certain animosity towards applied science, those aims had been achieved through industrial backing bestowed preferentially on ventures linked to industrial requirements.[181] It was the insight of Rutherford and his contacts into private industry, for example, which allowed Appleton to impart courses on wireless, including some laboratory research along with Balthazar Van der Pol, from 1919.[182] The investigation was a fundamental aspect,

Technological Change in Germany, the United Kingdom, and the United States'. In W. Murray & A.R. Millett, *Military Innovation in the Interwar Period.* Cambridge: CUP, 265–99, pp. 275–87.

[180]The issue of education and industry is reflected in numerous studies. Although published long time ago, by far the best work is still by Michael Sanderson, 1972. *The Universities and British Industry, 1850–1970.* London: Routledge & Kegan Paul, pp. 243–313. See also Sanderson (1999, pp. 55–73); Wickenden (1929, pp. 130–97); Austen Albu, 'British Attitudes to Engineering Education: A Historical Perspective', in Pavitt (1980, pp. 19–87); and Moseley (1977).

[181]Financial support came from Austin and Mond for physics, and Sir John Siddeley for aeronautics. J.P.C. Roach, 1959. 'The University of Cambridge'. In *Victoria County History of Cambridgeshire.* Vol. 3. Oxford: OUP.

[182]The link between Rutherford and A.P.M. Fleming, the director of research of Metropolitan-Vickers, facilitated the experiment of artificial disintegration in 1932. See Allibone (1984, pp. 151–73).

and enrolments and specialized courses rose progressively. In the 1920s, both the lectures and the research on radio waves propagation and electrical oscillations were integrated in the curriculum of the Natural Sciences Tripos;[183] and from 1927 onwards, the young John Ashworth Ratcliffe devised a radio ionosphere research programme which, financed by an annual grant of £500,[184] attracted a dazzlingly international stream of graduates, chiefly from Australia but also from New Zealand and Canada.[185]

Oxford, by contrast, barely showed any disposition to cultivate applied science or to create a narrow training school for technicians. Lindemann had resuscitated the Clarendon Laboratory from its lethargic state, but Oxford's particular academic structure acted as a constraint.[186] The improvement in practical training which came with the opening, in 1910, of the Electrical Laboratory, under the direction of J.S.E. Townsend, did much to alleviate the situation.[187] In fact, it was after World War I, when high-vacuum tubes were applied for the first time for wireless communication, that significant results in high-frequency waves and ionization of gases were achieved.[188]

The University and Colleges of London more typically found their strengths in applied physics, aeronautics, and engineering than in the pure sciences of Cambridge. London was one of the bastions of Britain's radio industry in the 1920s. Its territory housed a high concentration of commercial firms, which in some cases exceeded that of the rest of the country.[189] Density on this scale was rivalled only by those of the great metropolises in Eastern America. Radio propagation and thermionic vacuum tubes became a forte of King's College under Appleton's supervision, especially in the elucidation of the physical structure of the ionosphere.[190] The emphasis of the activities of the college

[183]Appleton's lecture course on 'Electrical oscillations and radio telegraphy' especially appealed to students, partly by its breadth, 'impinging so closely on practical engineering matters', in Ratcliffe's words. See R. Clark (1971, p. 33).

[184]J.G. Crowther, 1974. *The Cavendish Laboratory, 1874–1974*. London: MacMillan, p. 267.

[185]Budden (1988, p. 677). Again, since 1921, Cambridge University Wireless Society (which had a membership of 80 in that year) organized weekly lectures and activities on the subject. See 'Wireless Club Reports', *The Wireless World*, 1921, 9, 805; and 10, 86.

[186]Morrell (1992, p. 268).

[187]Engel (1957, p. 261).

[188]Bleaney (1994, p. 255).

[189]In 1924, 125 of London's wireless apparatus makers competed against 77 British and Irish firms; 114 wireless components manufacturers against 82; and more than one in four radio engineers had his workplace in the capital. See *Kelly's Directory of the Electrical Industry, Wireless and Allied Trades*. London, 1924.

[190]Ratcliffe (1966a, pp. 3–4).

was on strictly technical training in what the promoters defined as an excellent venue for men entering industry.[191] The establishment of the Halley Stewart Laboratory in 1932 was typical of a private patronage bestowed on an influential personality, such as Appleton, associated with the British RRB and with growing links with radio industry and the BBC.[192] Beyond its scientific results, the importance of the Laboratory lay in the creation of the intellectual seed corn of radio scientists for the war and postwar years.[193]

Although London University earned much of its prestige in technical training for the teaching of chemical engineering, there were other minor educational institutions with which the local radio industry, in diverse ways albeit with a clear industrial orientation, interwove links of common interests.[194] These included the City and Guilds of London Institute (without a doubt the most significant),[195] the Woolwich Polytechnic, the Northampton Polytechnic Institute, the London Telegraph Training College, and the Radio Engineering College. Generally, the object of their courses was to provide instruction for selected apprentices to become highly trained technicians, with national diplomas and certificates and starting from the base of a solid preparation in electrical engineering.[196] It would be too ingenuous, and profoundly simplistic, for these colleges to be labelled as second-rate or as a mere refuge for inept students. In fact, this network of schools was usually organized by arrangement and cooperation with local firms in those fields in which the big universities offered no service.[197] One might see evident similarities, for example, in

[191]Cromer (1933, p. 53).

[192]R. Clark (1971, p. 69).

[193]White (1975, pp. 41–3); and AAAS. [111/4/5], 'F.W.G. White's Biographical Notes, vol. 1', pp. 23–4.

[194]Millis (1932, pp. 58–9).

[195]As Professor of Applied Physics and Electrical Engineering from 1916 to 1926, William Henry Eccles was a staunch advocate of any attempt to extend the emphasis of the institute on wireless studies and to direct its very limited teaching towards industry. During that time, he wrote two books which proved to be of great value to students: *Wireless, Telegraphy and Telephony: a Handbook of Formulae, Data and Information* (1915); and *Continuous Wave Wireless Telegraphy* (1921). See Ratcliffe (1971, p. 204) and City and Guilds of London Institute, 1993. *A Short History, 1878–1992*. London, pp. 63–4.

[196]See IEEA, [UK0108 NAEST 065/A/5], 'Education, apprenticeship and employment', 1936–37.

[197]Whereas 37 students from the City and Guilds of London were awarded a national diploma for their good marks in Radio-communication in 1937, only one London University student obtained the PhD in communication. See *Woolwich Polytechnic. Distribution of Prizes and Medals* (London, 1937), p. 4.

the methods and procedures for instruction in different schools.[198] Just as the Northampton Polytechnic offered a scheme of 'sandwich' engineering training through cooperative firms such as Pye Radio, GEC, and British Thomson-Huston Co. (normally 250 students attended the day courses in 1935),[199] so the Woolwich Polytechnic offered similar courses and the establishment of the Electrical Communication Laboratory was oriented towards the service of the Johnson and Phillips Ltd Company.[200]

By virtue of local traditions and cultures, the universities of the Midlands and the North of England more easily adapted to the demands of industrial employment. Manchester University had valuable vitalizing effects on developing the new sectors of the Lancashire economy, in particular that of electrical industry.[201] The Metropolitan-Vickers Electrical Co. of Trafford Park, one of the 'Big-Six' British radio firms, required of its school apprentices to attend the College of Technology, Manchester, or the Royal Technical College, Salford, for part-time technical instruction.[202] A sign that practical apprenticeship was the bridge linking the academic world and private industry was the fact that many of them were eligible to compete for scholarships leading to a degree course of engineering at the university.[203] Close bonds with new industries were also developed at Liverpool University, notably in electrical engineering and communications, with British Insulated Cables and the Automatic Telephone Company. The university had a flourishing electrical engineering department which coordinated all experimental work in wireless for the RRB.[204] Moreover, the town being of particular maritime importance, it is not surprising that Liverpool

[198] As an innovative method in the 1930s, the so-called 'sandwich system' was intended to alternate academic studies and work experience (usually six months of pupilage in any radio firm in the second year). See 'University Training of Engineers: Timing of Practical Works Experience', *JC*, 1937, 16, 567–8.

[199] See 'A Sandwich Engineering Training', *JC*, 1935, 14(150), 176.

[200] In the second year, as part of the specialization in Electrical Engineering, Woolwich Polytechnic provided the subject of Electrical Communication, whose syllabus included: the propagation and reception of waves, wireless direction-finding, general principles of the theory, design, and construction of instruments and apparatus employed in telegraphic and telephonic working. See IEEA, [UK0108 NAEST 065/A/4], 'Education and apprenticeship', *Woolwich Polytechnic, School of Engineering* (London, 1935–36), p. 11.

[201] Sanderson (1972a, p. 253).

[202] 'Personnel Training in the Metropolitan-Vickers Electrical Co. Ltd'., in R.W. Ferguson ed., 1935. *Training in industry: a report embodying the results of enquiries conducted between 1931 and 1934 by the Association for Education in Industry and Commerce.* London, 111.

[203] 'Practical Training in Engineering: Metropolitan-Vickers Electrical Co. Ltd'., *JC*, 1926, 5(6), 8–11, p. 10.

[204] *Liverpool Courier*, 20 January 1920 and 4 June 1920.

Wireless College (opened in 1901) became the main training centre for operators from local firms, such as Marconi International Marine Communication, British Wireless Marine Service, and Radio Communication.[205]

It is hard to encapsulate the factors which led to an increasingly high demand of graduates, as well as the qualities that were required of them. The inclusion of PhD degrees in the university curricula (owing to the impossibility of realizing them in Germany during the war), and the proliferation of research laboratories in both industry and the DSIR were trends that the entrepreneurs could not ignore, as we will see in the next section.[206] Perhaps the fact that, among the leading firms of the nascent radio industry, there were electrical companies which opted for restructuring, and consequently were forced to change their techniques—in such a way that the distinction between electrical and wireless company rendered practically meaningless—, was a determining factor. But it also seems that, for the managers of these firms who were now coming to prominence, the actual performance of the job was much more demanding than it had been in the eyes of their predecessors. This was in part an inevitable consequence of the severe depression of the period, which had deified competition and magnified the dangers of failure.

There were, however, other more ethereal reasons which the erstwhile industrialists had conspicuously overlooked. Perhaps owing to the harshness of these difficult years, the preferences of radio entrepreneurs tended to less definable qualities, which in theory would appear more frequently in the university graduate, not so much for his talent, as for his spirit and broad outlook. A.P.M. Fleming, the Director of the Metro-Vickers, mentioned a 'particular temperamental quality',[207] 'an independence of thought', 'a sense of responsibility and self-reliance', and, above all, 'an outstanding personality';[208] the editorial column of the *Journal of Careers* especially emphasized 'the importance of personality',[209] but also referred to 'energy and tact',[210] and the British Science

[205]NAUK, [Post 30/4495], 'Liverpool Wireless College. Wireless Telegraphy Establishment'.

[206]H.R. Lang, 1934. 'Physics in Industry: a Developing Field of Work'. *JC*, 13, 527–31. As the Secretary of the Institute of Physics noted, the demand for physicists for research and industrial posts in particular exceeded the supply.

[207]A.P.M. Fleming, 1929. 'Engineering and Other Openings in the Wireless Industry'. *JC*, 8, 9–12, p. 10.

[208]A.P.M. Fleming, 1932. 'Looking Ahead in Electrical Engineering. Future Openings for Trained Engineers'. *JC*, 11, 7–14, p. 12.

[209]'Physicists for Industry. Where They Find Scope and How They Should Be Trained', *JC*, 1936, 15, 435–6, p. 436.

[210]'The Wireless Industry. Requirements and Prospects of Technical Appointments', *JC*, 1932, 11, 7–11, p. 10.

Guild 1934 report to 'pluck, self-reliance, and character'.[211] In all events, these were somewhat subtle and incorporeal attributes, which expressed an implied disenchantment with the narrowness of old apprenticeship systems, rather than an attachment to university values. The effect of this was the idealization of the archetypal graduate, or what at that time was popularized with the cliché of 'right type' in industrial and careers literature, superseding—in the words of M. Sanderson—the 'right stamp' and 'right calibre' metaphors of earlier periods.[212]

It is tempting to conclude that the myriad of educational offers catalysed an immediate positive reaction for the virtues of university–industry linkage, so permitting the numerous competitors in Britain to feed on skilful human resources.[213] In its enquiry into the state of technical training in 1937, however, the Board of Education warned of the excessive dependence on evening courses (in marked contrast with that of the Continent),[214] and to the inhibited response that it could generate in employers.[215] But even if many firms sought alternative solutions, it is not clear to what extent the limited organizational capacity of schools restrained entrepreneurial stimulus to investment in the radio sector. Perhaps these other comments help us to understand the imperfections of linkage; in a memorandum to the Board of Education in 1935, the British Institution of Radio Engineers drew attention to the shortage of trained radio personnel and to the obsolete equipment of technical schools.[216] In the light of these criticisms, it seems clear that the levels of output of

[211]'Careers in Industrial Research: Importance of Suitable University Training', *JC*, 1934, 13, 507.

[212]Sanderson (1972a, p. 247).

[213]Reviewing the benefits of education for the radio industry, Sir Ambrose Fleming predicted a brilliant future for radio-communication engineering. See A. Fleming, 1930. 'Electro-Communication Engineering as a Career'. *JC*, 9(99), 7–12, p. 8.

[214]Board of Education, 1937. *Annual Report*. London, p. 10.

[215]The list of curricula of the technical schools and colleges of the UK is, in this respect, instructive. The high presence of evening attendance is consistent with the fears expressed by the Board. According to the *Handbook to the Technical and Art Schools and Colleges of the U.K.* (London, 1925), pp. 7, 39, 61, 94, 104, 111, 124, 129, 130, 132, and 154, wireless evening courses were imparted at Croydon Polytechnic Central, Lauder Technical School (Dunfermline), Greenock Technical School, and Municipal Technical Institute (Limerick), while wireless telegraphy courses were taught at the Technical College (Bradford), Municipal School of Science (Great Yarmouth), Technical School and Fielden Art Classes, Royal Technical College (Glasgow), and Nautical College (Leith); only the Municipal College (Portsmouth) and the Marine School of South Shields offered day courses.

[216]The British Institution of Radio Engineers, 1944. *Post-War Development in Radio Engineering*. London, p. 7.

qualified personnel from universities and colleges to industry were, at least, insufficient, and their preparation, in some cases, deficient.[217]

The criticism of radio engineers appealing for their own schemes of technical training contrasts powerfully with the enthusiasm of broadcasting leaders for the educational system emanating from universities. For example, R.T.B. Wynn, the head of the Engineering Information Department of the BBC, acknowledged that the Company 'required a recognised university or engineering college training as a condition of appointment'.[218] Likewise P.P. Eckersley, perhaps the most dynamic figure in the BBC, thought that Oxford and Cambridge were ideal for those inclining 'towards an eventual research job, and the provincial universities for those who wanted to become practical engineers'.[219] The explanation for this warm welcome is simple. For although the access to non-graduates was not refused, the Company's management was becoming more complex and demanding. Broadcasting industry was new and scientifically very sophisticated; many of its leaders had been intellectually forged in academic classrooms (mostly as electrical engineers), and demanded a high degree of scientific and technical expertise.[220] The new chief engineers of the BBC could have understood the general shift towards the specialization of radio engineering that the British Institution was intended to promote, but at the same time they could have felt a profound sympathy for curricula which led to posts as diverse as radio technicians, directors of programmes, announcers, and superintending engineers.[221] It is therefore hardly surprising that, given the particular idiosyncrasy of this industry, selection of graduates at the BBC increased in geometric progression during the first years: from the first engineering appointment on February 1923 (P.P. Eckersley), to 46 on December 1923, 179 by the end of 1925, and 213 in 1926.[222]

[217]In the US the position of radio engineer had a certain distinction. For an analysis of the role of American radio engineers in broadcasting policy, see Slotten (1995).

[218] 'Twentieth-Century Careers. II—Wireless Engineering', *JC*, 1935, 14, 658–64, p. 662.

[219]P.P. Eckersley, 1933. 'The Future of Wireless Engineering. Considerable Expansion Inevitable'. *JC*, 12, 7–10, p. 9.

[220]In fact, there were suspicions that with respect to BBC appointments candidates with Oxford or Cambridge degrees enjoyed an undue preference. In 1936 non-engineering staff consisted of 395 graduates (76 from Oxford, 40 from Cambridge, and 75 from other universities) and 204 with no degree. See 'Employment in Broadcasting: Broadcasting Committee's views on Staff Selection at the BBC', *JC*, 1936, 15, 290–2, p. 291.

[221]A.P.M. Fleming, 1932b. 'Looking Ahead in Electrical Engineering. Future Openings for Trained Engineers'. *JC*, 11, 7–14, p. 11.

[222]For the list of the first Engineers-in-Charge and the organization chart of the Engineering Department, see Edward Pawley, 1972. *BBC Engineering, 1922–1972*. London: BBC Publications, pp. 71–7.

Education in the industrial context

Paradoxically, the main significance of the university–industry linkage lies in its contribution to the preservation of an alternative educational system within the industry. Let us see how. Like broadcasting leaders, most radio entrepreneurs perceived that universities, by introducing assorted forms of engineering, were certainly contributing to the rise of new industries and to the change of industrial structures.[223] Their assistance and expertise were widely recognized. Sir Hugo Hirst, the Chairman of the General Electric Co., deemed their participation as highly 'satisfactory';[224] Sir Ambrose Fleming stated that the possession of a university honours degree was 'a passport generally to a responsible and interesting post';[225] whereas the heads of the Metro-Vickers Co. saw no obstacle to admitting that 'a university course or advanced technical college course is the most desirable preparation for the profession'.[226] The rapid formation of research departments after World War I, many of which were the result of the shock wave generated by the needs of the conflict, was, in this respect, an important spring of fresh openings.[227] However, the flow of talent into radio industry was moderate, albeit effective, due to constraints in the educational system.[228] By and large, in spite of the rate of absorption of graduates gradually increasing, the radio industry always remained eager for certain skills. The imbalance between working needs and output of graduates was certainly not negligible.[229]

Satisfied, on the one hand, with an increasingly skilful workforce, but overwhelmed, on the other, by a galloping expansion of the sector, the entrepreneurs resorted to technical training within the company. Earlier in the century, the stimulation of premium apprenticeship, its attendant specialization via pupilage, preliminary workshop courses, and piecemeal practical teaching had

[223] 'The Wireless Industry. Requirements and Prospects', op. cit., pp. 7–11.

[224] In H. Hirst's words, 'as a rule, they are given experience in all departments, with a view to promotion to posts of minor and major responsibility as soon as they are competent'. See 'Opportunities in the electrical industry. An interview with Sir Hugo Hirst', *JC*, 1928, 7, 29–30, p. 29.

[225] J.A. Fleming (1930, p. 8).

[226] 'Practical Training in Engineering: II Metropolitan-Vickers Electrical Co., Ltd'., *JC*, 1926, 5, 8–11, on p. 9.

[227] Sir Frank Smith (1933, pp. 19–22), the Secretary of the DSIR, emphasized the importance of industrial research and the cooperation with universities for the development of new industries.

[228] For our present purposes, it would take a long time to determine why universities did not fulfil in general the appetite of employers. On this point, see Sanderson (1972a, pp. 243–313).

[229] On this question, see J.J. Thomson, 1931. 'Training and Careers in Physics'. *JC*, 10, 18–20.

been appropriate instruments not only of instruction but also of control.[230] But by the 1920s, with a galloping augmentation of radio exports (from 300,000 sets in 1923 to over 1,200,000 in 1927)[231] and with a flourishing broadcasting (36,000 wireless licences in 1922 as opposed to almost nine million in 1939)[232] beginning to play an important role in developing the radio industry, and vice versa, the response must be necessarily of a greater magnitude. Now, in a design that was initiated by the Marconi Wireless Telegraph Company twenty years before, the serious shortages that might have arisen if they had depended solely on university supply were remedied with home-grown training schemes, close to some of the main technical schools, and which were not exclusive to the wireless sector.[233] It was a scheme upon which the big firms, GEC, Metro-Vickers, BTH, and Western Electric in part built up their empire and from which they received much credit, though the conception, at least in the wireless industry, without doubt emanated from Marconi Company and its School of Wireless Communication, the first of its kind in the world.[234]

Of the several wireless firms and cable companies in which the tactic of training of premium apprentices was pursued, the Marconi School was the pioneer in providing a teaching centre for qualified engineers.[235] From its foundation in 1901 (first at Frinton and then at Chelmsford) until the expansion in 1935 (with the creation of an experimental laboratory), the school served as a plant for the output of operators and a way of profitably pouring the talents of young students into radio practice. The company's primary objective was to provide a basic instruction to probationary engineers (in the absence of high-training industrial institutions),[236] but after World War I it concentrated on openings in wireless operating.[237] In both respects, it was considerably

[230]Wickenden (1929, pp. 140–4).

[231]A.J. Smith (1928, p. 589).

[232]See M. Pegg, 1983. *Broadcasting and Society, 1918–1939*. London: Croom Helm, p. 7.

[233]Employers of high growth industries like motor cars adopted training schemes in their own way to help satisfy their labour requirements. See Thoms & Donnelly (1985).

[234]For the prospects of radio engineering at the Marconi Co., see R.N. Vyvyan, 'Wireless as a Career', in Vyvyan (1933, pp. 219–31).

[235]The standard history of the Marconi School in its early years is H.M. Dowsett, 1937. 'The Marconi School of Wireless Communication'. *The Marconi Review*, 66, 1–14. See also W.J. Baker (1970, pp. 274–7).

[236]'The Wireless Operator. Qualifications and Prospects', *The Wireless World*, 1913, 1(2), 102–4.

[237]By virtue of the Merchant Shipping Acts of 1919 and 1932, merchant vessels and fishing fleets of 1,600 tons and over were required to be equipped with apparatus for radiotelegraphy and to carry wireless operators. See 'Training and Prospects in Wireless Operating', *JC*, 1927, 6, 21–2.

successful. Specialized engineers in telegraphic technique, marine operators for ship and shore stations, and the prestigious 'Marconi Apprenticeship Scheme', designed in cooperation with the Mid-Essex Technical College,[238] contributed to showcase the company as a brand with an aura of inventiveness and freshness that soon had emulators.[239] The new subsidiaries of the company, deployed all over the world, soon became avid recipients of operators, and the scheme was admirably transplanted to the Amalgamated Wireless Company, Sydney, but also to Madrid, Montreal, and Babylon, Long Island, New York.[240]

Although the company's training scheme was inseparable from its industrial roots, there were numerous signs that, especially during the war, the heads of the company felt easy, not to say complaisant, with somewhat awkward, thorny affairs related to defence. In this sense, even the training scheme might not be understood in isolation from the imperial motives and hidden links by which the Company's policy was characterized. When in 1914 the Army had shown an interest in the organization of a large-scale wireless training school at the Crystal Palace, the Marconi School soon facilitated its instructors and lecturer engineers to promote the cause.[241] A year later, it proffered its experimental section at Brookland for the creation of a wireless training school for pilots;[242] and in 1917 it would lend its teaching staff to the Royal Navy's Marine Observers School at Eastchurch.[243] Fifteen years later, industrial training would move to the commercial and imperial realm. After the inauguration of a school for pilots and wireless operators in 1935 by the Imperial Airways Ltd. with Marconi Company's assistance, it was seduced by the lure of commercial aviation that beguiled many other radio firms. By 1936, it had already been capable of instructing a body of pilots and radio officers for the Empire's air forces in the newly formed Marconi's Air Radio Training School

[238]NAUK. [ED 82/18]. 'Chelmsford, Mid-Essex Technical College, 1931–1935'.

[239]Other reputable training centres for wireless operators were Siemens Brothers & Co., Radio Communication Company, and the British Marine Wireless Service. See 'Wireless Operating: a Career or a Training Ground for Other Careers?', *JC*, 1934, 13, 618–22.

[240]'Wireless Engineers in Training. Apparatus and Methods at the Chelmsford Wireless College of the Marconi Company', *The Wireless World*, 1927, 20, 221–3.

[241]W.J. Baker (1970, p. 164); and Hancock (1950, pp. 88–95).

[242]NAUK, [AIR 1/754/204/4/77]. 'War establishment for R.F.C. Wireless School' (1915 Nov.), 'Establishment of R.F.C. Wireless School at Brooklands' (1915 Dec.), and 'Training of Increased Number of Observers and Status of Training at Wireless School, Brooklands' (1916 Apr.–1917 Jan.).

[243]W.J. Baker (1970, pp. 169–70).

at Croydon.[244] This strategy of deliberate and meticulously calculated alliances fostered not only the dissemination of industrial training but also, albeit perhaps more subtly, the confluence of interests transcending such frail boundaries as existed between Marconi and the fighting services.

Another significant feature of the *specificità Marconiana* was apparent in the development of a cultural milieu. Here, the priorities for training and instruction gave the Company a marked distinctive character. In characteristic manner, its policy had been directed to encourage and collect the reports of its engineers and operators throughout the world for then disseminating to the public in general. That aim was channelled through the foundation of 'The Marconi Press Agency Ltd.' in October 1910 (rechristened 'The Wireless Press' in 1914),[245] an event to which W.J. Baker referred vividly as 'an instance of tall oaks growing from little acorns'.[246] The publication of *The Marconigraph* in 1911, the first periodical in the world dealing exclusively with wireless science (retitled *The Wireless World* in 1913), undoubtedly contributed to the high status and distinction associated with the profession as it was cultivated by the Company.[247] To be sure, periodical series such as *The Year Book of Wireless Telegraphy* (printed annually from 1913 until 1925) and various textbooks such as *The Elementary Principles of Wireless Telegraphy* and *Handbook of Technical Instruction for Wireless Telegraphists* were considerably more advanced than those of their competitors.[248] They were indicative, nonetheless, of a substantial engagement by the senior engineers in the values of the company's educational milieu. Given this promising atmosphere, it is not surprising that engineers and physicists from everywhere responded so effectually to the call for participation launched from Marconi House in the Strand, London (over a hundred books appeared before it was sold to Messrs. Iliffe and Sons in 1925).[249] Nor is it surprising that the heads themselves of the

[244]Ibid., p. 243.

[245]In changing name, the company sought to reflect more accurately its multiple activities. See 'The Wireless Press, Limited', *The Wireless World*, 1914, 2(17), 300; and NAUK, [J 13/7146]. 'Wireless Press, Ltd., 1914'.

[246]W.J. Baker (1970, p. 277).

[247]Hugo S. Pocock, 1971. 'Sixty years', *Wireless World*, 77, 153–5.

[248]Raymond Dorrington Bangay & Oscar Frank Brown, *The Elementary Principles of Wireless Telegraphy* (1914); and John Clayborough Hawkhead, *Handbook of Technical Instruction for Wireless Telegraphists* (1913). The former was translated into Spanish in 1917 and French in 1918.

[249]Apart from the literature already mentioned, copies from the early period of 'The Wireless Press' are almost exclusively in the form of, and limited to, textbooks, technical handbooks, and compendia. Such are, for example, the classic treatises by John Ambrose Fleming, *The Thermionic Valve and Its Developments in Radio-Telegraphy and Telephony* (1919) and *Fifty Years of Electricity* (1921); William Henry Eccles, *Continuous Wave Wireless Telegraphy* (1921); R. Keen, *Direction and Position*

company sought the motivation, publicity, and justification for its own research activities in the publication of specialist technical journals, first with *Radio Review* (1919–20) and then with the illustrious *The Marconi Review* (from October 1928 onwards). As result of all this, the intellectual production of the company was exemplary but it was so in a manner that left leading its role in precisely the hands where it had always been.[250]

In reality, the amplitude of Marconi Company's training experience was but the reflection of the power of Britain's radio industry as a whole. Economically, the sector witnessed an accelerated and vibrant growth, despite a period of depression that hardly had any repercussions on the sale of radio receivers. Understandably, electrical entrepreneurs, so proud of their long training tradition, showed an overt disposition to the educational adaptation. They were so minded, yet somewhat preoccupied about the industrial transformation as was Marconi itself. The big companies feared the competition of an industry that in its first three years had generated 200 new setmaker firms and that in one year consumed as many sets as in the whole of the years preceding it. Radio set manufacture, however, was for them a relatively minor diversification that had to adapt as quickly as it could to an essentially 'heavy electrical culture'.[251] In this respect, the adoption of a variety of formulas of apprenticeship served to train qualified electrical engineers and technicians, rather than radio engineers. But even so, it both widened the pyramidal base of radio industry and convinced employers that to invest in industrial education was something on which there could be no hesitancy.[252]

Finder (1922); and the practical handbooks by Phillip R. Coursey, *The Radio Experimenter's Handbook* (1922); Marconi's Wireless Telegraph Company, *Instructions for the Use of the Direction Finder* (1915); Harold Ward, *Pocket Dictionary of Technical Terms Used in Wireless Telegraphy* (1918); Phillip R. Coursey, *The Radio Experimenter's Handbook* (1922); Bertram Hoyle, *Standard Tables and Equations in Radio-Telegraphy* (1919); and Elmer Eustice Bucher, *Practical Wireless Telegraphy: a Complete Text Book for Students of Radio Communication* (1917).

[250]It is hard to understand radio development in Britain in isolation from Marconi's editorial empire. In fact, The Wireless Press was even stronger than we indicate in the text, for another publishing subsidiary was open in New York, and one of its flagships, *The Wireless World*, became the official organ of the Wireless Society of London (later known as the Radio Society of Great Britain).

[251]Keith Geddes & Gordon Bussey, 1991. *The Setmakers: A History of the Radio and Television Industry.* London: Brema, 29.

[252]The electrical and radio industries were among the most important of the industrial economies that remained loyal to the motto 'qualified training for industry'. For the response of different firms in a local context, see David Thoms, 'Technical Education and the Transformation of Coventry's Industrial Economy, 1900–1939', in Summerfield & Evans (1990, pp. 37–54).

The conviction that technical training should abandon the traditional system of external evening school and opt unequivocally for domestic schemes—either part-time or full day courses—guided industrial education policy in wireless industry. As early as 1916, the management of the American subsidiary British Westinghouse Company had reversed the then controversial practice of study after the working day by instating a department of education and a technical college for the purposes of selecting and instructing its apprentices.[253] Some years later, now under the control of its British successor, Metropolitan-Vickers Electrical Co., the assortment of educational offers was extraordinarily rich.[254] The courses were of outstanding quality, and, above all, benefited by addressing wide labour sectors, such as training for artisans and professional employment and assistance for engineers and postgraduates.[255] Whereas the executives of the old firm had regarded industrial education as an almost paternal duty in the formative period of 14-years-old boys, by the late 1920s the new Director of Research and Education, A.P.M. Fleming, could recruit, in appreciable amounts, university students for his purposes.[256]

Electrical industry entrepreneurs adopted a variety of training schemes to adapt to the changing nature of the relations between electrical and wireless demands. These ranged from the three-year courses of practical experience for university graduates to engineering courses of five years duration, including night classes in associated colleges. GEC, for instance, became well known in the 1920s for the emphasis it gave to the recruitment of young men in order to provide them with a full engineering course at the main works at Witton, Birmingham,[257] while BTH was justly famous for its five types of apprenticeship training courses in the Rugby works, all available without payment of premium.[258]

[253]IEEA [NA ESST 70]. 'Some features of the educational and research work of the British Westinghouse Company', p. 1.

[254] 'New Apprenticeship Course of the Metropolitan-Vickers Electrical Co. Ltd'., *JC*, 1929, 8, 45–6; 'Practical Training in Engineering: Metropolitan-Vickers Electrical Co. Ltd'., *JC*, 1926, 5(6), 8–11.

[255]Metropolitan-Vickers Electrical Co., 1921. *Education for Industry. A description of the Apprenticeship Courses and Methods of Training of the Metropolitan-Vickers Electrical Co.* Manchester, pp. 4–33.

[256]In the late 1930s, for example, hundreds of engineering graduates entered the works of the firm each year for practical training, most of which took the course as a postgraduate one and held posts of responsibility. On the history of the company, see Read (1998) and Dummelow (1949).

[257]'Practical Training in Engineering: XXV. General Electric Co., Ltd., Witton Engineering Works, Birmingham', *JC*, 1928, 7, 33–7; 'Commerce and Industry: Pioneers in the British Electrical Industry,' *JC*, 1926, 5, 29–30.

[258]'The Profession of Electrical Engineering. An Outstanding Training Scheme', *JC*, 1932, 11, 33–6; and 'Practical Training in Engineering: VII. The British Thomson-Houston Co'., *JC*, 1926, 5, 16–9.

Another highly popular strategy, practised above all in very specialized and diversified small firms, was that of Ferranti Ltd. who, having no facilities for the general training of electrical engineers, imposed the policy that executive positions were exclusively filled by men trained in their own organization.[259] As part of a general tendency towards expansion and volatility that extended after the Depression, the labour market in the radio industry became a coveted sector for which the few surviving major companies outbidded each other.[260] It was within this context of an increasingly competitive labour market that other leading firms bordering on the field of radio, such as Standard Telephones and Cables Ltd. and the Indo-European Telegraph Co., promoted the development of electrical communication engineering in their own companies.[261]

However, in electrical firms as in technical colleges, this training was rarely understood as synonymous with radio specialization. *De facto*, even the body par excellence that was a purveyor of certificates on the subject, the Institute of Electrical Engineers, bestowed only a 'National Certificate in Electrical Engineering (Radio)' in 1920. Plainly, the autonomy of radio engineering in Britain occurred more slowly than it seems to have done in the USA and Germany. While the issue of national qualifications was an important one, one should not underestimate the sustained engagement by other parallel institutions in the realization of examinations on wireless technology. Establishments such as the British Institute of Radio Engineers (created in 1925), the Institute of Wireless Technology (1931), and the City and Guilds had a preponderant role in the awarding of the official diplomas which were so 'generously' used in the retail trade. It explains the apparent paradox that, while the sector had a substantial deficit of superior academic degrees, the editors of a *distingué* journal of the trade bemoaned the fact that anyone could place on his shop front the much-abused sign 'Radio Engineer'.

Clearly, craftiness was a common practice. In reality, the official distinction between electrical and radio specialists came late to Britain.[262] In the early 1920s, the gap between an electrical engineer who specialized in power supply or in the use of heavy electrical plant (and who might have been trained within

[259]'Opportunities in Commerce and Industry: XXII. Ferranti, Ltd., Electrical and General Engineers, Hollinwood. Methods of Training Sales and Technical Staff', *JC*, 1932, 11, 30–2.

[260]Geddes & Bussey (1991, pp. 103–12).

[261]'Practical Training in Engineering: XV. Standard Telephones and Cables Ltd'., *JC*, 1927, 6, 20–3; 'The Indo-European Telegraph Service', *JC*, 1928, 7, 37–9.

[262]On the contrary, some American universities were granting bachelor's degrees in radio engineering from the early 1920s, which corresponded to the equivalent degrees in electrical engineering. See Jome (1925, pp. 78–82); and C. Jansky, 1926. 'Collegiate Training for the Radio Engineering Field'. *Proc. IRE*, 14, 431–45.

a radio setmakers firm, such as Ferranti and Cossor, but who might equally have been in one of the great railway workshops, such as Metro-Vickers) and the radio engineer, with his own exigencies, was substantial in practice, albeit minuscule for legal purposes.[263] But in the next decade, that gap widened even more, at least in terms of complexity and specialization, if not of sophistication, with the arrival of mass production. And yet the provision of official certificates was not modified.[264] That widening was due in part to the erection of their own technology: the design of aerials and the use of thermionic valves required their own language, for example. But it was also due, especially in the imperial context, to the diversification of services: broadcasting and point-to-point communications demanded more than a six-month module of a general engineering course, as was the general norm.[265] That lack of distinctive approach was vital, because not only did it determine the procedures of selection in the electrical firms themselves but, perhaps much more important, it tended to shape the 'approach, scope and direction' of the technical college education policy.[266] As a result, it disproportionately tipped the scales in the electrical engineer's favour.[267] The shortage of qualified personnel in radio is reflected in the recruitment of technicians during World War II. Whereas between 1941 and 1942 war needs demanded a total of 3,050 radio officers, the British Government was able to supply only 1,600, among them 200 from Canada and 100 from Australia and New Zealand.[268]

Radio industry: the catalyst of ionospheric research

At this stage, it is needless to say that, in the interwar period, the British radio industry, as in many other countries, experienced rapid scientific development. Government research organizations, like the RRB, benefited from this impetus; the universities and industry strengthened bonds; and the radio industry as a

[263]On the differences in qualifications and expectations, see 'The Wireless Engineer. Qualifications and Prospects', *The Wireless World*, 1913, 1, 16–17; A.P.M. Fleming (1927, 1930).

[264]The extent of British backwardness in providing national certificates in radio engineering is evident in the fact that the first examination of the newly founded British Institute of Radio Engineers based on radio principles and applications was held in 1929. See Graham D. Clifford & Frank W. Sharp, 1989. *A 20th Century Professional Institution. The Story of the I.E.R.E., 1925–1988.* London, pp. 3–4.

[265]W.H. Eccles, 1925. 'The Work of a Wireless Engineer'. *The Wireless World*, 16, 649–51.

[266]The British Institution of Radio Engineers (1944, p. 9).

[267]On the discontent of wireless engineers with the official discrimination, see W.J. Reader, 1987. *A History of the Institution of Electrical Engineers, 1871–1971.* London: Peter Peregrinus Ltd., pp. 111–15.

[268]See NAUK [AVIA 42/42], 'Radio Training Scheme, 1940–42', Report 3.

whole, and in particular the firms themselves, witnessed a very considerable rise in research and development.[269] Yet this apparently categorical asseveration has to be read with caution. Here, a clarification is necessary. Radio was not a single entity. As diverse applications (on paper) as navigation, point-to-point communication, and broadcasting—the three realms of radio industry—resulted from specific procedures and developments. Each device differed in its methods and forms in accordance with its application. Furthermore, each piece of the device was not a simple unity but a composition of valves, resistors, coils, condensers, knobs, etc., assembled on a chassis, each of which had been subject to its own rhythm of innovation and development. In sum, these applications proceeded at different velocities within the radio industry, and affected atmospheric research in various degrees and ways.[270]

From its inception, the British radio industry became primarily the affair of the Marconi organization, concerned with the development of marine radio.[271] And it remained so until World War I, although point-to-point communication—notably through the trans-atlantic radio service—became increasingly important. An aggressive monopolistic policy tending to preclude any inter-communication between Marconi's own system and that of its rivals, led it to an unchallengeable position.[272] The Navy provided a market for the nascent radio industry, and it did as profusely as the Dominions and colonies had done in other areas of trade for generations past; battle fleets and coastal stations were equipped as sophisticatedly as imperial defence would require; and industrial expansion was consolidated (and to some extent was sustained) by fruitful fecund cooperation with the Admiralty.[273]

[269]Industrial research promoted privately within the firm has often been underestimated, disparaged and even ignored by historians of the interwar years. For an illuminating study concerned with the traditions of scientific research in private firms, see Michael Anderson, 1972b. 'Research and the Firm in British Industry, 1919–39'. *Science Studies*, 2, 107–51. Precisely this lack of industrial research has been attributed as an important cause of the British economic 'decline'. For a view against this interpretation, see Edgerton (1994).

[270]The best study of the aspects of the radio industry that are treated in this chapter is Sturmey (1958). See also Gordon Bussey, 1990. *Wireless, the Crucial Decade. History of the British Wireless Industry, 1924–34*. London: Peter Peregrinus Ltd.; Geddes & Bussey (1991); and Pegg (1983, pp. 49–54).

[271]It should be noted that the Post Office's contribution to radio development during these early years was practically insignificant. See Pocock (1988, pp. 1–13).

[272]David Read, 1998a. 'The Radio Communication Company'. *Bull. BVWS*, 23(3), 4–9, p. 6.

[273]Here we cannot examine in detail this cooperation. See NAUK, [ADM 1/29126], 'Wireless Telegraphy: Collaboration between Marconi's Wireless Co. and the Admiralty' (including 'Relations of Admiralty and Marconi Company', 19 February 1925).

The third realm of the radio industry, broadcasting, did not appear on the scene until the early 1920s. During World War I, the high military demand for valves for radio telephony and defensive priorities permitted the refashioning of the industrial strategies not only of the valve manufacturers themselves (Marconi's, and Cossor and Ediswan Companies) but also of a wider sector, that of the electric-lamp firms.[274] Three emblematic companies, BTH, GEC, and Metrovick, initiated the mass production of R-valves (a British version of the successful French triode valve), and after the war large surplus stocks of these valves became the mainstay of radio amateur's receivers.[275] Between 1918 and 1921, broadcasting (at that time far more advanced in the US) began to be organized on a factory basis and within a consensus framework. A policy of compromise between the Post Office and amateur wireless societies, on the one hand, and the reconciliation of conflicting interests between the six big firms (Marconi's, Metrovick, GEC, BTH, Western Electric, and Radio Communication) and the minor companies, on the other, gave rise to the birth of the British Broadcasting Company in December 1922.[276]

Clearly, broadcasting as an entertainment resulted from the engagement of the 'big six' (rather than the entertainment trades'), who expected to benefit from the sale of radio sets. In this respect, the regulation of broadcasting (as a result, largely, of Post Office pressure) generated in the first years indirect rewards thanks to the steadiness of turnovers.[277] But it is also true that it provoked, as if by domino effect, a burgeoning, if excessive competition obligating the sharing of profits: the setmaking industry, in particular, was so prolific that in 1925, only three years after the BBC's foundation, it was estimated that there was 1,600 manufacturers employing more than 40,000 people.[278] As a result, the 'big six', being somewhat inexpert in consumer products, found setmaking overwhelmingly competitive and scarcely profitable.[279]

A rather significant facet of this expansion was reflected in another deeply ingrained national tradition stemming from demonstrations in the private

[274]Ramsbottom (1995, pp. 114–18).

[275]Read (1998b, p. 17).

[276]The process of the BBC's foundation has been oversimplified here. For more details, see Bussey (1990, pp. 21–5), Pawley (1972, pp. 19–21), Briggs (1961, pp. 3–17), and Walker (1992, pp. 14–25).

[277]The annual turnover of the radio industry increased from 7,800,000 in 1926 to 30 million in 1931. See Hugh J. Schonfield, ed., 1933. *The Book of British Industry*. London: Denis Archer, 292.

[278]Sturmey (1958, p. 160).

[279]The only firm of the big six which maintained a great interest in the receiver market was GEC.

wireless societies. The network of specialist magazines was an important focus for radio amateurism. Here, avid and somewhat nostalgic home constructors of crystal and valve sets contrived their own accoutrements and proclaimed themselves as the legitimate heirs to the mythical pioneers of electromagnetism.[280] Especially in the mid-1920s, news-stands and appliance shops witnessed a flood of wireless periodicals (up to 30 between 1924 and 1926). Equipped with step-by-step guides, notably on home-made assemblies, they conferred an even more popular appeal on craftsmanship. Publishing houses such as the respectable 'The Wireless Press' and its rival 'Radio Press' emerged as the patrons of ventures which served as a banner of consumerist material culture fostered by radio firms.[281] As a result, magazines with variable circulation, some enduring but the majority ephemeral, such as *Wireless*, *Popular Wireless*, and *Wireless Constructor*, were able to channel the enthusiasm of wireless audiences, and, through them, especially in urban areas of modest presence, low-income listeners fashioned their own radio receivers.[282]

Although the BBC's formation was an incentive for competitiveness, the uncontrolled proliferation of small, weaker firms and the lack of research from the setmakers eventually took a heavy toll on trade.[283] In particular, the reduction of royalties and unfavourable conditions for exportation deprived the radio industry of its most lucrative business, the trade in kits of spare parts. The home-made receiver became a relative rarity. And so what until 1930 had been a highly prosperous ground for the electrical firms and assemblers now became so disadvantageous and volatile a field that in the mid-1930s the radio industrialists had no realistic alternative but specialization and concentration into a few hands.[284] The drastic diminution in the number of firms at radio exhibitions (from 72 in 1931 to 29 in 1939) clearly illustrates the new adaptation of manufacturers precisely at a time in which the receiving licences doubled (from 4.3 to 8.9 million pounds).[285] And the appearance of specialized firms—such as Erie for resistors and Dubilier for capacitors, to mention only two examples—ilustrates the atomization of manufacture. With the increment of competitiveness, the centre of production in the setmaking industry shifted unmistakably from the once thriving electrical sector to the components industry, and thereby to electronics.

[280]G. Bussey, 'Home Construction and Kit Sets', in Bussey (1990, pp. 91–119).

[281]G. Bussey, 1976. *Vintage Crystal Sets, 1922–1927*. London: IPC Press, 11–12.

[282]Pegg (1983, pp. 44–7).

[283]For a list of radio set and components companies, 1922–27, see Bussey (1976, pp. 43–69).

[284]*The Economist*, 22 August, 1936, 359.

[285]Sturmey (1958, p. 166).

Table 2 List of the most important wireless magazines in Britain in the interwar years

British Journals	Publishing House	Periodicity
Amateur Wireless & Electrics	Cassell & Co.	Weekly
Broadcaster (The)		Monthly
Broadcaster Radio & Gramophone Trade		Annual
Broadcaster & Wireless Retailer		Weekly
Electrical Review (The)	The Electrical Review Ltd.	Weekly
Experimental Wireless	Percival & Marshall Co.	Monthly
Gramophone (The)		
Irish Radio News		Weekly
Journal of the Institution of Electrical Engineers		Annual
Listener (The)	Broadcasting House	
Listener In	United Press Pty. Ltd.	Weekly
Modern Wireless	Radio Press Ltd.	Monthly
Popular Wireless	Amalgamated Press Ltd.	Weekly
Post Office Electrical Engineers' Journal	The Electrical Review Ltd.	Quarterly
Practical Wireless		
Radio Pictorial		Weekly
Radio Retailing (of New York)	McGraw-Hill Publishing	Monthly
Radio Times	George Newness Ltd.	Weekly
Scottish Radio Dealer	McNaughtan & Sinclair	Monthly
Scottish Radio Directory & Year Book (The)	McNaughtan & Sinclair	Annual
Sound Wave	Sound Wave Publ. Ltd.	Monthly
Talking Machine & Wireless Trade News		Monthly
Wireless & Allied Trades' Review	Wireless Press Ltd.	Twice-weekly
Wireless Constructor	Amalgamated Press Ltd.	Monthly
Wireless Engineer	Iliffe & Sons Ltd.	Monthly
Wireless & Gramophone Export Trader & Buyer Guide (The)	The Trader Publishing Co.	Quarterly
Wireless & Gramophone Trader (The)	The Trader Publishing Co.	Annual
Wireless & Gramophone Trader Year Book & Diary (The)	The Trader Publishing Co.	Annual
Wireless Magazine		Monthly
Wireless Review	Amalgamated Press Ltd.	Weekly
Wireless Trader (The)	The Trader Publishing Co.	Monthly
Wireless Weekly	Radio Press Ltd.	Weekly
Wireless World & Radio Review	Wireless Press Ltd.	Twice-weekly
World Radio	Broadcasting House	Weekly

Source: *Kelly's Directory of the Electrical Industry, Wireless and Allied Trades* (London, 1924), p. 523; and *Kelly's Directory of the Wireless and Allied Trades* (London, 1934), pp. 320–1, 447.

The mixture of over-expansion and restructuring had several consequences. Firstly, in response to American competition, and in a context of depression which affected it much less than other economic sectors, the radio industry largely opted for diversification and subcontracting. The setmaking fabrics penetrated the domains of electrical industry with ease. Most of the major manufacturers (GEC, Ferranti, and Cossor are good examples) directed their capitals to expanding production areas into domestic receivers, electric cleaners, electronic equipment, and cooking appliances, with the aim of achieving self-sufficiency.[286] Secondly, the certainty that radio set production irreversibly abandoned the new demand trade and progressively moved towards that of replacement and renewal guided the new industrial trend.[287] Replacement became a symptom of saturation and diminution of interest. And, thirdly, it was through the effect of the Great Depression, when the need for extra surplus had become urgent, that the advantages of export markets began to be fully realized.[288] By the mid-1930s, with all these elements (diversification, export growth, and an increasingly favourable patent policy), there emerged a new breed of specialists more responsive to commercial innovation than their predecessors had been in the 1920s. Given the adamant determination of these manufacturers to resist any offensive from American imports, it was no coincidence, but rather the commencement of a regular pattern of research and development, that leading companies such as Frank Murphy's,[289] EMI,[290] Ekco,[291] and Pye Radio,[292] established in these years teams of young researchers within their laboratories.

Although zeal for research and development reached magnitudes hitherto unknown, it cannot be understood in isolation from the high level of competitiveness and consumerism which distinguished these two decades of commercial success. Much talent and expertise was devoted to design (circuits, cabinets, and accessories), operation, and performance (such as the calibration of tuning dials). And it was concentrated on valve makers and to a lesser

[286]See Robert Clayton & Joan Algar, 1989. *The GEC Research Laboratories: 1919–1984*. London: Peter Peregrinus Ltd., 18, 40–1.

[287]By 1931, one half of the total radio trade was made up of replacements and renewals, and the difference was to grow in the following years. See Sturmey (1958, p. 188).

[288]Although always poor, the export performance of the British radio set industry went from 0.5% of the annual world trade in the early 1930s to 9.6% in 1936. For a comparison of radio exports in the late 1920s, see A.J. Smith (1928, pp. 589–90). See also Sturmey (1958, p. 183).

[289]Long (1985) and Geddes & Bussey (1991, pp. 157–66).

[290]James A. Lodge, 1987. 'THORN EMI Central Research Laboratories: An Anecdotal History'. *Physics in Technology*, 18(6), 258–68. See also Martin (1986, pp. 1–3).

[291]Geddes & Bussey (1991, pp. 114–15).

[292]Gordon Bussey, 1979. *The Story of Pye Wireless*. Pye Ltd.

extent on setmakers, rather than on assemblers, whose firms had assumed the role of adopting standard components and imitating imported products. The keen sense of innovativeness that pervaded the advanced American radio industry produced their receivers, at lower prices, and pervaded the British market with an aura of quality and sophistication which threatened the enthusiasm of the new specialists, while fulfilling the interests of the subsidiaries of the large American firms.[293]

It is tempting to suggest that the buoyant radio industry meant an accelerating momentum for atmospheric research in Britain, so allowing radiophysicists to edge ahead. In his presidential address to the Institute of Physics in 1930, W.H. Eccles exalted the virtues of wireless technology for research and physicists' role in its development.[294] But even if it seems clear that the radio industry provided radiophysicists with all the technology they needed, it is not true that all of this served to catalyse the investigation of the ionosphere. Here, therefore, it is necessary to qualify what is meant. In an analogy that helps to understand the complexity of the radio industry, the editor of the *Journal of Careers* compared it to a four-tiered pyramid in which competition occurred within, rather than between, the tiers.[295] On a wide and solid base lay the large electrical firms with their laboratories, which produced an extensive range of products and for which radio setmaking was only a sideline (it had not been so in the early 1920s). On the next level were the specialists, engaged in the manufacture of accessories and components, which emerged during the 1930s and whose research was associated with electronic developments of a short-term nature. Following these lay the assemblers and dealers, which encouraged neither innovation nor invention. And at the top of the pyramid, the valve-making industry.[296] In this last tier the furtherance of investigation always remained implicit, camouflaged beneath a commitment to activities of commercial and military interest which gained credence after World War I. The

[293]Since World War I, the British government had imposed protective import duties upon items of electrical goods, such as wireless valves, vacuum tubes, and rectifiers. These duties had, without doubt, benefited national production, but not by any means prevented American importation. See Alfred Plummer, 1937. *New British Industries in the Twentieth Century*. London: Isaac Pitman, 45.

[294]W.E. Eccles, 1930. *The Influence of Physical Research on the Development of Wireless*. London: Institute of Physics. See also 'Editorial: Physical Research and the Wireless Industry', *EW & WE*, 1930, 7(82), 359–60.

[295] 'The Wireless Industry. Requirements and Prospects of Technical Appointments', *JC*, 1932, 11, 7–11, pp. 7–8.

[296]Sturmey (1958, pp. 264–6), the author of *The Economic Development of Radio*, proposes a three-tiered structure, even if somewhat different, to explain the electronics industry.

Table 3 Directory of the wireless industry in 1924 and 1934

	1924		1934	
	London	GB & Ireland (ex. London)	London	GB (ex. London)
Radio engineers	57	200	49	176
Radio gramophone manufacturers			21	18
Wireless amplifiers			16	28
Wireless apparatus makers	125	77	40	45
Wireless cabinet makers			23	55
Components and acces sories manufacturers	114	82	43	46
Wireless equipment (ship's) manufacturers	3	2	2	3
Wireless receiving set wholesalers			10	37
Wireless receiving set dealers—retail			517	4,865*
Wireless receiving set manufacturers			79	69
Wireless relay services			16	116
Wireless supplies wholesalers			97	352
Wireless supplies dealers—retail			1,539	12,362*
Wireless instrument makers	51	58		
Wireless masts and accessories	9	24		
Wireless telegraph contractors	3	12	1	4
Wireless telephone receiving sets	44	46		
Valve manufacturers	6	21	22	9

* Approximate figures.

Source: *Kelly's Directory of the Electrical Industry, Wireless and Allied Trades* (London, 1924); and *Kelly's Directory of the Wireless and Allied Trades* (London, 1934).

sustained sobriety of this select group can be regarded as a feature reinforcing the solidity of the industrial pyramid; but it also produced, as will be seen in the next section, long-lasting and opulent benefits for ionosphere research and the Admiralty.

Thermionic valves and the strategies of research success

The radio valve industry was one of the flugships of Britain's new industries. Its production grew from 2.5 million devices at the time broadcasting was organized in 1924 to 5.6 in 1930, and up to 11.8 after the Depression in 1935, 90% of which were used in broadcast receivers.[297] In the period between 1930 and 1939, the annual average of exports were almost twice that of imports.[298] Expansion on this scale was proportionally comparable to that of motor vehicles, pneumatic tyres, or electric lamps, among the newer industries. In contrast to the relative lack of innovativeness of setmakers, the valve industry was technologically far more self-contained.[299] According to historian S.G. Sturmey, the different conditions in industry and broadcasting ensured valve-making developments were generally quicker in Britain than in the US.[300]

The prominent position of the sector of thermionic valves had immediate consequences for the investigations of the laboratories of experimental physics. It would be very odd if this had not been the care. After World War I, the main radio research centres—RRS at Slough, NPL at Teddington, Cavendish Laboratory—jealously preserved the very few existing valves and their versatile properties. Reminiscing about the milestones in ionosphere research, Appleton expressly mentioned valves by having 'made the production of waves of all wavelengths, long and short, a relatively easy matter'.[301] Unlike the older spark transmitters, the new valves which were especially designed for short-wave generators, allowed a precise tuning of the wavelength. But not only in radio science; in other fields of physics, they also had tenacious defenders. Artifices such as 'the amplification of ionisation currents by valves', this 'highly spccialized product of the radio industry', as mentioned by Patrick Blackett, an old Cavendish man, made 'atomic physics progress'.[302] And he went even further in his argument: the technique of experimental physics witnessed rapid changes, and each new technical advance broadened the knowledge of

[297]K.R. Thrower, 1992. *History of the British Radio Valve to 1940*. London: MMA, 2. See also J.A. Fleming (1934).

[298]In value, exports were £388,000 while imports were £216,000. See H.W. Richardson, 1961. 'The New Industries between the Wars'. *Oxford Economic Papers*, 1961, 13, 360–83, p. 363.

[299]See T.C.H. Going, 1955. 'The Growth of the Electron Tube Industry'. In Newcomen Society, *History of Thermionic Devices*, Conference Proceedings. London, 41–56; P.R. Morris, 1993–94. 'A Review of the Development of the Thermionic Valve Industry'. *Trans. Newcomen Soc.*, 65, 57–73. On valve development in general, see Tyne (1977) and Stokes (1982).

[300]Sturmey (1958, p. 265).

[301]E.V. Appleton, 'The Romance of Short Waves,' *World-Radio*, 1930, November 7th, 710.

[302]P.M.S. Blackett, 'The Craft of Experimental Physics,' in Wright (1933, pp. 68, 86–7).

the physical world. 'In part these changes come from within the laboratories themselves...but to an important extent...is influenced by the technical achievements of industry'.[303] This correlation between science and industry is particularly manifest in the case of thermionic valves. Provided we move out from the narrow focus of the manufacturing sector and look at valve research as a whole, then it is interrelationship rather than separateness, the intersection of links rather than unconnected developments, that emerges as the dominant feature in this technology not only at an institutional level but also at a personal one. It is the purpose of this section, therefore, to examine the interplay of industry, the fighting services, government laboratories, and the universities in these conditions, and to demonstrate how this formed part of the research strategies.[304]

Before World War I, valve was one technology in which Britain, through Marconi's and the Admiralty, was predominately the pacemaker in commercial applications and in which self-sufficiency was a reality. The company had cemented its dominant position not so much through fundamental research as by securing the exclusive right to sell and loan some of the most valuable patents (in particular that of Fleming's pioneering valve and De Forest's triode or three-electrode valve from 1911 onwards).[305] Likewise, it had reinforced its market through a contract signed with the Admiralty expiring in 1914, by means of which the Royal Navy bought or borrowed Marconi's radio equipment, whether for experiment or routine use.[306] During the war, despite being previously developed in France, R-valve was a credit to the malleability of the British manufacturing fabric. In scarcely two years, it had become the vanguard of another young industry, that of the electric lamp.[307] It incarnated the metamorphosis from the hand-made, laboratory product into the large-scale, production-line manufacture.[308]

[303]Ibid., p. 68.

[304]Jeff Hughes (1998) shows how similar links between universities and the electrical firms in the 1930s became crucial to the experiments on the artificial disintegration of atomic nuclei at the Cavendish Laboratory.

[305]MacLaurin (1949, pp. 43–4). For a discussion on how the granting of patent monopolies affected technical, social, and economic progress in radio, see S.G. Sturmey, 1960. 'Patents and Progress in Radio'. *The Manchester School*, 28, 19–36. Compare with the American radio industry: L.S. Reich, 1977. 'Research, Patents, and the Struggle to Control Radio: A Study of Big Business and the Uses of Industrial Research'. *Business History Review*, 51, 208–34.

[306]Pocock (1988, p. 170); and NAUK. [ADM 1/29126]. 'Wireless Telegraphy: Collaboration between Marconi's Wireless Company Ltd and the Admiralty', 1912–1933.

[307]A.A. Bright, 1949. *The Electric Lamp Industry*. New York: MacMillan, 304–5.

[308]S. Wood, 1989. 'The Valve: Industrial Aspects before 1925'. *Bull. BVWS*, 14, 12–17, p. 16.

The war was without doubt a stimulus to the rejuvenation of valve. The new R-valve was small, high-vacuum, and robust, of innovative design, and, above all, benefited from a technical expertise developed by two new Navy laboratories, those of the base ship HMS Vernon and HM Signal School at Portsmouth.[309] Whereas the sole three-valve companies of the early years of the century (Marconi's, Cossor, and Ediswan) had concentrated on the mode of operating soft valves, by 1918 the Signal School's staff was to specialize in the theory of the hard triode and in the design of special valves for particular purposes.[310] What before was skill and mystery, now was to become physical research: its performance as a detector, amplifier, and oscillation generator relied upon its internal physical dimensions and structure.[311] The bonds between Signal School and some of the most distinguished physicists of the time—such as Sir J.J. Thomson at the Cavendish—show how benefits of cooperation flowed in both directions. Thomson, who was interested in the technicalities of cathode ray tubes, worked for them in the quest for the optimum form for electrodes; his pupil, B.S. Gossling, was typical of a university generation in that he had begun his career as a designer of large valves at Signal School for the imperatives of war, becoming then head of the division and ascending eventually to the top of his profession as leading valve specialist at the GEC Research Laboratory.

The interwoven character of the relations between the Admiralty and private firms is demonstrated clearly by a case history whose leading protagonist is a skilful apprentice by the name of Stanley Mullard.[312] It was Mullard, endowed with great practical experience of lampmaking, who during his career service at Signal School persuaded officers to use fused silica instead of glass—less corrosive to seawater—for the envelopes of high-power transmitter valves.[313] In 1920, after dutifully earning high repute as manufacturer of high-quality products, Mullard put forward to be the founder of a company specializing in valves with the Admiralty's assistance and RCC's financial support.

[309]For the Navy's contribution to the development of valves, see Malcolm Foley, 'Industrial and High Power Thermionic Devices', in Newcomen Society (1995, pp. 89–99).

[310]B.S. Gossling, 1920. 'The Development of Thermionic Valves for Naval Uses'. *Jour. IEE*, 58, 670–703, esp. 670–1. For the early valves, see also Smith-Rose (1918).

[311]E.V. Appleton, 1955. 'Thermionic Devices from the Development of the Triode up to 1939'. In Institution of Electrical Engineers, *Thermionic Valves, 1904–1945: The First Fifty Years*. London: IEE, 17–25, p. 22.

[312]Geddes & Bussey (1991, pp. 49–57); Sturmey (1958, p. 35).

[313]M. Foley, 'Industrial and High-Power Thermionic Devices', in Newcomen Society (1995, p. 95) and Roberts (1923).

Until World War II, the new Mullard Company supplied the Royal Navy with silica valves vital for communications, and served as a platform of production for a great number of high-power silica triodes and rectifiers developed in Signal School.[314] Another important route by which the Admiralty stimulated radio engineering was through the secret radar programme. Between 1935, when the plan was devised, and 1939, the Signal School produced experimental silica valves for Watson Watt and his team. The importance of valves was beyond doubt: 'the Air Ministry and the War Office', noted an Admiralty minute, 'are almost totally dependent on Signal School for research and development of silica valves...There is a very real and grave risk that the whole RDF defence system...may be brought to a virtual standstill in the moment of greatest urgency.'[315] As in Mullard's case, the Admiralty quickly created bonds with manufacturing firms. An arrangement of 1936 whereby the GEC developed valves for very short-wave communications (below 60 cm.) with Admiralty's assistance soon evolved into one of far greater magnitude, as the technical supply to the RAF.[316] And another agreement with Marconi's permitted Watson Watt to examine the reliability of direction-finding on short waves, using magnetron valve oscillators developed by the Signal School.[317] It is a mark of the success of Admiralty's perseverance in weaving multifarious links that, when soon afterwards resources and production on a much larger scale were solicited for, the main valve firms and laboratories—Mullard, GEC, Marconi-Osram, Cossor, and the NPL—manufactured a record-breaking 44 million, and, perhaps most importantly, responded to the confidence that had been placed in them.

The inter-institutional nature of the initiatives was also clearly visible in the directives which marked the RRB's programmes. In a variety of ways the imperial and defence requirements that had consumed enormous energy during World War I had their prolongation in the postwar. The formation of a committee on thermionic valves (the largest within the RRB) was a sign of the special attention paid to applied science in conformity with the new circumstances. Not only did RRB's radiophysicists pursue aspects associated with the reception and transmission of radio telephony (exhorted during the Imperial

[314]'The Manufacture of Radio Valves', *Electrical Review*, 1926, 23 April, 648–50.

[315]Foley (1995, p. 97).

[316]NAUK. [AVIA 46/47]. 'Radio Industry: Interviews and Papers', Interview with Dr Peterson and B.S. Gossling at the GEC. Research Laboratories on 27 August , 1945.

[317]*Report of the Committee of the Privy Council for Scientific and Industrial Research for the Year 1935–36*. London, 1936, p. 84; *Report of the Committee of the Privy Council for Scientific and Industrial Research for the Year 1936–37*. London, 1937, p. 84; Watson Watt (1936).

Conferences of 1921 and of 1923); they also solved problems of valve production and design within areas which hitherto had been totally the preserve of manufacturers.[318] This was to a large extent the result of the pressure exerted from the fighting services: their three representatives in the committee sought the promotion of activities in congruence with the policy of their departments. But it was also due in part to the growing prominence of new firms solely specialized in valves with a technical proficiency able to satisfy radiophysicists' conditions.[319] As a RRB Report noted, Cossor Co.'s competence in producing 'an oscillograph superior to any of the types abroad' was determinant for the advances of Watson Watt's team in atmospherics and radar.[320] High Vacuum Valve Co. and Marconi-Osram (a firm shared by GEC and Marconi's) were representative of this new breed.

These interrelations would have not been so productive to the interests of radiophysicists if they had not been accompanied by a peculiar characteristic of valves, their versatility.[321] There can be no doubt that versatility did much to improve electronic instrumentation upon which ionospheric physics was to progress notably in the 1930s. For the thermionic valve was to transform imaging and electronic counting technology, replacing mechanical devices wherever a delicate and nimble indicator was required: Watson Watt's group, for example, contrived and developed over 20 different applications of cathode ray oscillograph (varying valves and circuits) for short-wave direction-finding, atmospherics, and the interpretation of layers on screen images.[322] Researcher's expertise, too, played its part: Appleton's squegger circuit (based on a triode oscillator) gave his team an unprecedented control on wave modulation and a new way of distinguishing the individual pulses from their echoes, which the pulse system of Breit and Tuve was unable to do.[323] In the 1930s, this oscillatory circuit together with cathode ray oscillograph served to visualize and photograph the distribution of echoes (and, thereby, the equivalent

[318] *Report of the Committee of the Privy Council for Scientific and Industrial Research for the Year 1919–20.* London, 1920, p. 56.

[319] For a detailed description of the major and secondary British valve manufacturers, see Thrower (1992, pp. 2–6).

[320] *Report of the Committee of the Privy Council for Scientific and Industrial Research for the Year 1932–33.* London, 1933, p. 72; R.A. Watson Watt et al., 1933. *Applications of the Cathode Ray Oscillograph in Radio Research.* London: H.M. Stationery Office, p. 4.

[321] For the main applications of valves, see D.P. Leggatt, 'Applications and the Growth of New Industries', in Newcomen Society (1995, 63–8, p. 63).

[322] Watson Watt et al. (1933).

[323] For Appleton's squegger circuit and radar research, see Bradley, Excell, & Rowlands (2000–01, p. 121).

heights of layers) in a way that would have been inconceivable only ten years before.[324] By mid-1930s, the valve amplifiers abounding in the laboratories of nuclear physics and physiology were to become, in Appleton's words, 'primary tools of scientific research'.[325] Hence, with the benefit of hindsight, it does not seem surprising that since the RRB designated a special committee to deal with valve amplifiers in 1925 (intuiting their burgeoning role),[326] the desire for propitiating areas of confluence between government laboratories and industry was a common denominator in the agendas of all committees.[327] In 1931, the NPL drew up a résumé of the literature on thermionics for the RRB with more than a hundred references,[328] and, four years later, when valve production reached a prewar maximum of 11.4 million units, every laboratory in Britain was valve-equipped.

At the Cavendish as at King's College, London, radio physics and style coalesced into one concept. Ratcliffe alluded to a 'style' in which simple equipment was synonymous with elegance,[329] 'a feeling that the best science was that done in the simplest way';[330] Appleton, for his part, invoked 'imagination' and 'inspiration' as the electromotive forces for investigation.[331] In both cases there was an apologia for small-scale science, for heroism before paucity, which was designated as 'sealing wax and string'. How far this was the result of a certain scarcity characteristic of much of prewar physics,[332] how far a recurrent cliché evincing a caricature of reality,[333] or how far the deification of a practice

[324]Watson Watt et al. (1933, pp. 111–19).

[325]Appleton (1955, p. 25). It is worth noting that Appleton's early research on valves had led to the conferring of his MSc and DSc degrees. Furthermore, his only textbook, *Thermionic Vacuum Tubes* (London, 1932), became part of the respected *Methuen Monographs on Physical Subjects*. Other early papers are Appleton (1919) and Appleton & Van der Pol (1921, 1922).

[326]*Report of the Committee of the Privy Council for Scientific and Industrial Research for the Year 1925–26*. London, 1926, p. 87.

[327]Examples of exchange of views and cooperation with manufacturers include the *Reports for the Years 1926–27*, p. 79; *1927–28*, p.102; *1931–32*, p. 83; *1932–33*, p. 72; and *1933–34*, p. 86.

[328]NAUK. [DSIR 36/2231]. 'Thermionic Valves: Résumé of Literature on Thermionics by NPL for RRB, 1927, 1931'.

[329]'Just as a mathematical proof derived neatly was better than one involving laborious calculations'. J.A. Ratcliffe, 1975. 'Physics in a University Laboratory before and after World War II'. *Proc. Roy. Soc. Lond.*, A342, 457–64, p. 464.

[330]Ibid., p. 464.

[331]R.W. Clark (1971, p. 34).

[332]J. Hughes (1998, pp. 58–9).

[333]A former Cavendish man, J.A. Crowther (1926, p. 58), stated that the 'oft-repeated legend' had little foundation.

resulting from over-sentimental reminiscences[334] is not clear. It is probable that each element was in part true. But it is also certain that a price was paid for the idealization and the emphasis on straightforwardness. Physicists often tended to oversimplify the interactions between institutions and industry, and the social complexities of day-to-day practice; to overemphasize their distinctness and ingeniousness; and to reify what were in reality only perceptions and ideals.

Some features, however, are worth considering. The maintenance of a style was of much more than symbolic importance. In essence, the following motto encapsulated Appleton's research method: 'the easiest way to make a name for yourself is to take a technique which has been fully developed in another subject and then apply it to your own.'[335] In practice, this meant disassembling equipments and analysing components in order to redesign new circuits and devices in accordance with needs, all this within a context of shortage and austerity. Understandably, Appleton was equally preoccupied by the material sustenance as he was interested by the simplicity. And his team of physicists, were as beguiled by style as avid for creation and innovation. Appleton thought of science, technology, industry, and government as systems with interfaces, characterized by a high degree of intersection and overlap. In this respect, the post of industrial consultant that he held in the early 1920s—acting as adviser to, *inter alia*, the Post Office, the Pye group, Trippe and Philips, and from 1927 to *The Times* on wireless subjects—and his numerous contacts—an intimate friendship with the brothers Eckersley, who were closely linked to the BBC and Marconi's[336]— provided him not only with remunerative benefits but also material ones (various companies presented him with equipment with the hope of influencing his decisions).[337] It was scarcely less the case for Ratcliffe and the Cavendish radio group, although here the ambit of influence was considerably less. From its formation in 1928, the realization of contacts was a *sine qua non* for materializing any research and a way of profitably accessing large-scale facilities, especially those of powerful transmitters. Special transmissions for the study of wave interaction from the BBC, assistance for the study of long waves from the Post Office and Cable & Wireless, and aerials installed by

[334]For an argument in favour of the myth, see Derek J. de Solla Prize, 1984. 'Of Sealing Wax and String'. *Natural History*, 93, 49–57.

[335]As Appleton reminded his pupils. See W.R. Piggott, 1994. 'Some Reminiscences of Work with Sir Edward Appleton'. *Jour. Atm. Terr. Phys.*, 56(6), 727–31, p. 727.

[336]R.W. Clark (1971, pp. 31, 68–9).

[337]Piggott (1994, p. 727).

the RAF. facilitated investigations that refuted any notion of separateness and intellectual isolation in prewar physics.[338]

Conclusion

In its editorial of 30 April 1930, *The Wireless World*, the most acclaimed radio journal with a wide acceptance among engineers and physicists, bemoaned the 'apparent lack of centralisation' in the field of radio research and development. 'Public money', it asserted, 'is expended in providing facilities and staff for radio research for the Post Office, the Army, the Navy, the Air Force, the BBC, and other Government services.' And it advanced the following intimation: 'If it were possible to obtain the figures it would be extremely interesting to know... the average annual expenditure on radio research financed by the Government. We believe that such figures would show that the Radio Research Board receives only a very small proportion of the total amount.'[339]

The insinuation of the editorial article about the prevalence of the influence of the fighting services is worthy of our attention, for in this period the state-funding of R&D was concentrated overwhelmingly in defence—for example, in the radar programme, with the Admiralty, the War Office, the Air Ministry, and the RAF as main beneficiaries—and scarcely at all in civil ministries.[340] The editorial's opinion points to the specific identification of the multinstitutional endeavour as the dominant characteristic of the investigative nature, even if it adduces no evidence for the business-funded R&D.

This very omission, which affects one of our theses—that the nascency and maturation of the British radio physicist community correlates with the economic blossoming of the radio industry—, reminds us that the global panorama depicted by the editorial cannot be the whole picture. The datum that in 1922 the RRB's total budget for the implementation of radio activities amounted to about £6,650, as opposed to £40,000 earmarked by Marconi's (almost seven times higher), indicates the necessity for amending such a picture.[341] Hence, in order to account for the particular appropriateness of a socio-economic environment to British ionospheric physics and those hypothetical correlations, we have felt obligated to invoke precisely those factors which the 'declinist' approaches generally ignore and underestimate—*inter alia*, industry-funded

[338] J.A. Ratcliffe, 1978. 'Wireless and the Upper Atmosphere', 1900–1935. *Contemporary Physics*, 19, 495–504.

[339] 'Centralising Radio Development', *The Wireless World*, 1930, 26(18), 47.

[340] Edgerton (1994, p. 52).

[341] That was the sum earmarked for further research at the Departments of Franklin and Round. See W.J. Baker (1970, p. 268).

R&D, business strategies related to the Empire, and industrialists' engagement with technical education.[342] The arguments advanced for the relative economic 'decline' by standard sources is thereby clearly inaccurate in our case. In the light of the foregoing, it seems hard to deny that the emergence of the British radio scientist community and the anticipated progress in the physical doctrine of the ionosphere shown in the preceding chapter were *in fact* nourished and sustained by the burgeoning of the incipient radio industry.

In sum, in the interwar years upper atmospheric research, and in particular ionospheric physics, blossomed outstandingly in Britain—*in hoc*, 'the world leader in that branch of knowledge' in the words of a prominent meteorologist. The reasons for this flowering are structural, not incidental: the existence of deep-rooted traditions (mathematical-physics Cambridge school, laboratory-based experimental physics, and Humboldtian-style terrestrial physics), and the funds purveyed by the government (through the RRB, but also through the fighting services) and by private firms (especially Marconi's). However, if the contribution of the industry was so determinant (by no means inferior to state-funded), it was not exclusively due to the pyramidal hierarchic structure (with valves as *avant-garde*) and its function as a purveyor of technology. Entrepreneurs' 'spirit for technical education and preparation' also played its part: as a nervous system whose functional cerebrum reposed in the firm and whose nerves ramified though the fibres of universities and engineering schools, the radio industry conflated R&D with education. But there was an additional ingredient. In this period radio industry (above all Marconi's) and state radio enterprise became so intensely coloured by geopolitics that every development of, or research related to, ionospheric physics would be, in retrospect, difficult to explain without bearing in mind that geopolitical imprint, and this was but the imprint of the Empire.

[342]The 'declinist' theses on the British technical and industrial failure and its connection with problems in technical education are transparently explained in C. Barnett (1986), Wiener (1981), and Mowery (1986).

3

FROM DOMINION TO NATION: ATMOSPHERIC SCIENCES AND RADIO RESEARCH IN AUSTRALIA

Introduction

Concluding an essay on the creative spirit in Australia, Geoffrey Serle appeals to the quotation of the architect and writer Robin Boyd on the philistine character of a country where, unlike other countries, the artist and man of ideas have always been required to justify their existence. In providing an answer to this hostility, Serle mentions the old Australian anti-intellectualism identifying art with amorality, vanity, and effeminacy, and academic intellect with contemplation and futility. And, being rejected, artists and academics felt impelled to believe that their true home was in Britain.[1] While it is unclear whether his conclusions may be extrapolated to our case, the question that he poses in his essay is an important one: did Australian radio scientists, as men of creative ethos, feel impelled to act in a similar way? And supposing that Britain shaped radio and atmospheric research in Australia, in which way and to what extent did it do so? In the light of the success they would achieve, it appears highly attractive to ascertain the factors which facilitated the cultivation of first-rate research in a country with deep-rooted anti-intellectual attitudes in which the values of the European migrants of business ethics, evangelical convictions and egalitarian assumptions prevailed above all.[2]

Between 1918 and 1939, Australia was neither a colony nor a completely sovereign nation but an autonomous Dominion within the British Empire–Commonwealth. It was an entity in a state of metamorphosis. Among Australian scientists this generated a combination of isolation and uncertainty, a protracted sense of dependency on Britain in scientific affairs, and almost no independent voice until after World War II. Nevertheless, it also generated,

[1]Geoffrey Serle, 1973. *From Deserts the Prophets Come: the Creative Spirit in Australia, 1788–1972.* Melbourne, on pp. 214–15.

[2]For important works on the idiosyncrasy of Australian science, see Roy M. MacLeod, 1988. 'The Practical Man: Myth and Metaphor in Anglo-Australian Science'. *Aus. Cul. Hist.*, 8, 24–49; Ann Moyal, 1986. *A Bright and Savage Land: Science in Colonial Australia.* Sydney: William Collins. See also D. Walker, 1997–98. 'Climate, Civilization and Character in Australia', 1880–1940. *Aus. Cul. Hist.*, 16, 77–95.

even if not frequently, a clear consciousness in some scientists of the benefits of the bond with Britain, and the privileges and obligations this entailed. Above all, the assertion of its status in the new world order, particularly in the Pacific, and the pursuit of Australia's interests within an imperial framework, were prevalent attitudes.[3]

One of the criticisms that have been often made of science—and politics in general—in Australia is that it was too ready to make itself captive to the tentacles of Britain. Thus, it is argued, when Australian scientists were tempted by innovation and original research, a relatively fragile academic infrastructure and a weak manufacturing tradition, although not in all the fields, repressed their impulses and made them look to Britain.[4] Like Australia's international identity, the identity of scientists themselves was subsumed within a larger whole, abstracted in a breeding ground for dependency, subservience, and delegation. The ambiguity of being in an intermediate stage between Empire and nation originated tensions, and these annihilated inspiration and creativity. However, closer inspection raises at least certain doubts in this reasoning. Firstly, it is reasonable to wonder what alternatives there were. But, moreover, the consolidation of a mining sector based upon its own resources, the significant advances in aviation and telecommunications, the development of a munitions manufacturing industry, and the pre-eminent position in 'autochthonous sciences' such as geology and zoology, all raise situations in which the alleged ambiguity concurs, in space and time, with ingenuity and resourcefulness. Perhaps if the case of Australian radiophysics (such as those mentioned) does not seem like that of its other facets such as its external political relations or defence policy, this is because the dilemmas of dependence and nation-building within the Empire were resolved in other ways.[5]

To be sure, the rapid development of Australian radio and ionospheric physics in the interwar years contrasts powerfully with the delayed development of Australian independence, for in both cases there was a scientific-cultural

[3]For a moderate view on Australia's commitment to empire, see Millar (1978). For a more passionate and vigorous nationalist approach, see, for example, McCarthy (1976). On Australia's status in the Pacific, see Thompson (1980) and the standard work by Meaney (1976).

[4]For indispensable works on Australian science at that time, see R.W. Home, 1988. 'The Physical Sciences: String, Sealing Wax and Self-Sufficiency'. In Roy Macleod, ed. *The Commonwealth of Science. ANZAAS and the Scientific Enterprise in Australasia, 1888–1988*. Melbourne, 147–65; and R.W. Home, ed., 1988. *Australian Science in the Making*. Cambridge: CUP.

[5]For the ways in which Australians resolved these dilemmas in such disparate issues as security, diplomacy, trade, or immigration, see C. Bridge & B. Attard eds., 2000. *Between Empire and Nation: Australia's External Relations from Federation to the Second World War*. Melbourne: Australian Scholarly Publishing. For defence policy, see J. McCarthy (1971).

isolation and an attachment to the Empire for geo-strategic and ideological reasons. Before nationhood was achieved in Australia, the modest community of radiophysicists managed to reconcile an emergent sense of nationality with the openness to international stimulus, beyond the imperial horizon. The transit from a dependent Dominion towards independence has, therefore, its scientific echoes.[6]

The development of radio and ionospheric physics in Australia is inseparable from that of the Empire. Specific circumstances of time and place, however, may colour local advances in such a way as to shape the features of that field and the organizational models that are transplanted from the mother country but also, and equally importantly, the way in which relations with other countries are conceived. To be more precise, and particularly in the case of ionospheric physics, the British influence reaches its zenith in the early stages of its institutionalization and then suffers a continued decline. But in this regression the Australian community shows a mixture of longing for rivalry and a certain traditional veneration for British scientific authority, in contrast to increasingly profound respect for American technological expertise. The scientific field is, therefore, the lens through which a process of social and political maturation is projected.[7]

Converging different research traditions

When, in 1926, the Executive Committee of the Council of Scientific and Industrial Research (CSIR) decided to establish a Radio Research Board (RRB), following the British precedent, ionospheric physics had just begun to reinforce the pillars of what was to be a prestigious and autonomous discipline. In a series of brilliant experiments, radiophysicists did succeed in convincing the scientific community of the existence of reflected layers in the upper atmosphere. Ionospheric layered structure was the atmospheric counterpart of quantum physicists' atomic structure, and chemists' periodic table. The timing could not have been more propitious. While Appleton and Barnett, in England, and Breit and Tuve, in the US, confirmed by means of radio waves the early hypothesis of

[6]For suggestive reflections on the use of science for national purposes, Roy M. MacLeod, 1982. 'On Visiting the *Moving Metropolis*: Reflections on the Architecture of Imperial Science'. *HRAS*, 5, 1–16; R. MacLeod, 1993. 'Passages in Imperial Science: From Empire to Commonwealth'. *Journal of World History*, 4, 117–50; R.M. MacLeod, 'From Imperial to National Science', in Macleod (1988, pp. 40–72).

[7]R.W. Home & S.G. Kohlstedt, eds., 1991. *International Science and National Scientific Identity*. London: Kluwer Academic Publishers.

Heaviside, Kennelly, and Eccles, Australian authorities decided to earmark their modest resources for radio research, and not only for agriculture.[8]

The Australian decision was neither arbitrary nor spontaneous, however. By the middle of the 1920s, two scientific traditions sharing common knowledge areas and benefiting each other converged in Australia. The first was Humboldt-style terrestrial physics, being deep-rooted in colonial science and in nineteenth-century magnetic explorations. From viewpoints both institutional and scientific, it was based upon observatory-style data recording and analysis of geophysical variables. As R. Home notes: 'the science of physics, in its evolution in Australia, included and for a time was dominated by a great deal of work of non-experimental, data-recording kind, much of it done in institutions other than laboratories.'[9] The second was the study of wave propagation in the atmosphere, which was much more practical in character, and centred largely on laboratory-based experimental work. Unlike the terrestrial physics tradition, the latter was to extend to universities, government institutions, and private enterprises, following the prevailing pattern in the major industrial countries of the Northern Hemisphere.[10]

How geophysics in the service of technology favoured the expansion of colonialism is revealed in the erection of magnetic observatories. Magnetic research flourished at several Australian colonial institutions. The Rossbank Observatory in Hobart rose to international prominence as part of an international 'magnetic crusade', coordinated and patronized by Edward Sabine, an Irish Fellow of the Royal Society. Understanding the magnetic conditions of the earth and its possible liaison with cycles of sunspot activity absorbed most of the time of observers. Behind the recordings of magnetic variations lay the Humboldtian vision of cosmic relations, in which forces operating in the sun and the earth appeared to be interconnected. The new magnetic and meteorological Flagstaff Observatory in Melbourne, opened under the direction of the German geophysicist Georg Neumayer, followed the same

[8]On atmospheric and radio research in Australia, see the well-documented, if rather chaotic work of W.F. Evans, 1973. *History of Radio Research* Board, 1926–1945. Melbourne. See also Gillmor (1991), White & Huxley (1974), and White (1975).

[9]R.W. Home, 1994. 'Defining the Boundaries of the Field: Early Stages of the Physics Discipline in Australia'. In R. McLeod & R. Jarrell, eds. *Dominions Apart: Reflections on the Culture of Science and Technology in Canada and Australia, 1850–1945*, Special Issue of *Scientia Canadensis*, 17, 53–70, p. 70.

[10]For a general and little-known overview on Australia's geophysics with copious bibliography, see A.A. Day (1966).

traces.[11] Financed by metropolitan capital, these institutions were but convenient platforms erected for of geo-strategic reasons on which colonial scientists could develop their observational programmes. International regard resulted from such scientific enterprises. Although technical equipment might have impressed local communities, one wonders if these initiatives were but an ephemeral illusion due to foreign intrusions on the making of the intellectual landscape of Australia.[12]

Scientists from Europe had, of course, brought with them the measuring instruments, but also an imported scientific programme which, unfortunately, had fomented no local geophysical research. Such a mode of operation was far removed from that which took place at the Watheroo Observatory during the 1920s. This observatory was organized and financed by the Carnegie Institution of Washington, through its Department of Terrestrial Magnetism (DTM). The study of the atmosphere and the magnetic and electrical properties of the earth became a central task for the two outposts established by the DTM: Watheroo, in the southeast Indian Ocean region, and Huancayo, Peru, close to the magnetic equator.[13] The case of Australia is particularly striking, because the heads of the observatory adopted the strategy of recruiting assistants locally, which gave young graduates in physics one of the few grounds for further training outside the teaching field. Furthermore, Watheroo staff became completely integrated into the local scientific community, participating at the meetings of the Australian Association for the Advancement of Science (Perth, 1926, and Hobart, 1928), and developing links with the University of Western Australia.

The concern of young researchers revolved around understanding how solar action sustained or inhibited the electrical variations in the atmosphere of South Hemispheric regions, such as that of Watheroo. Under the supervision of the American observer-in-charge, H.F. Johnston, routine measurements of atmospheric electrical conductivity began in 1922 and of atmospheric potential gradients in 1924.[14] As early as 1926, in parallel with the emergence of interest in improving communications with headquarters in Washington, the DTM encouraged research on radio propagation. The task fell to J.E.I. Cairns,

[11]For the magnetic explorations in the colonial age, see Cawood (1979). For Australian cases, Savours & McConnell (1982) and Home (1991, pp. 40–53).

[12]See R.W. Home, 1987. 'The Beginnings of an Australian Physics Community'. In Nathan Reingold & Marc Rothenberg eds. *Scientific Colonialism: A Cross-Cultural Comparison*. Washington, D.C.: Smithsonian Institution Press, 3–34, pp. 4–5.

[13]Tuve (1947).

[14]H.F. Johnston (1926).

a graduate of the University of Western Australia. His enterprise related not so much to upper atmospheric physics as to links between radio fading and atmospheric potential-gradient recorded at the observatory.[15]

The power of radio waves to facilitate metropolitan communication was also investigated, and remarkably so by other laboratory-based experimental groups. For reasons both political and technical, the burgeoning field of radio research was considered a national and international business rather than a scientific enterprise. The most emblematic case in Australia, as we will see in the next chapter, was the firm known as Amalgamated Wireless Australasia, Ltd. (AWA), one of Marconi Company's subsidiaries.

Nevertheless, not all radio research was associated with lucrative practices. The Department of Natural Philosophy at Melbourne University and its new professor, Thomas Howell Laby, exemplified the investigation based on scientific parameters rather than public utility ones.'[16] Laby had acquired a high level of experience at the Cavendish Laboratory as an 1851 Exhibition Scholar with J.J. Thomson. When he returned to Australia in 1915, to the chair left by his predecessor, T.R. Lyle, he combined an extraordinary dexterity in laboratory technique with a far-sighted vision of new radiations as active agents in the ionization of gases. At his department, he consolidated several parallel areas of physical research—X-rays, thermal conductivity, and, around the mid-1920s, radio propagation in the atmosphere—, which enabled young researchers to familiarize themselves with the leading issues of experimental physics. One might assert that he firmly applied his fame as a scientist of exceptional independence, originality, and spirit of critical exactitude (acquired at Cavendish) to the training of research students. 'My experience with these men', he said to A.D Ross in 1926, 'makes me consider that experience of research work and of thesis writing is an essential part of the training of anyone who calls himself a physicist. Lectures and examinations do not train originality of mind, independence of thought, capacity to take responsibility, or (in Australia) the candidate's power of expression.'[17]

Laby's faith in physical research was as intense and passionate as it was uncommon in Australia. Both his adroitness and ability to be a 'good

[15]On the Watheroo Observatory and the appearance of Australian-American cooperative links, see R.W. Home, 1994. 'To Watheroo and Back: The DTM in Australia, 1911–1947'. In Gregory Good ed. *The Earth, the Heavens and the Carnegie Institution of Washington.* Washington, D.C.: American Geophysical Union, 149–60. For the early activities, see Wait (1923).

[16]MUA. 'Laby's papers and correspondence'. Box 85/144. On Laby's early research, see Muirhead (1996, pp. 9–32).

[17]Home (1987, p. 17).

if dominant administrator' contributed, without a doubt, to its success.[18] To support his activities, he sought both university funds and external sources. In this regard, it is understandable that he sensed in the boom of commercial broadcasting an attractive financial resource. In October 1926, he elicited from the Broadcasting Company of Australia (3LO Melbourne) a promise of a three-year research fellowship (£500 per year) for his department. In return Laby committed himself to investigate physical problems arising in broadcasting.[19] One of his young researchers, R.O. Cherry, examined the practicalities of transmission and reception through field strength measurements; this was rapidly and favourably received by the company.[20] By 1926, when the RRB was launched, Laby's team was one of the focal points of Australian research.

Paving the way

Laby was not the only instigator of a series of projects which would modernize Australia's telecommunications system. The Postmaster General's (PMG) Department was another important agent as organizer and regulator of Australian radio.[21] Part of its high level of attainment was due to its head, Harry Percy Brown (nicknamed 'Pooh-Bah' Brown), a small corpulent Englishman, who had in 1923 solicited one year's extended leave of absence from the British Post Office for planning and assisting in the reform of the PMG Department. An expert member of the British Institution of Electrical Engineers, Brown put into practice much of knowledge acquired on telegraphic and telephonic networks in the British service. To be sure, he came accompanied by a great reputation—indeed, he was considered by his colleagues as 'the most capable man' in Britain—, but his reputation grew even more during the time he was in charge of the PMG, from 1923 to 1939, becoming 'one of Australia's public service *giants*'.[22]

Brown, as an experienced engineer, showed a special vision of the interaction between investigation, telecommunication techniques, and the British Empire. Immediately after taking over the post of permanent head, he

[18]H.C. Bolton, 1990. 'Optical Instruments in Australia in the 1939–45 War: Successes and Lost Opportunities'. *Australian Physicist*, 27 (3), 32.

[19]MUA. Laby's papers: 'Static and fading tests conducted by the Broadcasting Company of Australia in conjunction with the Wireless Institute of Australia, February and March, 1928'.

[20]Richard O. Cherry, 1927?. *Signal Strength Measurements of 3LO Melbourne*. Melbourne: Broadcasting Company of Australia, *Paper II*, 1928?.

[21]A classic and indispensable work on the PMG and communication systems in Australia is Ann Moyal, 1984. *Clear Across Australia. A History of Telecommunications*. Victoria: Thomas Nelson

[22]Ibid., p. 118.

advocated the establishment of the PMG research laboratory in Melbourne to guarantee an advanced telecommunication system abroad, and especially with the rest of the Empire. Radio research with such aims was favourably viewed by the Australian scientific authorities. The chief engineer in charge of the laboratory was S.H. Witt, a skilful self-taught engineer trained on the department's internal courses.[23] In the following years, large long-term projects emerged steadily. The introduction of high-frequency carrier systems, the technical planning of the National Broadcasting System, and the design and construction of broadcasting stations, were to call for the creative and intuitive grasp of radio science of men like Brown and Witt.[24]

Sydney University and its Electrical Engineering Department was another seedbed of radio practice. From among its professors, the long figure of John Madsen was to stand out, an engineer-physicist of unimpeachable character, a very down-to-earth and approachable person who was regarded with affection by students and colleagues due to his flexibility and understanding.[25] Madsen's success in organizing science—for which he always had a special gift, as he had demonstrated in the organization of the RRB and the National Standards Laboratory—came from his long absorption in similar overseas experiences and the benefits of a slow maturing of ideas. He regarded himself as a person influenced by his stay at Adelaide University (and by his apprenticeship with the older Bragg, William Henry Bragg). There he had organized the School of Electrical Engineering, after visiting the principal universities in England and US, and meeting J.J. Thomson and Lord Kelvin at Cavendish in 1903. Yet, he was as much engineer as physical expert.[26] Judicious in the strict sense of the word, Madsen's particular effectiveness lay in his ability to apply, and even achieve, the inseparability of science and engineering while professing that the

[23]Moyal (2002).

[24]Although the staff was made up initially of only one man, the chief engineer, by 1925 a total of five researchers were working at the place, becoming by 1938 a staff of 57 persons distributed throughout a three-storey building. The bulk of activities were not recorded except in internal laboratory and test internal reports. See S.H. Witt, 1938. 'The Research Laboratories of the Postmaster General's Department'. *Proc. Wor. Rad. Con.*, 3–7.

[25]For students and colleagues' opinion of Madsen, see AAAS. Box 72/1; and SUA, 'John Madsen, 1879–1969', G3/158 Box 35.

[26]Madsen's thesis at Adelaide University, entitled 'The Ionisation of Gases after their Removal from the Influence of the Ionising Agent', was in the front line of experimental physics at that time. He proceeded with his investigations into radioactivity and on the electron. Apart from their estimable intrinsic value, Madsen's works were highly meritorious because of his isolation from the active centres of England. See Home (1981).

progress of electrical engineering had its origin in physics, a view not always shared by engineers at that time.[27]

Radio science was a fertile ground to put into practice that particular conception. Like Laby, Madsen negotiated a three-year grant (also £500 per year) from a private firm, the Farmer's Broadcasting Company in Sydney, to carry out wireless research. But unlike Laby, he planned a basic rather than pragmatic line of research. As a result of that policy, W.G. Baker, a young graduate returning after two years' research overseas, investigated the refraction of short waves in the upper atmosphere, a pioneering topic that had recently received attention from British and American physicists.[28] Madsen aimed to fuse physics and engineering.

The different nuances in laboratory-type research between Sydney and Melbourne, along with the differences between the heads of PMG and AWA, as a result of commercial interests and radio communication lobbies, would colour in varying degree of light and shade the naissance of the Australian RRB.[29]

Organizing radio and atmospheric research in the 1920s

Throughout its history, Australia has had a dependent economy. In the interwar years there existed a clear relationship between Australia's dependence on the British market and its specialization as a primary producer. For most Australians, it was an indisputable reality, and to a certain degree logical, that Britain possessed the laboratories and factories and produced the secondary products, whereas Australia provided the raw materials and markets for the manufactured goods. As it was, Australian scientists occupied themselves in areas related to agriculture and mining while those of the metropolis specialized in more 'avant-garde' but less mundane fields like physics.[30] The fact that radio research was the only field of investigation not related to agriculture to be subsidised by CSIR might reinforce this conviction.[31]

[27]AAAS. [Box 72/1], 'J.P.V. Madsen's Papers and Correspondence', C.G. MacDonald, 5 May 1954; and W.N. Christiansen, 'Emeritus Professor Sir John Madsen'.

[28]W.G. Baker, 1926. 'Refraction of Short Waves in the Upper Atmosphere'. *Tran. Am. Ins. Ele. Eng.*, 45, 302–33.

[29]For the differences between Sydney and Melbourne's research, see Home (1987, pp. 20–1).

[30]For Australia and its place in the international economy in the interwar years, see Dyster & Meredith (1990, esp. chapters 5 and 6). For the different dependent economic models, such as that of Australia, see Sinclair (1976) and Maddock & McLean (1987).

[31]G. Currie & J. Graham, 1974. CSIR 1926–1939. *Public Administration*, 33(3), 230–52, pp. 235–6.

Clearly, science and economy are intertwined. However, to establish a link between economic dependence on Britain and Australia's dependent scientific relationship with Britain is an arduous task. What is certain is that throughout the 1920s most Australian scientists took for granted the status of the country, and felt comfortable within an imperial framework. The acceptance of this dependence was similar to that proceeding from governing political circles, especially from the conservative parties, for which loyalty to Australia was identified as synonymous with loyalty to Britain and the imperial link.[32] Perhaps one of the most attractive aspects of radio research is to examine how the apparent contradiction between nationhood and the willingness of Australia's scientific authorities to embrace a dependent relationship within the Empire was reconciled. Here, unlike in other fields of science, the reaction of physicists involved in radio and atmospheric sciences was not characterized by indifference or indolence, and their predilection towards Britain had a rationale deriving from the socio-political realities of the day and from cultural historical traditions.[33]

After World War I, imperial science was redesigned to suit the political ideals of Empire. In the language of scientific boards, that meant cooperation, not metropolitanism. This was exemplified in the spirit of the first meetings of the British RRB, in which the members of the Board aspired to strengthen 'communication with the Government officers engaged in radio telegraphic work in the Dominions, the Crown Colonies, and in India'. They also expressed their hope that the 'willing and helpful co-operation' with Australia's authorities could be intensified.[34] The message was well heeded in the Antipodes. In 1921, the Deputy Director of Radio Service, Melbourne, requested information in order to establish observations on atmospherics in Australia.[35]

The project did not prosper but the spirit of 'cooperation' survived. Scientific self-sufficiency in the Dominions was the end pursued of which the visits of Australian scientific leaders to Britain, paying from time to time, were the means of rapprochement. In one of Laby's travels to England, in 1925, Admiral Sir Henry Jackson, chairman of the British RRB, suggested the

[32]Attard (1999, p. 24).

[33]George Currie & John Graham, 1968. 'Growth of Scientific Research in Australia: The Council for Scientific and Industrial Research and the Empire Marketing Board'. *RAAS*, 1(3), 25–35.

[34]*Report of the Committee of the Privy Council for Scientific and Industrial Research for the Year 1920–21.* London, 1921, p. 55.

[35]DPA. Minutes of the 11th Meeting of the Sub-Committee B (Atmospherics) held on 17 November 1921. DSIR, Radio Research Board.

advisability of organizing radio research in Australia in order, among other reasons, to solve the problem of long-distance transmission, a matter concerning both countries.[36] The adopted strategy came in the form of slow but steady encouragement to Australian scientists, often as an invitation to adopt British scientific organizations as a model. From the point of view of Australian interests, while sharing the need for stimulating scientific research, the authorities also put much emphasis on including radio manufacturing.

Imperial self-sufficiency through science needed leaders able to rise to the occasion. In this regard, Madsen would demonstrate to be 'the right man, in the right place, at the right time'.[37] In the months before the constitution of the RRB, he persuaded those concerned with wireless, and especially George Julius, the newly appointed Chairman of CSIR, of the need for ameliorating local facilities and equipments for radio research; he also persuaded them that Australia, as a nation, had the external responsibility to play its role in cooperative projects on a worldwide basis.[38]

Madsen's sense of duty and deliberateness of action made him scientifically a reliable leader, even the architect, of the RRB. Between June and November 1926, he canvassed all possible support to convince CSIR to foster research beyond the field of agriculture. He contacted Ernest Fisk, the manager of AWA, who received the proposal enthusiastically; sounded out the opinion of the influential A.C.D. Rivett, the executive officer of CSIR, who subscribed to almost all aspects of the initiative; and requested background information concerning the British parent board from Julius.[39]

At this stage, the fixations of governing leaders might be cited as significant impediments. The CSIR had decided to concentrate on research for primary industry, rather than secondary industry research. Behind this decision, there was a mix of prejudices and economic and political reasons. First, both the Hughes and the Bruce Conservative governments believed that the economic prosperity of the country rested on the efficient production of primary industries, as well as on the associated exchange trade, largely with Britain. Second, although it was recognized that secondary industry was vital for the creation of employment, it was considered that it would be difficult to compete with

[36]MUA. 'Laby's papers and correspondence'. Manuscript 'Origins of Australian Research Board, by T.H. Laby'. Box 85/144. See also Evans (1973, p. 12).

[37]Evans (1973, p. 2).

[38]G. Currie & J. Graham, 1971. 'G.A. Julius and Research for Secondary Industry'. *RAAS*, 2(1), 10–28.

[39]For the 'Summary of Information' sent by the British RRB, see W Evans (1973, Appendix 1). For Madsen's early contacts, see AAAS. 'General', Madsen to A.J. Gibson, 9 July 1926.

overseas rivals, for the scale of most secondary industries in Australia was severely constrained by its small population. Australia was too distant from the main international industrial markets, and its population was not big enough to generate a large-scale competitive economy. Finally, the burden of debts contracted by Australia forced it to earmark its limited financial funds for primary industry scientific research. To invest in science was accepted, but this must be applied to the huge and rich natural resources of Australia.[40]

CSIR's policy focus on the primary sector did not allow much room for manoeuvre. Madsen's first arrangements led to organization of a Radio Conference on 17 November. The success or failure of the meeting depended on the interplay of personalities taking up the appointment.[41] Madsen advocated clearly confining radio research entirely to the domain of science, where it would be free not only 'from the day-to-day governmental functions of regulation and administration of broadcasting, but more specifically, still, from the restrictions of secrecy associated with research for the armed services'.[42] He invoked the maxim of being firm in aim but flexible in method, when revealing to Brown that it was necessary to achieve a balance between what he termed the 'immediate practical difficulties' of broadcasting and 'the purely scientific side of the question'.[43] No doubt Madsen's well-balanced and sensible attitude, given the numerous commercial interests at stake, was necessary to eschew mercantilism in radio research.[44]

In 1927, Madsen's contacts bore their first fruits. The board was approved in June, and Madsen was appointed chairman. Among the other members, there were some well-known faces (Laby and Brown) and little-known names (F.G. Cresswell, the representative of Defence, and G.A. Cook, as secretary). The Navy, represented by Cresswell, had had an active interest in radio since

[40]There are a considerable number of studies on the institutional development of science in Australia in the twentieth century, but most of them tend to underscore the dominant role of the CSIR. See, for example, George Currie & John Graham, 1966. *The Origins of CSIRO: Science and the Commonwealth Government, 1901–1926*. Melbourne: CSIRO; and C.B. Schedvin, 1987. *Shaping Science and Industry: A History of Australia's Council for Scientific and Industrial Research, 1926–1949*. Sydney: Unwin. For a critical reading of this approach see A.T. Ross (1987).

[41]Representatives of the Navy, Army, Air Force, Munitions, Wireless Institute, AWA, Post Office and broadcasting stations attended the Conference. See NZNA. [SIR 1, 26/10/2], 'Radio Conference, 17th November 1926', and 'Explanatory Notes to Accompanying Report of the Radio Conference, by T.H. Laby'.

[42]Evans (1973, p. 16).

[43]Ibid., p. 17.

[44]The even balance between free-running basic research and applied research was a characteristic feature of what Schedvin defined as 'the culture of CSIRO'. See Schedvin (1982).

World War I. Fisk's absence deserves an additional explanation. AWA's monopoly in royalty payments over all broadcastings activities was a continuous source of tension and irritation to PMG, radio manufacturers, and broadcasting stations. Brown had grave misgivings about the government's share agreement with AWA in 1922—as a result of which, the government had acquired half- AWA's shares plus one. With these precedents, it is not difficult to see Brown's influential hand in the composition of the board. The fact that 70% of the funds for the RRB were supplied by the PMG reinforced his authority even more.[45]

At this stage, the exclusion of AWA might be seen as a significant setback. Madsen strove to include Fisk: 'This Company is a semi-government Department' responsible for 'almost all the constructive work in regard to Radio Telegraphy in Australia'. So, 'it will mean a tremendous amount to the Research Board if it can obtain the active cooperation of Amalgamated Wireless'.[46] Beyond his friendship with Fisk, this avowal was in keeping with his conception of the application of science to industrial needs. Madsen, rather than any one, knew Australia still suffered from a weak manufacturing base and a relatively little-developed university sector.

Transplanting the British research model

Traffic in ideas on cooperation and programme exchange could not be maintained only by correspondence. Immediately after the formation of the board, Madsen made an overseas visit to Britain, the USA, and several European countries. In Britain he visited the British Post Office, the BBC, and spent a long time at the National Physical Laboratory at Teddington and the Radio Research Station at Slough, accumulating information not only on radio research but also on testing basic primary standards.[47] In those days, the degree of concord was so intense that he hardly needed to explain his intentions to obtain help; and like his British colleagues from DSIR, he was convinced that

[45]AAAS. 'General', Fisk to Madsen, 12 November 1926; Madsen to Julius, 30 September 1930. See also Evans (1973, pp. 9–19).

[46]Madsen to Lightfoot, March 1927, in Evans (1973, p. 25).

[47]Despite these early contacts, the National Standards Laboratory was to be established a decade later. See R.W. Home, 'Science on Service, 1939–1945', in Home (1988b, 220–251, pp. 222–3). See also AAAS. [Box 72/1], J.P.V. Madsen, 'The Need for Standardisation Laboratories in Australia', [Presidential Address] *Proceedings of the Electrical Association of New South Wales*, 1913–14, 75–98.

practical problems had to be tackled, not just pragmatically, but as fundamentally as was necessary.[48]

One might consider that the sense of cooperation surrounding those appointments was not problematic for British authorities, given the deference showed to Madsen. By reading the minutes of the British RRB meetings, one observes that the organic language of cooperation gave way to a species of delegation, in which leaders in London seemed to feel almost obliged to help in Australia's scientific development and self-determination.[49] That is how the ready assistance of Tizard and Appleton in vetting the applicants for the Australian positions, the offer of training the new recruits at Slough—more than 90% of the Australian radio researchers would spend a period in Slough, Cavendish, or King's College—, and the selection and procurement of technical equipment for the Australian programme can be understood. The exquisite welcome was to stimulate the loyalty of Australian radiophysicists towards the British Empire. At least it did so until the mid-1930s.[50]

A further venture was Watson Watt's invitation to mutual collaboration and free equipment for atmospherics observations. This proposal was of a piece with his sense of tradition and Empire, and with the quest for a world-wide scheme—negotiated through URSI—for atmospherics research using Spectrum Atmospherics Recorders.[51] In a letter addressed to Stuart Smith, from the Australian embassy in London, Watson Watt employed a language exuding a somewhat paternalist pitch: 'I have provisionally drafted a scheme of allocation [of spectrum apparatus] which appears to me to give satisfactory sampling throughout the whole world. Unfortunately, however, the British Empire, which is so admirably distributed for such sampling, is somewhat heavily represented in the allocation, and it is just possible that some concessions may have to be made to national sentiments at points where they clash

[48]The importance of transplanting the British research model was voiced in many letters. See, for example, AAAS. 'General', Madsen to A.J. Gibson, 9 July 1926.

[49]See, for instance, the *Reports of the Committee of the Privy Council for Scientific and Industrial Research for the Years 1920–21.* London, p. 55; *1921–22*, p. 39; *1930–31*, p. 78.

[50]For the vetting of applicants, see Madsen to Cook, 10 and 16 April 1929; Application for Appointment, May 1929; F.L. McDougall to Madsen, 4 June 1929; Cook to Madsen, 21 June 1929. For the procurement of technical equipment, O.F. Brown, 7 August 1929; Cook to Madsen, 19 October 1929; Madsen to Cook, 23 October 1929; McDougall's Memorandum, 14 November 1929, NAA. 'Madsen Papers and Correspondence', G6/20 General, A10762/1.

[51]Some apparatus designed by British DSIR for recording the direction of arrival of atmospherics had already been supplied to the Watheroo Magnetic Observatory; one new cathode ray direction-finder, available at a reduced price, was destined to go to Mount Stromlo Observatory.

with abstract sampling theory. Even assuming, however, that my proposals are adopted integrally, it would not be possible to allocate more than two sets to the whole of Australia, and it would be necessary that these should be installed one in the East and one in the extreme West.'[52]

Despite Watson Watt's tone, cooperation without compulsion, free acceptance of programmes, compromise, and concord—among other features—characterized the relationship between the British and the Australian RRB. In the process, debts and favours were imposed on both sides. To be sure, it was not about a formal union but an organic relation. Imperial cooperation in radio research had exemplary precedents in other directions, such as the Geophysical Prospecting Scheme.[53] At first Britain's generosity exceeded Australia's expectations. Arguably, the most prized treasure ceded by the British RRB was not the equipment—similar to that initially employed in England—,[54] but the research programme to which the Australian RRB would adhere, close to that being followed in the UK.[55] The programme was handed over to M. S. Lloyd, the High Commissioner for Australia, in 1928. It suggested Madsen cooperate actively along the following five lines: (a) comparison of atmospherics and potential gradient—by the method developed by Watson Watt; (b) repetition of Appleton's experiments on polarization—specifically proposed by the Admiral H. Jackson; (c) attenuation of waves and coastal refraction; (d) ultra-short-wave beams—to be transmitted from Slough; and (e) development of the rotating beacon transmitter—as an aid to marine navigation. The instructions received from DSIR were referred to warmly and with gratitude by Australian researchers: 'We are beginning to appreciate the tremendous amount of time and energy which has been necessary to bring the apparatus to its present state of development, and how fortunate we are to be able to start our investigations with all that experience and information at our disposal.'[56] Indeed it is highly probable that the relations between both boards reached their highest point at that period.

[52]Ibid., Watson Watt to Stuart Smith, 19 November 1929.

[53]See Barry W. Butcher, 1984. 'Science and the Imperial Vision: The Imperial Geophysical Experimental Survey, 1928–1930'. *HRAS*, 6(1), 31–43.

[54]For the technological support, see *Reports of the Committee of the Privy Council for Scientific and Industrial Research for the Years 1929–30*. London, p. 98; *1930–31*, p. 82.

[55]British RRB's Secretary to M. S. Lloyd, containing the 'Programme of Research Work for 1928–29'. In NZNA. [Fold: Department of Scientific and Industrial Research, SIR 1, File 26/10/2], 'Investigations Overseas: Australia'.

[56]Munro to Lightfoot of 16 October, 1929. In Evans (1973, p. 50).

The programme of the Australian Board included the bulk of recommendations from its British counterpart. At first, a similar programme with three research lines (field strength, atmospherics, and radio fading) was designed for Sydney and Melbourne Universities, under the supervision of Madsen and Laby, respectively.[57] In the following years, however, while the Sydney group specialized successfully in upper atmosphere studies, the Melbourne group faced up to a thorny but attractive research topic: the interference with radio reception of atmospherics and its relation to the meteorology of thunderstorms. It is not clear why Laby decided on atmospherics. Maybe the response lies in his marked temperament, and in the fact that people working under his command did not fully share Laby's research style, the so-called Cavendish style.[58]

Two styles and two personalities

The differences in style between the two professors, Madsen and Laby, lay behind the Australian RRB. Both of them had qualities of leadership and principles that shaped the character of the board's policy-making. The description of their singular, if often complementary temperaments and talents enriches the understanding of radio research, and colours the intellectual and scientific landscape of physics in Australia.

Madsen's style was that of an exceptional scientist, who was not strictly either an electrical engineer or a radiophysicist, but who was characterized by superb mastery of diplomacy, discretion in methods, determination in organizing purposes, and dexterity in public relations. Broad-minded, attentive to detail, extremely energetic, and warmly conscious of the diverse research opinions within the RRB, Madsen combined the shrewdness of discernment on fundamental aspects of physics with the best virtues of the forward-planning engineer. His few opponents feared his capacity and pleasure in analysis (in fact, he was a keen fisherman who believed that even beach fishing had to be subject to scientific analysis). This mixture of faculties instilled into researchers' minds the enthusiasm for efficient working, and created at Sydney University a healthy atmosphere in which normal life and research could be pursued.[59]

Madsen's analytical mind worked as a grinding sieve, reporting meticulously on progress, scrutinizing local particularities, and accumulating a pile of disseminated ideas and proposals. The cogwheels of the grinder pulverized not only organizing complexities but also allowed him to distinguish practical

[57]NAA. [File A9873/2], 'Minutes of 5th Meeting of the RRB, 30th October 1928'.

[58]For a contemporary account on the Cavendish style, see Ratcliffe (1975).

[59]On J.P.V. Madsen, see White (1970) and Myers (1983).

applications from abstract developments.[60] 'Excellent world class research', F.W.G. White would write half a century later, 'resulted from Madsen's insistence on 50% pure and applied work.'[61] A sensitive stance on finance combined with his balanced view on research. Although he practised the manual craft of electrical engineering, stripping down magnetic devices, inductances, and condensers, and conferring with mechanics or contractors, he preferred to purchase new apparatus from Britain rather than obtain components at reduced price and install them in his own workshops in Sydney. This viewpoint was different, but more effective than Laby's practice, for researchers could concentrate on the scientific aspects without being involved in instrumentalities of technical devices. 'In the end', Madsen argued, 'it turns out to be more expensive than if the set were ordered complete from England.'[62]

Madsen extended his capacity of weighing pros and cons into to the realm of overseas public relations. In spite of being one of the busiest men on the planet—in the opinion of his students, 'the hardest to find when you want him'[63]—he had time to inspect *in situ* several overseas research centres. Madsen was of the same international mould as Rivett, and that style made him strive for keeping close links not only with Britain but also with the US, Europe, Canada, and New Zealand. He was, above all, aware of the fact that the development of wireless in Australia demanded more than a slavish application of experience from British research.[64]

All this did not prevent Madsen from immersing himself mainly in British DSIR's practical policy. Like Appleton and Tizard, he imagined the ideal pattern of scientific research as an equilateral triangle, a triangle with three equal sides and with university science, industrial science, and government science at its three corners. Also, like them, he judged it dangerous that the Australian government should conduct research for the immediate benefit of industries

[60]See also SUA. 'J.P.V. Madsen's papers', W.N. Christiansen, 'Emeritus Professor Sir John Madsen'; 'Radio Research Work', *Age*, August 1933; 'Extracts on Madsen's Qualifications to Occupy the Chair of Electrical Engineering', 14 February 1922.

[61]AAAS. [Fold MS 111/4], F.W.G. White to C.B. Schedvin, 9 August 1982.

[62]Although this quotation refers to the couple of Cathode Recording Detector Findings (one for Sydney and another for Melbourne) requested from the British RRB in 1929, it is applicable on the whole to Madsen and Laby's two ways of understanding research policy. See NAA. [File A10762/1] 'Radio Research Board/ G6/20—General', Madsen to Cook, 9 December 1929.

[63]White (1970, p. 54).

[64]Madsen's perception is observed clearly in his two John Murtagh Macrossan lectures in 1935: 'Radio Research and Ionosphere. Important Work in Progress' and 'Science Aiding Broadcasting: Atmospherics and Radio Waves'. The first lecture was published in *The Ionosphere and its Influence upon the Propagation of* Radio *Waves*. Sydney: Simmons Ltd., 1935.

not under government management. His bid for AWA's inclusion within the RRB's structure matched this 'triangular' vision.[65]

Despite their exclusion from the RRB, the participation of private industries in research activities was pivotal. Madsen was in favour of galvanizing industrial research by means of grants in aid given under suitable conditions. Investigations in radio factories would elicit a positive reaction in radiophysicists. But apart from Fisk from AWA, it does not seem that Australian radio managers grasped the far-sighted dimension of research. One might blame them for being conservative and unadventurous, preoccupied with the creation of mere establishments for production. But they probably might not reproach Madsen for not endeavouring to construct a new model of cooperation that would be of maximum benefit to Australia's emerging radio manufacturing industry, a rare prewar example of cooperation between governmental agencies and universities.

Madsen's vision was along similar lines to those being followed by CSIR. 'Government organisations', stated Rivett in 1928, 'are not the right agencies for attack upon the specific problems of the manufacturer. The individual firm or company can solve most of them, if it has the will... The Council will not take up the individual problems of this firm or that... Thus in radio work we have a part to play which no manufacturers of wireless outfits can play—and our Radio Research Board is doing it.'[66] But this tactic was not properly interpreted by everybody. Whereas Madsen was prudently divesting himself of his professional engagements with AWA and 2FC, Laby's stance regarding the 3LO broadcasting company raised well-founded suspicions. 'The greatest fear that I have in regard to Professor Laby', Madsen wrote to Lightfoot, 'is on the question of the Board dealing with matters of policy. Laby, apparently, thinks that the Board should deal with such matters,' while the other members of the Board insisted that it should restrict itself entirely to scientific work. 'Any suggestion of such policy would mean a disaster for the Board.'[67] Laby's behaviour appearing publicly as 'an advocate for 3LO' prevented him from being appointed chairman of the RRB, and, what was still worse, it increased his reputation for being an obdurate and difficult man.[68]

[65]For Madsen's commitment to search for interfaces between university, industry and government, see AAAS. 'Madsen's Correspondence', Madsen to Lightfoot, 3 March 1927; and Evans (1973, pp. 8–35).

[66]Rivett's quotation is in A.T. Ross (1987, p. 376).

[67]Madsen to Lightfoot, 26 February 1927, in Evans (1973, p. 26).

[68]Rivett to Madsen, 22 April 1927 and 5 May 1927. Ibid., p. 28.

Laby, a man of fragile constitution, and one year younger than Madsen, came from a humble farm family. He loved to say 'he had taken tuition in mathematics by correspondence'.[69] One might think that perhaps his keenness for precision, the confidence in his touch, and his originality and independence when tackling real problems lay in the lack of a good secondary education. The truth is that almost everything he knew was due to his self-taught background. Laby abhorred pretentious researchers, as well as slovenly work, whereas he adored accurate observation and a critical approach. His singular education had instilled a certain fear into him of being ignorant before people.[70] 'He developed highly scientific methods of safeguarding himself from adverse consequences of such ignorance', one of his best friends wrote.[71] He never had a natural gift for talking fluently before an audience, and even less before his students of physics. 'A terrible lecturer but superb with postgraduate students', remembered a pupil of his.[72] 'This fact was bound up with his lack of an ordinary general education, but also with a certain, almost physical, diffidence—which was strangely at variance with the quite exceptional confidence and assurance of this remarkable man's mind.'[73]

Laby's life changed completely in 1904, when he gained an '1851 Exhibition' scholarship for research at Cambridge. His objective in this case was purely chemical; a work of his on the possible change of weight in a chemical reaction had aroused the interest of the President of the Royal Society, Lord Rayleigh.[74] In the next four years, he devoted his talent and experimental ability to the field of physical chemistry, immersing himself with ease in empirical methods at Cavendish, and imbuing himself with J.J. Thomson's expertise. It was at this time that Laby became one of Rutherford's protégés, a relationship which would end in an intimate friendship. What united them was more than mutual affection. The two men shared the same picture of physical reality, inspired the same devotion for physics in their pupils, and engendered in them a desire to emulate their own attainments.[75]

[69]D.K. Picken, 1948. Thomas Howell Laby, 1880–1946. *Ob. Not. Fell. Roy. Soc*, 5, 733–55, p. 734.

[70]See also H.S.W. Massey, 1980. T.H. Laby F.R.S.—The Laby Memorial Lecture. *Aus. Phys.*, 17, 181–7; Cecily Close, 1983. Thomas Howell Laby (1880–1946), Physicist. *ADB*, 9, 640–1.

[71]Picken (1948, p. 734).

[72]Phillip Law's quotation is in Muirhead (1996, p.73).

[73]Picken (1948, p. 738).

[74]L.H. Martin, 1946. Obituary: Professor T.H. Laby, F.R.S. *Australian Journal of Science*, 9(2), 64–65.

[75]Katrina Dean, 2003. 'Inscribing Settler Science: Ernest Rutherford, Thomas Laby and the Making of Careers in Physics'. *History of Science*, 41, 217–40.

Laby held Rutherford in high regard, and vice versa. Rutherford advised him on the right person for filling a lectureship in Melbourne, and endeavoured to ensure a welcoming atmosphere in his laboratory for those of Laby's numerous students who gained an '1851 Exhibition' scholarship (13 did).[76] Laby's total confidence in the Cavendish, and Rutherford's emphasis on empirical research, often at the expense of discriminating against theoretical physics and industrial application, helped without doubt to define the method and scope of research in physics at Laby's department, and probably in all Australian universities.[77] The Rutherford–Cambridge tradition inculcated into researchers a 'string-and-ceiling-wax' mentality which persisted for decades, and, as in Cambridge, often led many physicists to adopt an ostentatious arrogance towards other disciplines considered as unhelpful and second-rate, such as engineering and applied physics.[78]

Laby was without doubt wedded to the British Empire. He was an active member of the 'Round Table Movement' for over 30 years.[79] This movement, founded by one of the leaders of the new Union of South Africa, Lord Alfred Milner, amalgamated the nineteenth-century beliefs of material progress and a series of racial convictions with the application of positivist science to imperial development.[80] After Laby's death in 1946, the secretary of the association, J.F. Foster, wrote to his family a long letter of condolence, expressing profound gratitude for the services provided by him from their inception in 1910, which were, in his opinion, 'among the most outstanding in the history of the whole Movement'.[81]

The Round Table united rigour with vision. Here Laby founds precisely the doctrine that suited his ideals and temperament. He embraced fervently the concept of 'organic Empire unity', for it allowed him to be an integral part of the Empire family while at the same time asserting his stand as a staunch

[76]On the role played by Laby in obtaining the 1851 scholarships, see ibid., pp. 232–4.

[77]In Prof. Oliphant's opinion, Laby was 'the most distinguished physicist working in the south hemisphere', and the man responsible for building 'the finest and most influential school of physics in the Southern Hemisphere'—quoted in Picken (1948, p. 753).

[78]See Home (1988a, pp. 147–165). The influence of Rutherford and Cavendish's style on Australian physics has been treated in many papers. For a critical approach, see S. Davies (1983), Jenkin (1983, 1990), and J.R. de Laeter (1989).

[79]Leonie Foster, 1986. *High Hopes: The Men and Motives of the Australian Round Table.* Melbourne, p. 218.

[80]John Kendle, 1975. *The Round Table Movement and Imperial Union.* Toronto: University of Toronto Press, 22–45.

[81]MUA. [Box 85/144], 'T.H. Laby's Papers and Correspondence', J.F. Foster to Laby's family.

nationalist.[82] Like many others, he spoke of the Empire as his 'family' and of Britain as his 'mother country'; he praised when matters went well, and commiserated when circumstances were unfavourable. The interests and hazards of Australia, of course, were by extension the interests and hazards of the Empire. He believed firmly that the safety of Australia could best be secured as a member of the Empire. The popular slogan 'Defence of the Empire through the development of the Empire' could be extended to his ideals on the development of imperial science. His great love was researching physical reality, the Cavendish modus operandi being the epitome of investigation. Nothing was more gratifying for him than seeing how his students won 1851 Exhibition awards, and elevated their training at Cambridge, as he had done.[83] 'He was prepared to do his utmost to promote and strengthen an effective union with Britain', read the anonymous obituary published in *The Australian Quarterly*.[84] Laby could not understand radio research without invoking his preferred threesome: Empire, criticism, and cooperation. He appreciated the skilful diplomacy of Madsen and Rivett in building up bridges with third countries, but he preferred to channel his efforts within the Empire, preferably with Britain via Rutherford and J.J. Thomson, and subsequently with New Zealand and South Africa.[85]

Bringing 'atmospherics' to Melbourne

Soon after the constitution of the RRB, Madsen and Rivett began to think about selecting radio researchers. The process of choice had been discussed at the British RRB meeting, in Slough, to which Madsen had been invited. By the spring of 1929, Madsen contacted Rutherford to inform him about the need for four radio scientists—two for Sydney and two for Melbourne. A panel—made up of Rutherford, Tizard, and Appleton—was immediately established to vet the merits of aspirants. It was especially important to gauge their technical skill and theoretical ability. Of eleven applicants who replied to the Australian advertisement, the panel recommended six, of whom the RRB chose four: George H. Munro, Leonard G.H. Huxley, David F. Martyn, and Alfred L. Green. All spent several months at Slough, under the supervision of Watson Watt, familiarizing themselves with the latest procedures and atmospherics equipments.

[82] Kendle (1975, p. 85).

[83] For the influence of '1851 Exhibition' awards on Australian science, see Wark (1977).

[84] 'T.H. Laby', *The Australian Quarterly*, 1946, 18 (3), 48–49.

[85] The copious correspondence (more than 300 letters) between Laby and Rutherford is deposited at the MUA [Box 85:144]. Laby was also in contact with, among many others, J.J. Thomson, P.W. Burbidge (University of Auckland), and B.F.J. Schonland (University of Johannesburg).

The first three choice candidates, together with J.L. Pawsey, who would join later, were assigned to Melbourne.[86]

A central element in the Australian programme of radio research involved employing in conjunction two cathode ray direction-finders (CRDF) installed in two points distant to each other so that the bearing of atmospherics could be studied through methods of triangulation. The basic components of CRDF were two square frame aerials, a pair of amplifiers, and a cathode ray oscilloscope. The oscilloscope provided the visual picture when showing the behaviour of the point of origin of an atmospheric from the course of the linear and luminous trace initiated by its recording. The mechanism, perfected by Bainbridge Bell in Slough, was to be known as the Narrow Sector Atmospherics Recorder. The Australian programme of atmospherics was conceived by Watson Watt (with URSI's support) as part of a coordinated study on a worldwide basis. Despite Watson Watt's indications, suggesting the setting up of two apparatuses in the east and the west of the country, they were installed in Melbourne and at Stromlo Observatory, Canberra, thus demarcating an observing base line of 300 miles.[87]

In seeking data that could supply global information on atmospherics, Watson Watt had persuaded Munro and Huxley to perform observations in their voyage to Australia on board the ship 'Baradine'. The fact that the itinerary crossed extensive geographical regions provided an excellent opportunity to locate, from many cross-bearings, the main active sources of atmospherics. The two young researchers realized that the most active sources were in the region of greatest thunderstorm activity in the world, in tropical central Africa. Their conclusions served to confirm the early suspicions of Watson Watt and his colleagues concerning lightning flashes as the chief origin of atmospherics.[88] Once they had arrived in Melbourne, in January 1930, Munro, the most acquainted with the practicalities of CRDF, was assigned to Laby's department to operate the same apparatus that he had employed in the 'Baradine'. Huxley, for his part, was posted to Stromlo, if for only a short time, since in May 1931 he was transferred to Sydney to fill the vacancy caused by Baker's resignation. Meanwhile, F.W. Wood continued recording systematic observations of

[86]A general description of the process of selection is given in Evans (1973, pp. 46–50). For an evaluation of the work on atmospherics carried out at Laby's department, see: White & Huxley (1974, pp. 15–27).

[87]For research on atmospherics in Melbourne, Evans (1973, pp. 186–214). For atmospherics at Stromlo, Allen (1978); Tom Frame & Don Faulkner, 2003. *Stromlo. An Australian Observatory.* New South Wales: Allen Unwin, 53–66.

[88]Munro & Huxley (1932).

atmospherics at Watheroo, with an early model directional recorder supplied by the British RRB.[89]

Munro, an adroit young New Zealander, who had received professional training in Slough, augured that investigation on atmospherics would find fertile ground in Australia. Individual atmospherics caused interferences with radio waves. Broadcasting executives requested help to optimize conditions for radio transmissions and to plan new commercial networks.[90] Furthermore, the possible relation of atmospherics with meteorology was for more evident in forecasting and aviation.[91] For more than a year, Munro was assisted by Joseph L. Pawsey, a physicist who would write his MSc thesis in 1931 on atmospherics while working at the RRB. Pawsey would be one of the first people to obtain postgraduate qualifications based on radio observations in Melbourne.[92] Thanks to Laby's influence, he would widen his training at Cambridge under J.A. Ratcliffe's supervision. Pawsey's resignation opened the way for Hugh C. Webster's appointment, another '1851 Exhibition' scholar from the University of Tasmania, who had taken a PhD at Cambridge.[93]

For Munro, atmospherics connoted not only the agent of interference with radio reception but also the way of discerning their fine structure. While needs for optimum frequency drove the first, reasons for physical knowledge stimulated the second. A set of uninterrupted observations at the range of wavelengths from 3,000 to 300 m. led him to conclude that interference was proportional to radio wavelength. Laby, who disagreed in private on the orientation given by Munro to his investigations, preferred to stress the possible link of atmospherics with meteorological features such as cold fronts. Munro acknowledged the value of such a link but was irritated at the fact that his results occasioned so many adverse comments. 'I am really pretty "fed up" with the conditions of working here', he wrote to Martyn, '[Laby's] attitude towards

[89]G.H. Munro & L.G.H. Huxley, 1932. 'Atmospherics in Australia'. *RRB Rep.*, 5(1); *CSIR Bull.*, 1932, 68.

[90]G.H. Munro, 'The Intensities of Atmospherics at Broadcasting Wavelengths'. Unpublished RRR Rep., 1932.

[91]G.H. Munro, 'The Utility of Observations of Atmospherics in Weather Forecasting in Australia'. Unpublished RRR Rep., 1932.

[92]Among the graduate students attached to RRB who achieved their MSc degrees in Melbourne, figure J.L. Pawsey, W.J. Wark, R.W. Boswell, F.J. Kerr, A.F.B. Nickson, F.G. Nicholls, and J.J. McNeill.

[93]For an excellent review of atmospherics and radio research in European countries at the early 1930s, see H.C. Webster, 'Report of Investigations of the State of Radio Research in Europe at the Present Time', 1933, in Evans (1973, appendix 8).

me and my work has had a pronounced damping effect on my natural enthusiasm and pleasure in what has otherwise been an interesting investigation.'[94]

But even Munro's own pattern of researching did not represent fully the reason for his resignation. Above all was Laby's mentality and his austere, thrifty policy tending in part to combat the difficulties caused by the Great Depression. Laby had been warned not to spend 'an undue proportion of their time on the construction of apparatus which could be made for them by lower paid mechanics'.[95] Munro felt that he did not fit well into Laby's research environment.[96] Laby had, as his highest aspiration, an emphatic, resolute, inflexible pattern of empirical research towards 'truth' in line with the tradition of the Cavendish.

It must have been discouraging for Laby to observe how two valuable researchers, Munro and Martyn—working on fading—abandoned his discipline to join Sydney's department. Both resignations did not alter Laby's firm resolve to establish definitive correlations between atmospherics and meteorological phenomena. This conception had much in common with the line defended by John T. Henderson of the National Research Council in Canada and by R. Bureau of the French Meteorological Service, who judged it indispensable to perform observations in collaboration with meteorologists themselves.[97]

Newly recruited student researchers, William J. Wark and Robert W. Boswell, along with Webster, knowing Laby's devotion to experimental precision and his tendency towards austerity, designed and constructed a second narrow sector recorder mechanically superior to Watson Watt's model.[98] This device, together with that being installed in Stromlo, under Arthur J. Higgs, and the detector operating in Toowoomba, Queensland, configured an efficient network of systematic and coordinated observations. As a result of this cooperation, it was possible to draw, by means of a gnomonic projection of Australia,

[94]Evans (1973, p. 83).

[95]NAA. [File A9873/2], 'Minutes of 9th Meeting of the Radio Research Board, 5th June 1930'.

[96]In 1933 Munro despaired of Laby: 'Of course there was no possibility of changing his mind. Webster, instead of being a help to me, is running round designing the damn thing and hunting up jam tins and alarm clocks'. NAA. [File A8520/11], 'Personal Correspondence of D.F. Martyn', Munro to Martyn, 18 October 1933.

[97]For Henderson's support, see NAA. [File A9873/2 (30)] 'Radio Research Board, Atmospherics', Henderson to Laby, 30 September 1937. Meanwhile Bureau encouraged the Melbourne group stating that he had 'never found an indubitable case when a centre of atmospherics did not coincide with a thunderstorm centre'. See NAA. 'Radio Research Board, Annual Reports, 1932–39', 10th Annual Report of the RRB, 1938.

[98]Boswell, Wark, & Webster (1936).

maps of isointensity lines showing the loci of thunderstorm points related to atmospherics.[99] These maps were for the geographical distribution of sources in atmospherics what maps of isobars and isotherms were for the localization of cold and hot fronts in meteorology. From these diagrams, estimations of the severity of atmospheric interference at different places could be calculated, and groups of thunderstorms of the frontal type could be detected.[100]

Laby's position in favour of atmospherics recording contrasts surprisingly with the pertinacious, obstinate resistance of a good number of Australian meteorologists towards their use in forecasting techniques. The disagreements between the officers of the RRB and of the Commonwealth Meteorological Bureau had been voiced in open dispute in meetings and the press during the previous years. Here is a taste of this friction, from the minutes of the 10th Meeting of the RRB, in 1931. 'The general feeling of the Board', the secretary wrote, 'is that the Meteorological Bureau must strengthen its scientific staff'.[101] The reproach also included the shortage of scientific publications. If, as Laby assured perceptively from his habitual critical standpoint, the reports were 'open to worldwide criticism, it would be a most useful inducement for accuracy. Had this policy been adopted in the case of the Weather Bureau, it would have been all to the good'.[102] Meteorologists, for their part, were extremely reluctant to accept the idea that the art of forecasting, requiring experience and intuition, could be turned into a laboratory science involving precision and talent.[103]

The tension between both bodies, however, decreased from 1935. After considerable reluctance, the Bureau—through the meteorologist H. Treolar—collaborated with the RRB to determine whether atmospherics were of potential value for weather forecasting. Although atmospherics could be associated, with some reservations, with cold fronts, Treolar concluded that their warning value for storms was minor. 'On the whole the system proved to be of less value for weather forecasting than for aviation', White would afterwards recall.[104] These conclusions virtually marked the end of the atmospheric programme of the RRB. One might understand Laby's frustration when he

[99]See, for instance, Higgs, Munro, & Webster (1935).

[100]For a valuable review of atmospherics research in Australia since its inception, see T.H. Laby, 1936. Contribution to 'Discussion on Thunderstorm Researches,. *Quart. Jour. Roy. Met.. Soc.*, 62, 507–16, 525–7.

[101]NAA. [File A9873/2], 'Radio Research Board, Board Minutes', Minutes of the 10th Meeting of the RRB, 12 March 1931.

[102]Ibid., NAA. Minutes of the 23rd Meeting of the RRB, 21 May 1935.

[103]For an account of the practice of forecasting at that time, see Harding (1984).

[104]White & Huxley (1974, p. 27).

found out that, while direction-finding on atmospherics was of little value to Australian meteorologists, the organizers of the International Meteorological Conference at Warsaw in 1935 recommended to continue radio investigations for weather forecasting, and active research was in progress in New Zealand, Canada, South Africa, the USA, and elsewhere.[105] Ten years after the commencement of the atmospherics programme, Laby had occasion to taste the most bitter side of research work.[106]

National Broadcasting Service: long versus medium waves

In the late 1920s, Australian broadcasting appeared to be given an unrepeatable opportunity to design a national network devised on a scientific basis. The dual system, which was introduced by Brown in July 1924, differentiated between the 'A' or 'national' stations (maintained by revenue from licences and based on the BBC model) and the 'B' or 'commercial' class (self-supported by advertising).[107] Stations had been licensed to operate on the medium frequency for five years, a period in which the number of licences increased from 38,000 in June 1925 to 310,000 in July 1929. This hybrid scheme—a combination of the British and American systems—favoured cities and small states like Victoria, while discriminating against the countryside and sparsely-populated states such as Western Australia.[108] The general public, especially in the bush, perceived the new medium as an instrument to alleviate isolation and loneliness.[109] The Bruce government enacted the nationalization of 'A' class stations at the expense of private companies as the most effective way out of social and state inequalities. In similar vein, it appointed a committee—chaired by Brown and including, among others, Madsen—to advice technically and scientifically on the planning of a national broadcasting service.[110]

Both Brown and Madsen imbued the planning with their keen vision. Basically, their recommendations would lead to, in 1932, the creation of a statutory instrumentality, the Australian Broadcasting Commission (ABC), the body in charge of controlling national broadcasting and based along BBC

[105]NAA. [File A9873/2], 'Radio Research Board, Board Minutes', Minutes of the 31st Meeting of the RRB, 23 November 1937.

[106]For the activities on atmospherics at RRB from 1935 onwards, see NAA. [File A9873/2(30)], 'Radio Research Board, Reports on Atmospherics'.

[107]Curnow (1963, p. 47).

[108]Lesley Johnson, 1988. *The Unseen Voice: A Cultural Study of Early Australian Radio*. London: Routledge, 31.

[109]Counihan (1982).

[110]Moyal (1984, pp. 117–51); and Evans (1973, pp. 38–41, 96–7, 215–25).

lines.[111] The recently nationalized 'A' stations would be operated by the PMG's department for three years. It would become the responsibility of Brown and his advisers not only to install and maintain the national service but also to decide the technical pattern on which both the 'A' and 'B' classes should be developed.[112] Although Brown was not formally compelled to solicit the board's opinion, he considered it his duty to seek their expert view, as a symbolic gesture in recognition of their services. The executive committee of CSIR was very receptive to Brown's goodwill, for most of the board's funds had been financed by the PMG. For the RRB board, it was a unique opportunity to gain credibility, an air of challenge, the occasion for self-affirmation.[113]

In the following months, the board expended its time and energy on studying the adaptation of broadcasting techniques to the rugged geography and meteorology in Australia. In a country of little scientific practice, frequency allocations, interferences, land lines versus short waves, atmospherics, and fading rapidly peopled radiophysicists' minds and contained the ambitions of many intrepid entrepreneurs. Listeners in general ignored the complexities pertaining to frequency selection for optimum performance. Audiences did not think about scientific-technical considerations but about the quality and entertainment of programmes. Readings from Dickens, jazz music, weather forecasts, and drama were their concerns.[114] But 'without such information', as declared Madsen, 'it would be a most unbusinesslike and unscientific manner of tackling' the service.[115] The issue resembled a mathematical problem. 'Given a roughly circular piece of country', wrote one observer, 'what frequencies and powers should most efficiently be employed in regional broadcasting transmitters to give an adequate service to the area?'[116]

In several reports on factors governing broadcasting, in the meetings of the board in 1932, in a widely circulated lecture delivered by Laby to the Institute of Radio Engineers in Melbourne in 1934, and again and again in the first half of the 1930s, the RRB's officers emphatically recommended the combination

[111]On the ABC's role as a 'cultural elevator', see Frank Dixon, 1975. *Inside the ABC: A Piece of Australian History*. Melbourne: Hawthorn Press, 65–9.

[112]The pattern of the PMG activities, which extended to most fields of radio communication, is described in *Broadcasting and the Australian Post Office 1923–73*. Australian Post Office, 1973.

[113]Moyal (1987, pp. 35–54).

[114]L. Johnson (1981).

[115]NAA. [File A10762/1], 'Madsen Papers and Correspondence', Madsen to Wright, 16 May 1928.

[116]Evans (1973, p. 96).

of long and medium waves in selected areas.[117] They particularly advocated for moving from rational criteria to commercial pressures. Quite apart from the question of the legitimacy of companies to impose their tenets in science, the decision of physicists' was in consonance with European procedure.[118]

Similarly, Fisk advocated long waves as ideal for Australian specificity.[119] Persuaded of their appropriateness, he invited P.P. Eckersley, the BBC's Chief Engineer, in 1932, who evaluated the frequency, power, and location of the Australian stations. In Eckersley's opinion, the current use of medium waves guaranteed a 'moderately good service' to urban listeners; but their extension, however calculated, to the entire geography could not emend the deficiencies of the exiguous, unsatisfactory, 'poor service' of rural listeners.[120] The introduction of long wave transmitters within the broadcasting system could eventuate in, or give rise to, changes with socially far-reaching repercussions.

From the technical and administrative viewpoint, Brown's and the PMG's part in the diffusion of broadcasting on the medium frequency was pivotal. From 1925 the Radio Research staff had been acquainted with field strength measurements within the international medium wave band (from wavelengths of 200 metres to 545 metres), as well as with the use of short wave for inland areas. The Conference on Broadcasting held in March 1930 in Melbourne was an unsuccessful attempt to conciliate the divergences—and sometimes hostility—existing between the PMG and RRB's staff.[121] What was debated is a matter of choice on a scientific basis, and what was chosen is a matter of expediency: the cold logic of physics was qualified by humans immersed in day-to-day reality, in lobbying stemming from commercial stations, and in

[117]See the following Reports of the RRB: R.O. Cherry, D.F. Martyn, G.H. Munro, and W.J. Wark, 'Factors Governing the Service Area of Broadcasting Stations in Australia', 1933; R.O. Cherry, 'Long versus Medium Waves for Australian Broadcasting', 1932; D.F. Martyn, 'Fading on Broadcast Frequencies', 1932; D.F. Martyn, 'Report on Broadcasting'; G.H. Munro and A.L. Green, 'Long versus Medium waves', 1933; G.H. Munro and A.L. Green, 'Measurements of Attenuation, Fading and Interference in South-Eastern Australia, at 200 Kilocycles per Second', *Jour. Inst. Eng. Aus.*, 1933, 5, 193–9.

[118]For both commercial and geo-strategic reasons, in Europe medium and long waves were favoured, whereas short wave was used for international broadcasting. Meanwhile, the American companies preferred short wave for domestic broadcasting. See Wood (1992, pp. 24–5) and Aitken (1985).

[119]AAAS. 'General', Fisk to Madsen, 12 November 1926.

[120]P.P. Eckersley, 1932. *A Planned System for Australia Broadcasting. A Report to Amalgamated Wireless Limited.* Institution of Radio Engineers, p. 4.

[121]NAA. [File A9873/2], Minutes of the 16th and 19th Meetings of the Radio Research Board, 14 March 1933 and 21 August 1933.

Table 4 Broadcasting stations and licences in force for broadcasting services, 1913–1938

Wireless Stations	1913	1925	1938	Broadcasting Stations	1924	1930	1932	1937	1938
Coast station	19	27	28	Stations	18	32	60	110	119
Ship	26	122	220	'A' or national			12		24
Aircraft			23	'B' or commercial			48		95
Broadcast		13	144						
Experimental		290	1737	Licensed listeners	38,336	312,192	350,000	,008,595	1,045,363

Source: A.S. McDonald, '1913–38. A quarter-century of Radio Engineering in Australia', in *Proceedings of the World Radio Convention*, Sydney, 4–14 April 1938 (Sydney, 1938), p. 8.

the radio technicalities particular to the Australian outback. The final decision for the development of national broadcasting rested with Brown, the ultimate authority. Between scientific ideals and pragmatism, Brown opted for the latter. The opportunity to erect a powerful long-wave station such as the successful Radio Luxembourg had vanished.[122]

At the time the Federal Government promulgated the Act of the ABC on 1 July 1932, frequencies had not yet been allocated.[123] Over the next six years the national network would increase by 100%, going from 60 stations (eight main, four regional and 48 commercial) in 1932 to 119 (24 national and 95 commercial) in 1938. By April 1938 the 1,020,000 licensed listeners were concentrated mainly in six metropolitan areas (60%), representing an audience of possibly two millions.[124] It was estimated that 13 out of every 20 households had wireless receiving sets (the greatest density being in Adelaide, 89%). The number of licenses had gone from 1% of Australian population in 1925 to 5% in 1932, and up to 15% in 1938, localizing mostly in urban areas. To compensate for this class differentiation among listeners and ameliorate the content of broadcasting, regional and capital city stations were interconnected through telephonic lines and began to transmit the same programmes.[125] Radio broadcasting would

[122]Evans (1973, pp. 215–26). Unlike Australia, American radio engineers played a key role in the policy of broadcasting. See Hugh Richard Slotten, 1995. 'Radio Engineers, the Federal Radio Commission, and the Social Shaping of Broadcast Technology'. *Technology and Culture*, 36, 950–86.

[123]The following data have been extracted from NSWA. [ML MSS 6275/ Box 25], 'Sir Ernest Fisk papers', Fold 'Facts and statistics on wireless broadcasting in Australia, 21st July 1938'.

[124]H.P. Brown, 1938. 'Broadcasting in Australia'. *Proc. Wor. Rad. Con.*, 1–15.

[125]H.P. Brown, 'The National Network', *The Listener-In*, 9 April 1932, p. 28.

participate gradually in the formation of a unique Australian culture. In the midst of radio practicalities and technicalities, medium-frequency radio transmission would prove an instigator of change in Australia's transition from a rural society to a consumerist and Americanized one.[126]

Ionospheric physics at Sydney

From 1925 onwards, the rate of production of experiments and theories on the ionosphere was so rapid that one might be tempted to discuss the beauty and originality of the new atmospheric picture with its layers and boundaries. Such a chronological and organized account would probably mutilate the limbs sustaining the body of atmospheric physics. Ionospheric physics developed on the heels of the expansion of observations to polar and Southern Hemisphere regions. As we have seen, just at the time that relations between Britain and the Dominions were being redefined, invisible bonds began to be displayed by consensus. Their objective was to facilitate Australia's development of radio research, but also the repetition of crucial experiments performed by British scientists.

The new relations demanded people who accepted the challenge on the basis of adherence and consent. Green, the only one of the four appointees assigned to Sydney, met all such requirements. An English student and graduate of King's College, London, Green had assisted Appleton and Ratcliffe in the study of the polarization of downcoming waves from the ionosphere. He was acquainted with the magneto-ionic theory developed by Appleton and Wilhelm Altar.[127] Appleton's experiments had proved that waves were elliptically polarized in the left-hand direction. Now, if his theoretical deductions were correct, 'similar measurements made in the Southern Hemisphere' should show right-handed polarization, and thus 'would yield evidence which would materially confirm' the validity of the magneto-ionic theory.[128]

Before leaving for Australia, Green, and Appleton himself discussed the importance of obtaining experimental data from Southern Hemisphere locations. In spite of employing improvised equipment, Green performed polarization measurements at Jervis Bay, New South Wales, in 1930, confirming Appleton's predictions. The Earth's magnetic field impinged upon the direction

[126]Scout Howden, 1997. 'Australian Radio and Cultural Formation'. *Access: History*, 1(1), 29–47, p. 40.

[127]Gillmor (1982).

[128]E.V. Appleton & J.A. Ratcliffe, 1928. 'On a Method of Determining the State of Polarization of Downcoming Wireless Waves'. *Proc. Roy. Soc. Lond.*, A 117, 576–88, p. 576.

of rotation of the plane of polarization of the radio waves. 'The results', Green telegraphed exultantly to Madsen, 'form the final link in the chain of proof of the Eccles–Larmor–Appleton magneto-ionic theory'.[129] Appleton, for his part, full of excitement, informed Ratcliffe of Green's achievement: 'We have now got what we consider conclusive evidence that the splitting is magneto-ionic.'[130] All 'this re-reading of things makes me think that our polarization paper is the best paper we have written.'[131] One year later, Green obtained the first estimations in the Southern Hemisphere of the height of both E and F layers, at the same levels as those found in the Northern Hemisphere.[132]

By 1933, Sydney's research would progress from surface propagation and atmospherics, through fading, to polarization and propagation of the upper atmosphere, to culminate in fundamental investigations of the ionosphere. The shift from local aspects of radio propagation to the elucidation of the structure and dynamics of the ionosphere was accompanied by international recognition. Through ionospheric physics came maturity (the sense of identifying with a group of Australian radiophysicists); through this maturity came adaptation (Empire bonds were modified). The spirit of cooperation nuanced gradually into a mixture of rivalry and emulation.

The slow, if progressive conversion was due to both external and internal factors. The external causes were associated with financial problems. The economic depression of the early 1930s had caused disparate working conditions. Officers' salaries, for instance, were lower than those of their counterparts at the British RRB and in private radio industries. In July 1931, Baker, who had specialized in field strength measurements and in radio broadcast transmission, joined AWA—supposedly for financial reasons—, becoming the manager of the Marconi School of Wireless. In September, Huxley accepted the post of lecturer in physics at the University College of Leicester—here also incomes were much higher—, abandoning his investigations on atmospherics and fading. These two losses were replaced by Martyn and Munro, transferring to Sydney from Melbourne. They provided the Sydney group with the combination of a solid theoretical basis and technical experience.[133]

The Sydney group improved its instrumental capacity substantially when Geoffrey Builder, a young Australian who had graduated from the University

[129]Green to Madsen, 25 June 1930. Evans (1973, p. 168).
[130]EUA. MS Gen. 1985, Appleton to Ratcliffe, 30 September 1931.
[131]Ibid., Appleton to Ratcliffe, 7 October 1930.
[132]A.L. Green (1932). On Green's polarization tests, see White & Huxley (1974, pp. 10–11).
[133]Evans (1973, pp. 83–5).

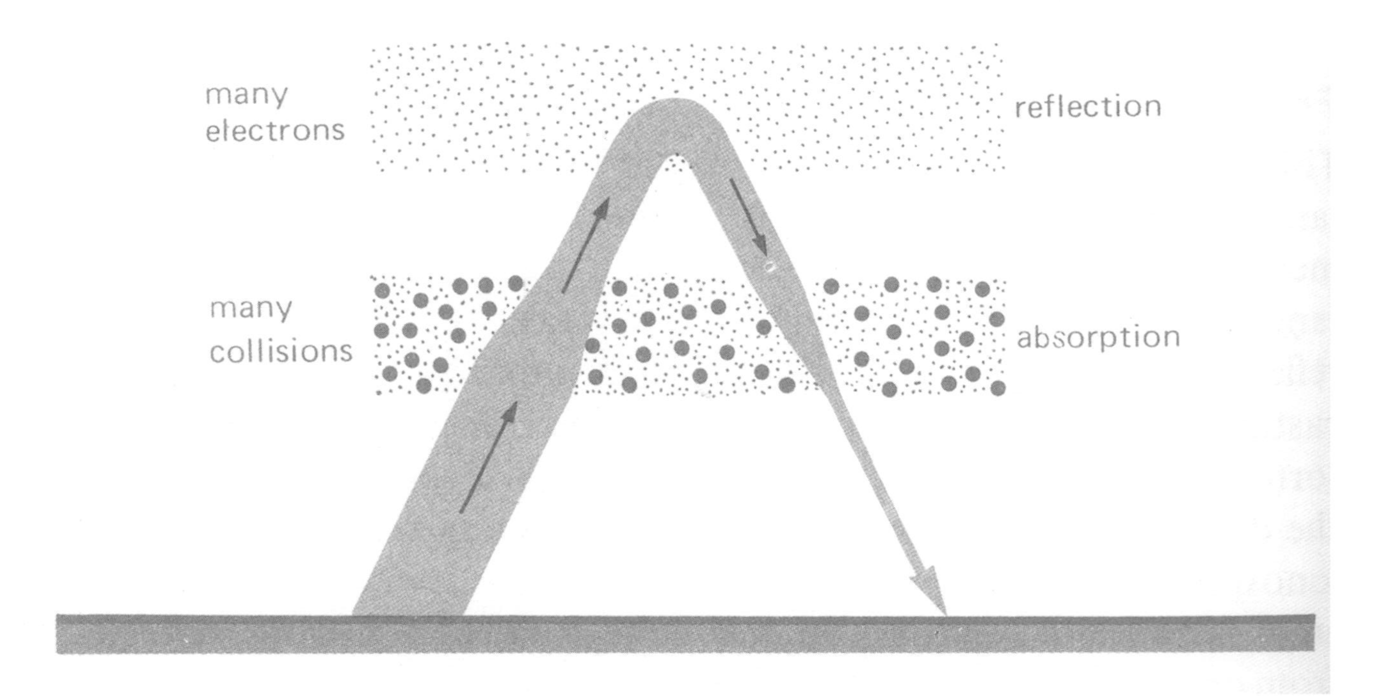

Fig. 16: If there are sufficient electrons high up, radio waves are reflected, and if there are too many electrons low down, they are absorbed. If the frequency is too great, there are not enough electrons to reflect; if it is too small, the absorption is too great. From Ratcliffe (1970), p. 206.*

*There are instances where we have been unable to trace or contact the copyright holder. If notified the publisher will be pleased to rectify any errors or omissions at the earliest opportunity.

of Western Australia, was contracted.[134] He had worked as a fitter at the railway workshops while attending evening classes at the Perth Technical School. He was attracted by radio research, just as the four young men recruited in Britain had been. While completing his M Sc, he took advantage of his post as assistant at the Watheroo Observatory to investigate atmospheric potential gradients. Skilfulness in the command of atmospherics derived from practice and familiarity.[135]

Builder's gifts for ameliorating techniques did not go unnoticed. Appleton was fortunate that Builder continued his postgraduate training at King's College. He earned Appleton's admiration after his contribution to the scientific expedition to Tromso, Norway, during the Polar Year 1932–33.[136] Builder's expertise persuaded Appleton to change his own frequency change method for the pulse-echo technique developed at the Carnegie Institution.[137] Builder displayed all his ingenuity and proficiency in constructing a semi-automatic virtual height recorder supplying the equivalent height (P′) – frequency (f) plot each 15 minutes. As happens in any quasi-continuous recording, it permitted observation of the evolution of the layer height. With the new device, monitoring the correlation of ionization with magnetic and solar activities appeared to be less cumbersome. 'I do not think there is anyone more familiar with it now than he is', Appleton confessed to Laby in 1933.[138]

Pulse modulation soon gravitated towards a standardized technique in England and elsewhere. Each P′f plot was like an Aladin's wonderful lamp: from each curve emerged as if by magic the equivalent height of layers, the number of ionized regions, and the gradient of ionization in each region. The main problem troubling physicists lay in recording observations in different terrestrial latitudes, rather than in technique.[139] Like Watson Watt and the 'Baradine', Builder provided Appleton with the figure of the equatorial ionization on his return by ship to Australia.[140]

[134]On G. Builder, see R.W. Home, 1993. Builder, Geoffrey (1906–1976), 'Physicist and Radio Engineer'. *ADB*, 13, 290–1.

[135]V.A. Bailey, 1960. 'Obituary: Geoffrey Builder'. *Australian Journal of Science*, 23 (5), 155–6.

[136]E.V. Appleton & G. Builder, 1932. 'Wireless Echoes of Short Delay'. *Proc. Phys. Soc. Lond.*, 44, 76–87.

[137]On the pros and cons of the different methods, see G. Builder, 1933. 'Wireless Apparatus for the Study of the Ionosphere'. *Jour. Inst. Elec. Eng.*, 73, 419–36.

[138]Appleton to Laby, 27 May 1933. Evans (1973, p. 228).

[139]Ibid., Builder to Cook, 7 June 1933, p. 228.

[140]Appleton to Laby, 27 May 1933. Evans (1973, p. 228).

In Sydney, Builder propounded the use of the British pulse-sounding set, but his offer was declined. Although rejection was due largely to lack of funds, there were other reasons. Martyn and Wood, for instance, were at this precise moment engaged in developing analogous automatic devices.[141] One wonders how the advance of the Sydney group might have changed with the adoption of pulse technique, or, indeed, what Builder would have contributed in the event of having felt completely identified with the board's scientific policy. What is certain is that systematic recording of observations was delayed for some years, and Builder remained only 12months with the RRB before joining AWA. There can be little doubt that in that short period of time he enriched the Sydney group with his new ideas on technique and instrumentation. Arguably, Builder's most valuable contribution was the design and construction of the pulsed echo transmitting equipment, the predecessor of the P′f ionosondes developed shortly afterwards in Sydney.[142]

Builder's broader ambitions would materialize by 1935, when the higher flexibility and efficiency of the pulse technique ousted the earlier frequency-change method. Most of the success was due to a classmate of his at King's College, O.O. Pulley, who had built the first manual P′f equipment for the British RRB.[143] Pulley, a graduate of the Department of Electrical Engineering, Sydney, was recruited again by Madsen for Australia in 1934. There he constructed a similar transmitter receiver. To observe both visually and photographically the continuous pattern of layer heights, he incorporated a cathode ray oscilloscope. By June 1935, Pulley's equipment provided the first routine ionograms in Australia.[144]

It would not be fair to give Pulley all the credit. Another graduate of Madsen's school, H.B. Wood—who had assisted Martyn since 1932—, developed the first fully automatic frequency sweep recorder at the Sydney workshops.[145] He obtained continuous recording of P′f ionograms from May 1936 onwards.[146] Madsen and his group were overcome with excitement and enthusiasm. He proudly informed the board that the new equipment might be instrumental in providing Martyn and Pulley with excellent material for theoretical

[141] David F. Martyn & Herbert B. Wood, 'A Frequency Recorder', *Jour. Inst. Elec. Eng.*, 1933, 5, 6–13; *CSIR Bull.*, 1935, 87; *RRB. Rep.*, 1935, 6, 49–58.

[142] Evans (1973, pp. 229–31).

[143] Pulley (1934a).

[144] Pulley (1934b).

[145] H.B. Wood (1936).

[146] Evans (1973, pp. 239–41).

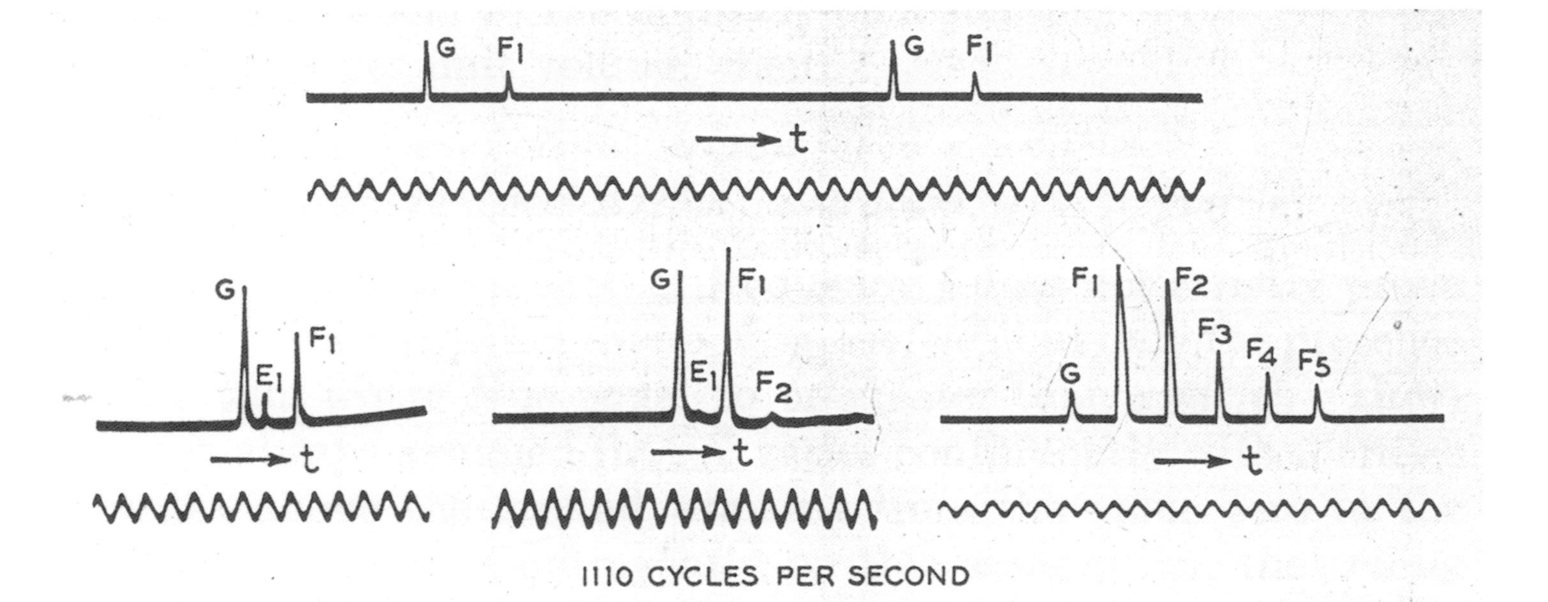

Fig. 17: Records of echoes. G, outgoing signal; E1, echo returned after one reflection from E-layer; F1, echo returned after one reflection from F-layer; F2…F5 echoes returned after two to five reflections from the F-layer, with intermediate reflections from the ground (Appleton and Builder). From Darrow (1940b), p. 431.

deductions on the ionosphere.[147] These advances might permit the monitoring of seasonal ionization and height changes, but also and most importantly, would alleviate investigators of all night observational work. Free time was essential for theoretical analysis. Madsen's thoughts appeared to be a premonition of what would occur some months later.

The genius of a complex man

Much of the reputation acquired by Australian radio research rests on the talent of a young Scotsman. Martyn's association with RRB yielded several major contributions of international dimension. Much of the success and team spirit that was forged at that time pivoted around him. Martyn's investigations of the ionosphere were, to a large degree, projections of his own personality.

David Forbes Martyn, the lead personality of the Australian community of radiophysicists, was a secretive and independently-minded man. One sees this in the trajectory of his life and his ideas. Son of an ophthalmic surgeon, at only 17 years of age, he abandoned his native Scotland to study physics at the Royal College of London. His professors soon perceived that his secretiveness and intelligence could fit in with the sharpness of mind required for radio investigations. With a first-class honours, he obtained a research scholarship at the University of Glasgow in 1926 while attending doctoral courses at the University of London. In Glasgow, he investigated the stability of the triode oscillator from both the theoretical and experimental viewpoint.[148] For the originality and inventiveness shown, the University of London awarded him a PhD in 1928. The position at the Australian RRB appeared when he most seemed in need of it. It was a unique opportunity for emancipating a theoretical physicist. Appleton probably noticed the singularity of Martyn's intuitions when he vetted the four applicants, for, among other working proposals, Martyn's included the modification of Appleton's frequency-change method for a new technique of continuous frequency sweeps.[149]

Martyn's revolutionary sense of physics, which appears in him from a young age, was the outcome of a defiant attitude on his part, even a remonstrance. It

[147] NAA. [File A9873/2], Minutes of the 25th Meeting of the RRB, 19 November 1935.

[148] Martyn (1927, 1929–30).

[149] On D.F. Martyn, see H. Massey, 1971. David Forbes Martyn, 1906–1970. *Biog. M. Fell. Roy. Soc.*, 17, 497–510; J.H. Piddington & M.L. Oliphant, 1971. David Forbes Martyn, 1906–1970. *RAAS*, 2(2), 47–60; R.W. Home, 2000. 'David Forbes Martyn (1906–1970), Physicist'. *ADB*, 15, 320–2; Gillmor (1991, pp. 188–9); and NAA. [File A8520/11], 'Personal Correspondence of David F. Martyn'.

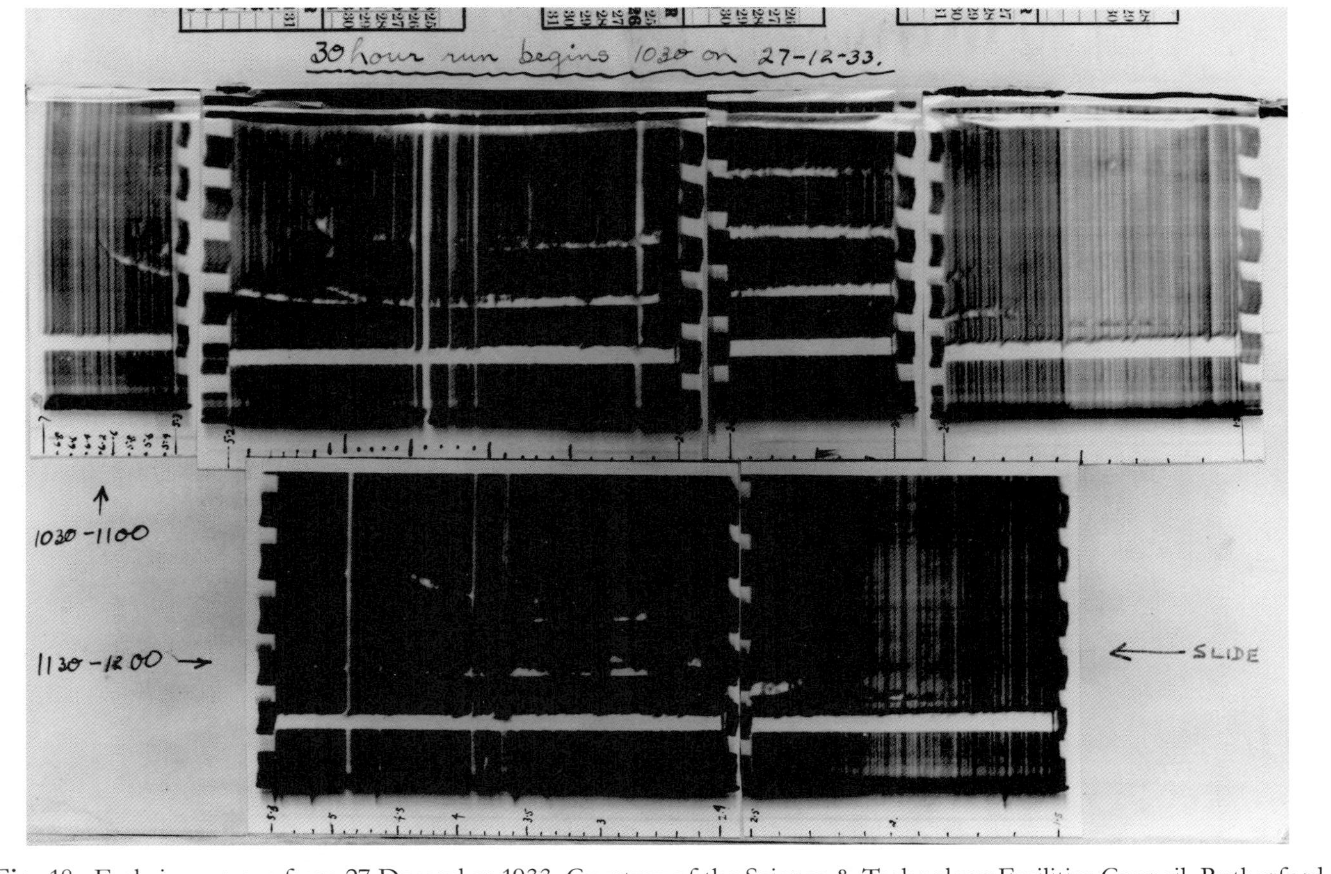

Fig. 18: Early ionograms from 27 December 1933. Courtesy of the Science & Technology Facilities Council, Rutherford Appleton Laboratory.

followed from the uncomfortable realization that he had an exceptional intuitive talent to formulate solutions, but that the problems were not going to be solved in the terms in which he had conceived them. The matter might seem scarcely worthy of attention, much less the nonconformity of a young man of his intelligence. But in Martyn, there was a well of tension accentuating his temper when he noticed incomprehension. As early as 1930, for instance, at the age of 24, Martyn wrote these prophetic words:

> I have investigated the practicability of receiving echoes [reflected from the Moon]. In attempting the experiment, the use of short-waves of only a few metres in length would be desirable [...] It is a relatively simple matter to construct a reflector to produce a concentration of energy or 'beam' in the required direction [...] This development appears to the writer to provide a new tool to physics and astronomy. Thus a wireless beam may be used as probe to investigate regions of the solar system, in much the same way that X-Rays have been used to examine the structure of matter. Observations on the time taken by the signal to return [...] will all yield information regarding the size and shape of the object encountered, its constitution and its velocity relative to the earth.[150]'

One might scarcely imagine in retrospect a more precise and marvellous prediction, not only of radar but also of the communication satellite or radio astronomy. This 'extravagant' project was but a mere anecdote at Laby's laboratory.

To Martyn, physics provided leadership and challenge. He conceived theoretical investigation of the ionosphere as a game in which he could exercise his mental faculties. As if it were an exciting adventure, he enjoyed finding fallacies and errors in the theories of his adversaries; he was proud when he or his Australian colleagues enunciated theorems and solved unfathomable enigmas.[151] However, teams—rather than individuals—could play that game, commanded as he thought by leaders of scientific communities. He was conscious of his role as leader in the Australian group, and of being the visible part of an enormous iceberg.[152]

Martyn's solid mathematical training permitted him to refute coherently and convincingly hypotheses put forward by overseas colleagues. When,

[150]D.F. Martyn, 1930. 'The Reflection of Wireless Signals from the Surface of the Moon'. Unpublished report. Melbourne: RRB.

[151]An example is Martyn to Piddington, 20 September 1937, in NAA. [File A9877/1], 'Personal Correspondence of David F. Martyn'.

[152]In Madsen's opinion, Martyn was well equipped to act as 'a leader of a team of investigators'. NAA. [File A9873/2], Minutes of the 24th Meeting of the Radio Research Board, 29th May 1935.

for instance, he developed in 1935 a theorem—now known as 'Martyn's theorem'—by which he mathematically related waves obliquely affecting the ionosphere to those of vertical incidence, he knew such a proposal would be essential for future work on ionospheric prediction. 'Recently English workers', he said euphorically, 'initially in ignorance of this conclusion, [...] discovered the relevant theorem and altered the plan of their work into a test of this theorem. The published account of their results provides an excellent verification.'[153] Here precisely lay the strength of a scientific community. Martyn could not understand investigation without inciting his workmates to cultivate the terrain of originality and creativity.[154] He appreciated the efforts being directed to putting Australia on the map, but appreciated above all a realm of freedom in which he could develop his scientific ideas. Some of his most outstanding contributions were serious attempts to inject originality and independence into the mathematics of theories of the ionosphere.

Like their British colleagues, Australian radiophysicists often worked in close collaboration with each other. In Martyn's case, the evidence allows one to say that he participated in the 1930s in more than 30 papers and unpublished reports on radio fading (in Melbourne), and on wave polarization, field strength, and radio propagation (in Sydney). Although from 1935 onwards he gravitated towards fundamental problems of the physical structure of the ionosphere, he paid special attention to the capacities of the pulse method, as well as to laying the cornerstone for the geophysical conception of the ionosphere. Martyn encouraged G.H. Godfrey and W.L. Price, two students at the Sydney Technical College, to research radiation and absorption in the upper atmosphere;[155] in 1935, he and Green established the first evidence for ionospheric moving disturbances;[156] and, in 1937, together with Munro and J.H. Piddington, he perfected the pulse-phase technique providing a continuous register of pulses reflected from the ionosphere.[157] Martyn as the theoretician and Munro as the experimentalist symbolized the articulation of two complementary pivots.

[153]Martyn's report on 'Radio Research and Related Investigations, 1st March 1937'. Evans (1973, p. 121).

[154]Some examples are Martyn to K. Kreielsheimer from the Physics Department, Auckland University, 13 March 1935; and Martyn to Higgs, 30 July 1935, in NAA. [File A9877/1], 'Personal Correspondence of David F. Martyn'.

[155]The paper was sent by Madsen to Rutherford in June 1937 and presented at the Royal Society of London in that year. Godfrey & Price (1937).

[156]Green & Martyn (1935).

[157]Martyn, Munro, & Piddington (1937).

In the classic treatment of radio propagation, it was tacitly assumed that the ionosphere was a linear transmitting medium; consequently, the bearing of a wave was unaffected by other waves traversing at that moment the same portion of space. Nevertheless, from Europe arrived news of observations not being consistent with these theoretical parameters. Radio Luxembourg, the first superpower broadcasting station, founded in 1933, had captured within scarcely two years a 50% share of the listening audience of Europe. The new station not only represented a commercial threat to for the BBC and French and German companies but also distorted (through interferences) waves emitted by less powerful stations.[158] The anomaly—known as 'cross modulation'—was first detected by Telligen in Switzerland, and later confirmed by Van der Pol in Holland. Observations had failed to establish rigorously their cause. Nor had other observers recognized how a broadcasting station, however powerful it was, could alter the physical state of the ionosphere. To explain the effect, Martyn and Bailey combined their knowledge of wave propagation with their notions on the motion of electrons in gases.[159] Bailey's familiarity with the conduction of electricity in gases, acquired as a result of his research experience at J.S. Townsend's electrical laboratory in Oxford, contributed to considering the effect from a different angle.[160]

In 1934, Bailey and Martyn argued that the current theory of interaction of radio waves was completely inappropriate: a powerful radio wave traversing the upper atmosphere could interact in a nonlinear manner with a weaker signal. To demonstrate it, they introduced a novel notion. A high-powered station would produce an appreciable change in the mean velocity of the electrons of that part of the ionosphere in the vicinity of the transmitter. To explain how this change could originate interferences, they appealed to the electromagnetic theory of waves. In an inhomogeneous atmosphere, the variation of their mean velocity produced a change in the frequency of collision of the electrons with molecules and hence in the absorbing power of the gas. As one of their major innovations, they demonstrated how the absorption power varied in accordance with the modulation frequency of the station. That would explain the fact that the modulation impinged upon any other carrier wave which could traverse the region. After Green's polarization tests, this theory was the most eulogized Australian achievement. 'Appleton made the *Luxembourg effect* the

[158]On the Radio Luxembourg broadcasting station and its effect on other European stations, see J. Wood (1992, pp. 43–6). On the interaction of radio waves, see Evans (1973, pp. 244–5).

[159]Bailey & Martyn (1934a, 1934b).

[160]R.W. Home, 1993. Victor Albert Bailey (1895–1964). *ADB*, 13, 90–1.

climax of his talk', would write Martyn to Bailey.[161] Appleton himself would confess to Bailey: 'Your theory accounts nicely for everything. The Australian Research Board has worked wonders; they have a fine team out there.'[162] The notion of a modulation of radio signal led to a rational interpretation of the empirically laboratory-analysed passage of charged particles through gases.[163]

The personal connection between Martyn and Madsen was to be as subtle and complex as the relations between the ionosphere and electromagnetic waves. Madsen had the very greatest admiration for Martyn's talent, which he labelled an 'excellent type of mind'.[164] Martyn at first found Madsen a ray of freedom compared to Laby's strictness. Madsen's circumspection and tolerance were a safeguard for Martyn's theoretical investigations. This was in 1933, after Martyn had sensed that ionospheric physics at his department in Sydney was the intellectual and physical environment which best suited his personality. In 1936, on the occasion of Martyn's journey to Europe, Madsen procured the best possible credentials for him so that he could visit the most prestigious scientific institutions. As an expression of gratitude, Martyn wrote to him: 'I feel that the probable success of this trip will be largely due to your foresight and wise advice.'[165]

During Martyn's overseas stay, communication between them was free-flowing and respectful. Martyn returned to Australia showered with international recognition and scientific admiration. Nevertheless, that circumstance itself altered the map of relations. Now, from the altar of celebrity, Martyn began to demand full control of his activities, as well as positions of responsibility. He began to distrust Madsen, but also Rivett and all his superiors, for exercising often paternal roles from their administrative posts. But, as Martyn conceded, paternalism itself might be a virtue if, as in Madsen's case, it arose from adherence to general principles and from his coherent view of investigation in state institutions as a driving force behind progress. 'No other institution exists in any part of the world in which it would be possible to

[161]NAA. [File A9877/1], 'Personal Correspondence of David F. Martyn', Martyn to Bailey, 16 June 1936.

[162]Ibid., Bailey to Martyn, 11 January 1935.

[163]It is significant to point out that the paper was communicated by J.S. Townsend, a world authority on interaction of waves in gases. The Luxembourg effect, such as was described by Martyn and Bailey, was to be confirmed in 1955 in the laboratory by Anderson and Goldstein. See Massey (1971, pp. 499–500).

[164]NAA. [File A9873/2], Minutes of the 24th Meeting of the Radio Research Board, 29 May 1935.

[165]Evans (1973, p. 399).

carry out pure research under more advantageous conditions', Martyn would say.[166]

Whereas Madsen had as his basic tenet the furtherance of postgraduate research, Martyn devoted his academic talent to gaining a brilliant curriculum as theoretician. Madsen wanted to emulate Henry Tizard's mastery of organization and stimulus; Martyn had as model a theoretician of distinction, Sydney Chapman, whose views he valued above all.[167] Around the mid-1930s, Madsen, then 35, was a family man having a special devotion to science, engineering, and his widespread family. In his opinion, scientific policy was comparable to a child's education: provided young graduates received sufficient support in their research endeavours, then one could expect them to obtain excellent results. Martyn, who was 30 in 1936, was a single man sharing a flat with Munro in Sydney. He appreciated team work, especially with Munro, Higgs, Kidson, and Piddington, but as an intellectual venture.[168] Madsen, prudent and administrative, was a family man both in science and at home; Martyn, intrepid and liberal, was a loner if very good company. Martyn, although sociable, was very reserved and could only enjoy himself with people of his own circle, with whom he might share his three preferred passions: wine, food, and cigars. Madsen, hospitable and cordial, turned his home into a focal point for his family and closest friends.

For a man of so many parts like Martyn, one might be surprised to observe that one of his greatest abilities was to distinguish the wheat from the chaff, to see the wood for the trees. Both in physics and life, he is characterized by his ability to discern the major factors operating behind very complex situations. His fine perception often led him into delicate situations. During his overseas visit in 1936, Martyn deduced that Watson Watt was undertaking highly secret research on radio location—an investigation that would turn into radar. Three years later, he would return to Britain to collaborate actively in the secret development of the UHF (ultra-high-frequency bands). After his travels, Martyn wielded a decisive influence in the creation of the new radiophysics laboratory at the University of Sydney, and in the application of the new technology of radar during the war.[169]

[166]Ibid., p. 399.

[167]Piddington & Oliphant (1971, p. 60).

[168]See, among others, NAA. 'Personal Correspondence of David F. Martyn'. Munro to Martyn, 18 October 1933; Martyn to Higgs, 30 July 1935; Martyn to Kidson, 6 February 1936; Martyn to Higgs, 27 April 1937; Martyn to Piddington, 20 September 1937.

[169]On the development of Radar in Australia and the Radiophysics Laboratory, see Schedvin (1987, Chapter 6, 'The Challenge of Radar'); Mellor (1958, Chapter 19, 'Radar'); W.F. Evans,

The word 'secretiveness' shrouded Martyn's life like an invisible veil. As he acted as a ferment and incentive for companions having full confidence in him, in the same way he was detached and touchy when having to share information with strangers. In the matter of the Luxembourg effect, Martyn never disclosed whether the basic idea had been the fruit of his inspiration or the result of Bailey's enquiries. Stealth and secretiveness came to him as a correlate of his complex character, and he extended it to all his life's facets. In the middle of the war, Martyn was spied on by the Australian secret service, supposedly because of his relationship with a German woman, Ella Horne, suspected of being involved in espionage related to radar. The affair did not become serious, but left an aftermath; in 1942, Martyn was replaced by F.W.G. White as Chief of the Division of Radiophysics. The incident was to aggravate his complex personal and institutional relations.[170]

Retouching the picture of the ionosphere

By the time that Builder and Pulley came to concentrate their technical skills on developing the P′t and P′f pulse-echo equipments, Martyn was in the middle of establishing a theory which would have a great impact on the physical knowledge of the ionosphere. In 1935, employing the method conceived by Pulley at King's College, Martyn and Pulley concluded that absolute temperatures between the E and F layers of the upper atmosphere were of the order of 1,200°K.[171] Beginning in the 1930s, thanks to Chapman and Hulburt's works, it was widely known that the ionosphere was much hotter than indirect observations had hitherto suggested. To obtain more precise conclusions, Martyn decided to use heights and ionization densities as indicators of the physical state of the ionosphere. Electron collision frequencies featured ionospheric temperature. Martyn had only to seek an agent responsible for increasing the temperature of the upper atmosphere up to unexpected limits. As Dobson would state later, that agent was much clearer in the low latitudes of the Southern Hemisphere than anywhere.[172]

1970. *History of the Radiophysics Advisory Board, 1939–45.* Melbourne: CSIRO; MacLeod (1999); and Marjorie Barnard, *One Single Weapon*, Manuscript deposited in the AAAS, Box 16, Chapter 11, 'Radiophysics Laboratory'.

[170]AAAS. [Box 4], 'Confidential enquiry in regard to the association of Dr. David Martyn with Mrs. Ella Horne of German Birth', April 1941; and 'Dr. D.F. Martyn and the Radiophysics Advisory Board', 1939–41, by F.W.G. White (Confidential).

[171]Martyn & Pulley (1936).

[172]Dobson's view is quoted in: NAA. [File A9877/1], 'Personal Correspondence of David F. Martyn', Martyn to Madsen, 10 August 1936.

In July, August, and September 1935, Pulley was recording data on ionization with the new pulse equipment. Although the range of observations was limited, Martyn concluded that the absorption of solar ultraviolet radiation by ozone was the cause of high temperatures.[173] He estimated that the atmosphere was almost completely mixed at those levels, consisting mainly of molecular nitrogen and atomic oxygen. Although the presence of ozone hardly reached the concentration of 1 in 1,000, it was enough to explain such an effect. Because of ozone, he postulated a correlation between ionization densities in the ionosphere and barometric pressure on ground level.[174] Meteorological changes modified ozone concentration and hence temperature variations. What was really notable in Martyn's picture was that for the first time there was suggested a fundamental connection between such apparently diverse atmospheric phenomena as ionization densities, ozone content, temperature of the stratosphere and ionosphere, noctilucent clouds, and the light of the night sky.[175] The approach, in the words of the Australian physicist H.S.W. Massey, 'was not only remarkable for its prescience but also as representing one of the earliest attempts to look at the atmosphere as a whole taking into account evidence available from all sources, including the ionosphere'.[176]

Appleton had recently postulated, albeit not convincingly, large temperature changes in the F region.[177] Madsen intuited the quality of Martyn's work. On 8 November 1935, he sent it to Rutherford for presenting before the Royal Society of London. When Laby read the paper before the European authorities on atmospheric physics, a heated discussion arose, at issue being the interrelation of physical and chemical phenomena.[178] The relevance of Martyn's deductions was acknowledged, for they would permit, *inter alia,* temperatures to be gauged far above the ground. However, in question were some features of physical mechanisms of the atmosphere. Ionospheric physicists increasingly questioned the nature of atomic reactions—whether, that is, the free electrons were removed through recombination between electrons and positive ions or through attachment to neutral molecules thereby forming negative ions, a key issue since 1930. Laby confessed to Martyn that Appleton had not

[173]For a discussion of Martyn's approach, see Massey (1971, pp. 500–2).

[174]NAA. [File A9877/1], 'Personal Correspondence of David F. Martyn', Martyn to E. Kidson, 6 February 1936; and Martyn to Higgs, 30 July 1935.

[175]Ibid., Martyn to F. Tate, 13 December 1935; and Martyn to E. Kidson, 6 February 1936.

[176]Massey (1971, p. 502).

[177]E.V. Appleton, 1935. 'Temperature Changes in the Higher Atmosphere'. *Nature*, 136, 52.

[178]NAA. [File A9877/1], 'Personal Correspondence of David F. Martyn', Laby to Madsen, 20th March 1936.

fully understood his paper, and that he 'tried to sidetrack the discussion with a red herring. The others ignored the red herring, however, after dismembering it'. Yet, he added, 'nobody wants to fight the Germans, whom they like, while they hate the French. It is a case of eat or be eaten'.[179] National undertones were given vent to in this atmosphere. In the end Appleton abandoned his position. Martyn's conclusions stood the test of time and were subsequently confirmed.[180] It is unnecessary to say that this was a matter of pride to Martyn and the Australian team.

Martyn continued his assault on the physical structure of the ionosphere by marshalling observational data and by obtaining reliable P′f and P′t curves, plots of virtual heights of layers against the frequency and the time needed for a signal to do the return journey. At Sydney graphs reflecting diurnal and seasonal height variations of layers were the order of the day; Martyn considered these graphs as the cornerstone of the new dynamic image of the ionosphere. Unravelling the behaviour of the ionosphere required continuous recordings and automatic equipment. By emitting a series of pulses to the sky, one could employ data collected by automatic P′f sets in different observatories to compare intensities of received echoes, from which could be calculated electron collision frequencies. If the structure of the ionosphere was composed of concentric if more or less discrete layers, as radiophysicists maintained, then one could use such frequencies to calculate ionization densities of layers. By 1936, a network of P′f equipment was established in Sydney, Liverpool, and Stromlo (the latter one being in the charge of Higgs) in which estimations of densities and temperatures were computed.[181]

Although P′f automatic technique had been used both in Britain and the USA since the early 1930s, unlike in Australia, measurements were mostly being completed manually. Furthermore, to extract accurate information from heights, it was necessary to employ a wide range of frequencies. Martyn was satisfied with the performance of the Australian equipment: 'Our P′f recorder', he wrote to Madsen, 'is a much better job than the Slough outfit, which is the only one in England. Their automatic range is only about a third of Wood's, and they have to do it by hand for satisfactory results.'[182] Over the next months the metropolis of ionospheric technique was laid down by Martyn's circle at

[179]Ibid., Martyn to Madsen, 1 June 1936.

[180]However, it is fair to state that the conclusions were arrived at on false premises. See Massey (1971, pp. 501–2).

[181]Evans (1973, pp. 239–42).

[182]NAA. [File A9877/1], 'Personal Correspondence of David F. Martyn', Martyn to Madsen, 29 June 1936.

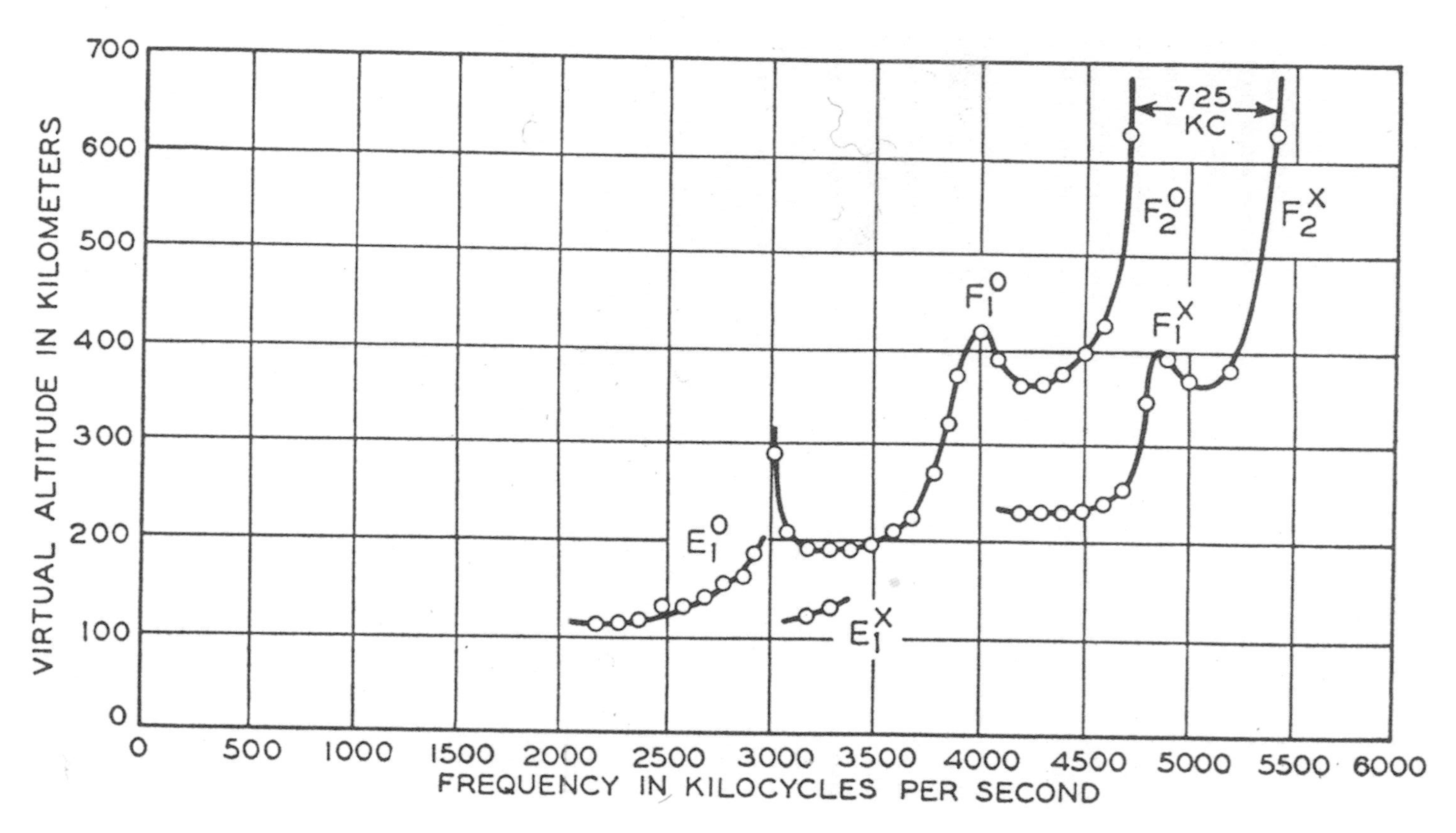

Fig. 19: Example of (h′, f) curve taken with sun high in the sky, showing E-branch, F-branch, gap between the branches, distortion near the gap, and crinkle indicating F_1 layer. From Darrow (1940b), p. 436.

Sydney. Setting out the objective of developing fully automatic P'f equipments, several leaders fixed their gaze on Australia. In 1936, Appleton requested a copy of the board's equipment for his use; similarly, F.W.G. White (who had worked with Appleton at King's College) purchased a P'f set (manufactured in Sydney) for use in New Zealand; and Seaton of the Watheroo Observatory, following directives from the Carnegie Institution, initiated negotiations to acquire an automatic model of the RRB.[183]

Regular and automatic recording would widen research fields. The big chain of investigation seemed to plot a linear sequence: from the early ionospheric works—which emerged as a logical development of studies of propagation, fading, and atmospherics—, to the physics and chemistry of the ionosphere, to conclude with solar physics. Martyn's broad-mindedness contributed, without a doubt, to this metamorphosis. From 1936 onwards, Martyn would concentrate on deciphering the relations between ionosphere, geomagnetism, and solar physics. 'The time is just ripe for a good discussion of radiation equilibrium in the high atmosphere', he allowed in 1936.[184]

The centre of operations was placed in the F region. With data supplied by Munro, Martyn and Higgs analysed and interpreted ionospheric disturbances, fade-outs and solar eruptions.[185] The level of rapport between the Sydney group and Stromlo Observatory was high; the agreement, profound. 'I have ordered the valves you request and they, together with the Cathode Ray Tube, should be sent up to you shortly', Martyn replied to Higgs. 'I shall be interested to hear if you observed the same effect...'[186] Ionospheric parameters began to emerge from the hourly, diurnal, seasonal, and sunspot cycle observations. Of as critical importance was location of ionospheric disturbances from the data associated with solar eruptions. In their extensive analysis of the behaviour of the solar disc during 1937, Martyn, Munro, and Higgs clearly identified bright hydrogen eruptions occurring almost simultaneously with changes in the ionosphere.[187] Subsequent observations would show that some ionospheric disturbances took place more frequently than solar flares.

[183]Evans (1973, pp. 240–2).

[184]NAA. [File A9877/1], 'Personal Correspondence of David F. Martyn', Martyn to Madsen, 14 September 1936.

[185]Bannon, Higgs, Martyn, & Munro (1940).

[186]NAA. [File A9877/1], 'Personal Correspondence of David F. Martyn', Martyn to Higgs, 27 April 1937.

[187]Higgs, Martyn, Munro, & Williams (1937).

Ozone and the holistic image of the atmosphere

Martyn and Pulley's work developed around the question whether ozone was essential for the ionosphere. Once answered affirmatively, the question obligated Australian radiophysicists to confront a momentous issue: is the atmospheric picture simplified and stratified by discrete layers, a gigantic natural laboratory with large-scale variables interdepending and interacting, or does it rather reflect a series of regions separated from one another with little interrelation? Moreover, does Australia represent a privileged platform to resolve such a question?

As for the first point, nothing seemed to be clear. Thanks to Chapman, Ferraro, and Stormer's investigations, atmospheric physics showed as many profiles as a polyhedron has sides: classical mechanics, atomic and molecular physics, optics, electromagnetism and geomagnetism. The sun, as the main radiation focus, acted on all these sides on the whole of the atmosphere. Yet, physicists preferred to divide the *corpus delicti* into two unconnected parts: the upper and lower atmosphere, separated by a thin layer of ozone. The former was the realm mainly of radiophysicists and geomagneticians, the latter, that of meteorologists. Australian physicists bet on a global conception of atmospheric sciences. Indeed, they said, ionization of the ionosphere was closely linked to meteorological conditions on the ground.[188] Adopting a position that might seem ambitious, they wagered on ozone as the bridge to traverse the apparently unbridgeable chasm between ionospheric and meteorological processes.[189]

We know today that Martyn believed firmly that Australia and ozone could play fundamental roles in the unifying process. He became convinced from his visit to Dobson, in Oxford, not only of the uniqueness of Australia but also of the sacrament of a global atmospheric picture: 'Our position in the Southern Hemisphere, our position on the *East* of a large continent (a new point to me), and our low latitude with its freedom from magnetic storms, put us in a unique position to solve the connection between the upper atmosphere changes and ground weather from our ionosphere results.'[190] With that extraordinary particularity it was only a question of awaiting the progressive

[188]D.F. Martyn, 1934. 'Atmospheric Pressure and the Ionisation of the Kennelly-Heaviside Layer'. *Nature*, 133, 294–5.

[189]On ozone research in Australia, see Evans (1973, pp. 252–7), Frame & Faulkner (2003, pp. 55–73), and Allen (1978, pp. 40–3).

[190]That was Dobson's opinion; NAA. [File A9877/1], 'Personal Correspondence of David F. Martyn', Martyn to Madsen, 10 August 1936.

elimination of variances and prejudices between meteorologists and physicists on the atmospheric front.[191]

The tensions, not to say animosities, between Australian radiophysicists and meteorologists did not help to close the profound cleavage in atmospheric physics. The precedent in atmospherics had shown that the predilections motivating and guiding their general lines of work were completely different. An example might help to clarify these points. At the Ozone Conference held in Oxford in September 1936, Martyn evinced envy at the degree of collaboration between Bureau, the head of the French Meteorological Bureau, and M. Vassy from the University of Sorbonne, Paris, who was researching into the possibility of ozone existing at high temperatures in the ionosphere. Dobson confessed to Martyn that he knew and collaborated with all the Commonwealth meteorologists except with those from Australia.[192] This Freudian slip, which in any other case would have been anecdotal, was but the confirmation of the failure of Australian meteorologists to understand the nature of physical research within the international community.[193]

Given the circumstances, one might no be surprised that Martyn sought support from Edward Kidson, the director of the New Zealand Meteorological Bureau, for whom he had a deep affection.[194] Not only did his support discredit Australian meteorologists, it also, in Martyn's opinion, was crucial to corroborate his ideas. Kidson had proved that meteorology in the Southern Hemisphere with its preponderant ocean surface was 'simpler than that in the Northern Hemisphere'.[195] He reminded Martyn that was the reason why

[191]Although it would be interesting to compare the development of geophysical disciplines, we may encourage comparisons between frontier disciplines, such as meteorology and the physics of the lower atmosphere, which often derived their impulse from the study of concomitant phenomena. The following references can be useful: Len Deacon, 'Turbulent Transfer in the Lower Atmosphere: Early Research', and Brian Ryan, 'Cloud Physics Research 1949–1984 and Beyond', both in Eric K. Webb, ed., 1997. *Windows on Meteorology: Australian perspective.* Melbourne: CSIRO, 162–79 and 142–61, respectively.

[192]NAA. [File A9877/1], 'Personal Correspondence of David F. Martyn', Martyn to Madsen, 10 August 1936. For the enigmatic Commonwealth meteorologist, see Nash (1990).

[193]For Dobson's attempts to organize regular daily surveys of ozone, see NAA. [File A10762/1], 'Madsen Papers and Correspondence', Dobson to Madsen, 10 February 1936; and Dobson to Madsen, 15 February 1936.

[194]Kidson had held brilliantly the post of assistant director of the Australian Commonwealth Meteorological Bureau from 1923 to 1927. See James W. Brodie, 'Kidson, Edward 1882–1939', *Dictionary of New Zealand Biography*, 1998, 4 [URL: http://www.dnzb.govt.nz/]. For Kidson's stay at Watheroo and Melbourne, see Kidson (1941, pp. 121–8).

[195]NAA. [File A9873/1], 'Societies, Institutes and Notes', 'Notes on a Suggested Programme of Investigation of Atmospheric Ozone in Australia, by D.F. Martyn'.

Fig. 20: The participants in the Conference on Atmospheric Ozone, Oxford, 9-11 September 1936.

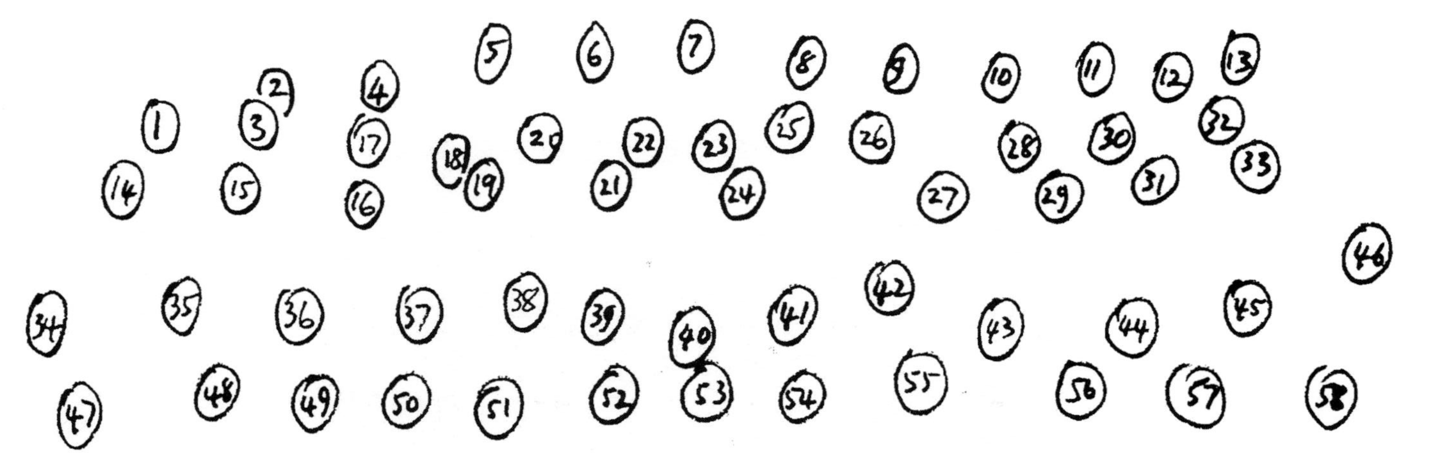

Fig. 20: *Continued.* 1 G.M.B. Dobson; **2** Capt. Heck; **3** J. Bjerknes; **4** M. Hessaby; **5** Prof. Cario; **6** T.W. Wormell; **7** Dr White; **8** D. Barbier; **9** F.J. Scrase; **10** F.J.W. Whipple; **11** E.H. Gowan; **12** M. Prettre; **13** R. Penndorf; **14** H.G. Kuhn; **15** W. Mörikofer; **16** A. Angström; **17** E. Vassy; **18** B. Lyot; **19** O.R. Wolf; **20** J.A. Ratcliffe; **21** R.H. Weightman; **22** R. Ladenburg; **23** Fr. L. Dumas; **24** F.W.P. Götz; **25** E. Stenz; **26** H. D. Harradon; **27** M. Dedebant; **28** J. Bartels; **29** C.L. Pekeris; **30** S. Chapman; **31** C.H. Collie; **32** F.A. Paneth; **33** A.R. Meetham; **34** Fr. P. Lejay, S.J.; **35** D. Chalonge; **36** J.A. Fleming; **37** R. Ayres; **38** D.F. Martyn; **39** S.K. Mitra; **40** B. Gutenberg; **41** E.V. Appleton; **42** Dr Hoelper; **43** Prof. Schmidt; **44** J. Cabannes; **45** J. Gauzit; **46** C.L. Godske; **47** V. Conrad; **48** Mme Michel; **49** Mme Gauzit; **50** Mme Vassy; **51** Mme Pettre; **52** Mme Lyot; **53** Mme Chalonge; **54** Mme Cabannes; **55** W. Gorczyski; **56** E. van Everdingen; **57** P. Wehrlé; **58** L. Weickmann. From G.M.B. Dobson (1968), pp. 402-3. Courtesy of the Optical Society of America.

a parallel variation of ionospheric ionization and ground barometric pressure had been discovered in Australia, and not in Europe and America.

The global picture, however beautiful it may be, was nothing but a conceptual hypothesis of work; ozone, a quasi-enigmatic element, whose vertical distribution was unknown; and the Australian Meteorological Bureau, a service for forecasting, rather than for research.[196] Martyn's goal was the conceptual unity of the atmosphere, but also a search for it by means of investigation within both the local and international scientific community. As a persuasive indicator that unification could succeed through cooperation, Martyn cited Dobson's proposal of organizing an international network for the study of ozone.[197] The proposal, approved at the 1935 International Meteorological Conference, provided for the installation of the new Dobson ozone spectrophotometer in different localities, so that an estimation of the total content of ozone could be obtained. And that was the point: with its help, Martyn stressed, 'the distribution of ozone and its connection with weather phenomena' could 'be studied intensively'.[198]

Madsen proceed to judge Martyn's ideas by their expectations. At the meeting of the board held in December 1935, he expressed the appropriateness of purchasing and installing three Dobson instruments in Canberra (at the Stromlo Observatory), Sydney (Riverview Observatory), and Melbourne (Meteorological Bureau).[199] The instruments, whose approximate cost was £400 each, needed to be calibrated by Dobson. Madsen persuaded W.S. Watt, the Commonwealth Meteorologist, of their usefulness for meteorology; and Martyn did the same with W.B. Rimmer, the director of the Stromlo Observatory. However, unlike W.S. Watt, Stromlo staff were acquainted with the measurement of atmospheric ozone. Assisted by the RRB, Higgs had been performing measurements since 1929 with the old Dobson spectrograph model, which for the fist time shed light on the amount of ozone above Australia.[200]

[196]See J.E. Gardner, 1997. *Stormy Weather: A History of Research in the Bureau of Meteorology*. Melbourne: Bureau of Meteorology, p. 740.

[197]NAA. [File A9877/1], 'Personal Correspondence of David F. Martyn', Martyn to Kidson, 6 February 1936; Martyn to Dobson, 23 March 1936; Martyn to Madsen, 10 August 1936.

[198]Report on 'Radio Research and Related Investigations', by D.L. Martyn, p. 13.

[199]NAA. [File A9873/2], 'Minutes of the 27th Meeting of the Radio Research Board, 17 December 1935.

[200]Higgs (1934). On ozone measurements at Stromlo, see Frame & Faulkner (2003, p. 64).

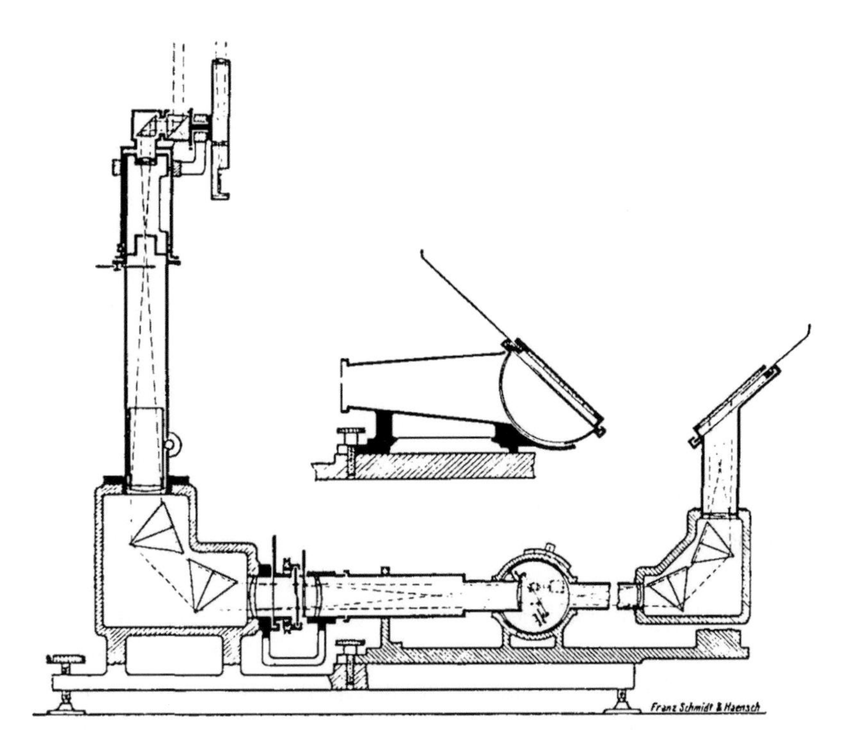

Fig. 21: The Fabry-Buisson ozone spectrophotometer gives two dispersions at right angles. The right-hand portion is shown rotated through 90°. From G.M.B. Dobson (1968), p. 388. Courtesy of the Optical Society of America.

Putting Australia on the map

There are key moments when a competent scientific community, such as that of the Australian radiophysicists, makes contributions of great magnitude to a scientific field. Martyn's overseas visit, from June to October 1936, is just one such moment. In April 1936, when the boat embarked for England, Martyn did not imagine that honours and distinctions would gravitate towards him in any remarkable way. Over the next months, he was elected to a Fellowship by the Institute of Physics and by the Royal Meteorological Society, London, and the University of London awarded him the degree of Doctor of Science. As an author of promising theoretical papers, he had to meet with the leaders of atmospheric physics. And they had something to gain from Martyn: the holistic and profound vision of a dynamic atmosphere of a theoretician of independent character and penetrating intuition, whose mathematical solidity admitted of no doubt. Martyn recognized that he was well placed to strike compromises for the sake of the Australian radio science. He had reason to believe that his venture, even if designed at first to satisfy his own interests, contributed to putting Australia on the map and to reinforce the sense of identity among the Australian radiophysics community.[201]

Martyn could scarcely have initiated overseas travel with better credentials than he received. No eulogy was spared by Rivett when depicting him as a most exceptional man, quite the best they had had on the board's staff. 'The man is a first-class worker, deserving of the designation brilliant. I am inclined to think that the radio work done by the C.S.I.R., though it does not receive much local recognition, is of a higher scientific standard than that attained in any of our other lines, and much of its excellence is due to Martyn.'[202] His commendations facilitated Martyn being awarded a grant by the Carnegie Institution for his European visit.

Madsen preferred to put emphasis on the benefits that Martyn's travel was to produce for the Australian community: 'The effect upon the further development of [Martyn's work] here would be very marked and must undoubtedly have a considerable influence upon the direction which these developments pursue.'[203] This is a crucial point in judging Martyn's behaviour. Madsen hoped that what the board deemed to be the main axis of its scientific policy—the call to international cooperation—would strengthen. He expected that the CSIR would adjust itself to what Martyn judged to be appropriate for the interests of the Australian community of physicists.

Madsen had compromised fruitfully in furthering and even extending the line of investigation which Martyn might pursue when returning. Rivett, albeit sharing Madsen's viewpoint, did not conceal his fears. 'My chief anxiety regarding him is that we may lose him before long, since efforts will be made to secure him for some post in England.'[204] Nevertheless, Martyn had adapted himself readily to life in the Antipodes, and, despite certain differences with Madsen, he identified with his new role of leader. Martyn's commitment and loyalty towards the Australian RRB appeared in his change of nationality while visiting Britain. 'It has caused a lot of amusement here', he wrote banteringly to Madsen, 'and I wouldn't be surprised if Appleton has had a lot of protests

[201]Evans (1973, pp. 127–36). See also Martyn's report: 'Radio Research and Related Investigations', ibid., appendix 9.

[202]NAA. [File A9877/1], 'Personal Correspondence of David F. Martyn', Rivett to F. Tate from the Carnegie Institution, Washington, 11 December 1935.

[203]Ibid., Madsen to F. Tate from the Carnegie Institution, Washington, 6 December 1935.

[204]Ibid., Rivett to F. Tate from the Carnegie Institution, Washington, DC, 11 December 1935.

from this side of the Tweed.'[205] Martyn's decision was perceived as an act of fidelity and adherence to science in Australia.

During the first weeks of his stay, Martyn took any opportunity he could find to discuss the implications in physics and chemistry of his recent work with Bailey. The features defining his faith in the validity of their theory and his determination to vie with British colleagues appear in his correspondence with Madsen, who became his mentor.[206] In these missives, Appleton was the rival to beat, but also the leader to emulate. Appleton and his fellows aroused contradictory sentiments in Martyn. On the one hand, he feared their capacity to assimilate ideas of Australian investigators to their own use; but, on the other, he admired the fact that they were able to stimulate cooperation with the Dominions without sacrificing their scientific leadership. Appleton was hospitable and affable with Martyn, albeit he felt somewhat aggrieved about the way in which Myers, the last recruited researcher, had been chosen by the board.[207] He saw in Martyn a rough diamond, a man of enormous qualities, but whose temperament it was necessary to refine. Martyn felt flattered: 'My own relations with Appleton are now most frank and cordial.'[208]

Confirmation that Martyn's desire to contrast and validate his ideas directed his actions comes from the General Assembly of the IUGG and the Conference on Atmospheric Ozone, held in Edinburgh and Oxford, respectively, in September 1936. Here he took the opportunity of exchanging impressions with Chapman, Appleton, Ratcliffe, Berkner, Bjerknes, Gutenberg, Mitra, Cabannes, White, Vassy, Gotz, etc., that is, the greatest authorities on specific subjects of the atmosphere. Martyn obtained their views on the applicability of his general theory of the ionosphere to each of the atmospheric areas. 'I know from them that Appleton tried every possible avenue of presenting his ideas anew...It took the form of a complete abandonment of his position, and full acknowledgements of ours.'[209] This sort of reaction and feeling and the gradual conviction of his originality increased Martyn's self-esteem and confidence. Here the bitter experience of his first stage in Melbourne gave way to euphoria. Martyn wrote to Madsen about the importance of

[205]Ibid., Martyn to Madsen, 16 June 1936.

[206]See, for example, ibid., Martyn to Madsen, 1 June 1936; Madsen to Martyn, 1 June 1936; Martyn to Madsen, 16 June 1936; Martyn to Madsen, 29 June 1936; Martyn to Madsen, 10 August 1936; Martyn to Madsen, 14 September 1936; Martyn to Madsen, 25 September 1936; Madsen to Martyn, 16 October 1936.

[207]Ibid., Martyn to Madsen, 29 June 1936.

[208]Ibid., Martyn to Madsen, 14 September 1936.

[209]Ibid., Martyn to Madsen, 14 September 1936.

the contacts established in Britain: 'The Board's work will receive prominent notice in all countries in the future, even to a greater extent than in the past.'[210] His personal successes were the catalysed force to attract attention from foreign experts.[211]

Martyn and Madsen began to be aware of the influence they wielded over the 'metropolis'.[212] One might be tempted to think that Australian scientists had been paying homage and gratuitously displaying deferential attitudes towards the British community of radiophysicists. Scientific conquests, however, demanded a redefinition of the relations within the Empire and with other countries. Martyn exhorted Madsen, in connection with the need for recording and examining magnetic data, to approach the Carnegie Institution. 'The Watheroo [Observatory] F regions results have put a spanner into Appleton's ideas of the latitude variations of ionisation, and it is very clear that our Southern Hemisphere observations put us in a key position for solving several difficulties.'[213]

Defining a new relationship with Britain

In radio science, the second half of the 1930s was critical in Australian's transition from a relationship of close cooperation with Britain towards the opening-up to other countries, especially the USA. The changeover was not the result of a premeditated programme but rather a pragmatic response to changing circumstances. Australian physicists believed that the country satisfied most conditions to be able to regard herself as the principal focus of radio science in the Pacific. With the credit gained by Green, Bailey, and Martyn's contributions a new situation emerged; with the new situation came ways to advance in her own diplomacy (new relations with the USA and New Zealand). For reasons both geographical and scientific, Australia surfaced as a strategic enclave for studying the ionosphere in the Southern Hemisphere. Science, again, was to play a role in Australia's march to nationhood.[214]

[210] Ibid., Martyn to Madsen, 14 September 1936.

[211] British leaders were unanimous in commending the originality of Australian works. See NAA. [File A10762/1], 'J.P.V. Madsen', Appleton to Madsen, 9 January 1936; Rutherford to Madsen, 3 December 1935, 14 June 1937, and 31 July 1937.

[212] See, for example, NAA. [File A9877/1], 'Personal Correspondence of David F. Martyn', Carl Störmer to Martyn, 18 June 1936; A.R. Boome (from Los Angeles Herald) to Martyn, 16 June 1936; Martyn to Madsen, 25 September 1936.

[213] Ibid., Martyn to Madsen, 25 September 1936.

[214] For a conceptual overview, see Roy MacLeod, 'From Imperial to National Science', in MacLeod (1988a, 40–72, esp. pp. 60–7).

For almost a decade, Australia's place within the map of relationships with Britain was a given fact rather than something to be justified. The members of the Australian Board, who were in some cases British or imperialists—as Brown or Laby—, did not feel the need for justifying their position. Nevertheless, technical development of ionosondes, success of theories and experiments, radio industry boom, distance and socio-economic juncture would fertilize the land in which Australians would begin to develop a separate and well-defined sense of national identity.[215]

Distinct interests and sometimes disparate approaches bred tensions pushing the Australian Board towards a more detached relationship with its British counterpart. From London, it was recognized that the amount of cooperation between the two bodies was not as satisfactory as it had been in the past. Watson Watt admitted that the staff of the British RRB 'had failed to inform the Australian workers of several important technical advances, and information in England concerning the work in progress in Australia had been similarly lacking.'[216] Despite this, Britain intended to retain leadership.

From Australia, it was also recognized that the old spirit of cooperation in the market of scientific ideas had faded away after a period of logical vacillations. What had hitherto been real cooperation was now mere exchange of information. This was exemplified particularly in the correspondence between Martyn and Piddington in 1937. Jack Piddington, a Walter & Eliza Hall Research Fellow researching in Cambridge, served as a good liaison link to obtain confidential information.[217] 'I cannot tell you anything that they [Appleton and Ratcliffe] are doing without first asking their permission, and when I do this they ask me what you are doing. I am quite sure you need not fear that they will commence any line of research when I tell them you are doing it.'[218] Martyn seemed to be affronted by the tone of the message. 'I would like first of all to make it quite clear that we do not want any details of Cambridge work', he replied to Piddington. 'All I would like to know is just in a general way what kind of problems are being attacked.' And he proceeded: 'As far as Appleton

[215]For a general study of Australian national identity, see Richard White, 1981. *Inventing Australia: Images and Identity, 1688–1980*. Sydney: Allen and Unwin. On the process of Australia's nationhood, see Stuart Macintyre, 1986. *The Oxford History of Australia, 1901–1942: The Succeeding Age*. Melbourne, pp. 222–324; and Manning Clark, 1993. *A History of Australia* (Melbourne, esp. Chapter 11, 'The Old Dead Tree and The Young Tree Green', 1916–1935, pp. 541–50.

[216]Evans (1973, p. 116).

[217]D.B. Melrose and H.C. Minett, 'Jack Hobart Piddington', *HRAS*, 1998, 12(2): 229–46.

[218]NAA. [File A9877/1], 'Personal Correspondence of David F. Martyn', Piddington to Martyn, 27 July 1937.

and Ratcliffe are concerned, we only wish to exchange quite general plans of attack', which meant the exchange of results with regard to the magneto-ionic theory and the critical gyro-magnetic frequency.[219] The dream of loyal and full cooperation seemed to have reached an end. A new stage, in which Britain no longer occupied centre, was to be inaugurated, and a new set of dependencies (fresh testimonies of partnership and commonwealth) would replace the old model of imperial centralism.

For the Australian Board, however, to displace Britain from its altar of leadership represented at least a source of dilemmas, but also an exercise of coherence and consistency in defence of their own interests. Different levels of dependency were interwoven in complex ways, binding the board's policy closely to Britain. Was the Australian decision correct? Retrospective judgement might be at least as difficult, although not as meditated, as the board's own decision making. At one pole, the Laby–Rutherford axe had generated juicy dividends. Nearly 40% of more than one hundred papers published by the RRB's officers from 1928 to 1939 were accepted by overseas journals, most of them through British channels (seven of them were discussed at the Royal Society of London, 13 were accepted for the issues of *Nature*, and many others by journals such as *Philosophical Magazine* and *Wireless Engineer*).[220] At the other, historians, especially in Australia, criticize the fact that Australian science looks quite so instinctively and automatically to England, rather than, say, France or Germany, for lines of investigation and leadership.[221]

The desideratum of initiating ultra-short-wave research dominated the board's scientific agenda around the mid-1930s. The echoes of Watson Watt's secret investigations, uncovered unofficially by Martyn and Piddington, on the detection of 'objects localised in the lower atmosphere' underlay these pretensions.[222] The opportunity came in 1936, and provides an exemplary instance of the caution with which the British authorities acted when cooperating on far-reaching issues. In a correspondence between Cook, the secretary of the Australian RRB, and R.L. Smith-Rose, his British counterpart, it was proposed to nominate ultra-short research lines in which Australians could 'cooperate

[219]Ibid., Martyn to Piddington, 20 September 1937.

[220]R.W. Home & P.J. Needham, 1990. *Physics in Australia to 1945: Bibliography and Biographical Register*. Melbourne.

[221]See, for example., R.W. Home, 'A World-Wide Scientific Network and Patronage System: Australian and Other *Colonia*l Fellows of the Royal Society of London', in Home & Kohlstedt (1991, pp. 151–79).

[222]NAA. [File A9877/1], 'Personal Correspondence of David F. Martyn', Martyn to Madsen, 29 June 1936; Piddington to Martyn, 22 July 1937.

and co-ordinate with similar projects in England'.[223] There was considerable scope for the study of wavelengths below 20 metres; Smith-Rose replied to Cook thanking him for the invitation, which would permit the improvement of transmissions between England and Australia. A similar proposal, he went on, was to be sent to New Zealand. This offer was doubtless disappointing to the Australian authorities, who clearly had in mind confidences concerning secret defence projects, rather than cooperation on long-distance propagation between Dominions. The disenchantment did not preclude Martyn and Munro formalizing the agreement.[224] If the British authorities had then shared their radar secrets, instead of in 1939, when hostilities in Europe were imminent, one might deem that there would have been reasons to preserve privileged treatment for Britain.

The commencement of an American research entente

An approach to the USA proved as ineluctable as the increasingly intense sense of Australian-ness. Like so many other aspects of prewar policy and society, the development of Australian science as an independent identity, and not as an appendage of the mother country, seemed a fact. The collaboration with the Carnegie Institution had several antecedents. On the one hand, A.L. Green of the RRB and J.W. Green, the new observer-in-charge at Watheroo, had expressed their interest in favouring a programme of 'Co-operative Ionosphere Research' including Washington, Huancayo, Alaska, and Watheroo.[225] On the other hand, Martyn and Berkner, as members of the committee formed by the URSI and IUGG (along with Appleton, Chapman, and Mitra), had recognized that close cooperation was the precondition to securing the plans of collecting worldwide ionospheric data.[226]

It was a competent trio who enabled a compromise between the two institutions in a way that the British authorities had not attained: Berkner, diligent and well-mannered, true to the behests of John A. Fleming, the Director of the DTM; Madsen, moderate and upright, acting for the good of Australian science as if it was his duty; and Martyn, renowned and audacious, seduced by amalgamating the quest for physical knowledge and the consolidation of Australian community. They were to represent the pieces of a game of chess

[223]Evans (1973, p. 271).

[224]D.F. Martyn & G.H. Munro, 'Report of Proposed Investigations of the Propagation of Short Waves between England and Australia', unpublished Report. Sydney: RRB, 1938.

[225]AAAS. [MS 86/1/15], J.W. Green to Ross, 8 April 1936.

[226]NAA. [File A9877/1], 'Personal Correspondence of David F. Martyn', Martyn to Madsen, 25 September 1936.

in which mutual respect and egalitarian treatment would prevail—a combination of pieces and personalities that, once coalescing, was to be of maximum benefit to both entities.[227]

Besides installing the automatic ionosonde in Watheroo, Berkner's visit to Australia in 1938 served to establish a basis for cooperation.[228] For Australians it was doubtless gratifying to observe the status they enjoyed within American academic circles. As Fleming wrote to Berkner before his departure, 'upon completion of your work at Watheroo you will proceed to Melbourne and Sydney and [...] work for a limited period with Australian radio research men associated with the organizations indicated above in order that the Department may derive the greatest possible benefit from the researches they are conducting'.[229]

The harmony between Martyn and Berkner probably helped to convince Madsen and Fleming to strengthen what had hitherto been isolated cases of cooperation. As the correspondence on the Lorentz polarization correction proves, Martyn and Berkner had fruitful discussions about the connections between ionization density in the F_2 region, radio fadeouts, and solar chromospheric eruptions. Comparisons of F_2 region data from Watheroo and from Stromlo showed that large fluctuations, apparently unrelated, occurred at each place. These results, they anticipated, demanded a revision of the ultraviolet theory of ionization production. Both leaders had embarked on a quest for links between ionosphere, geomagnetism, and solar radiation.[230] And both men agreed on their appreciation of geophysics as a global science dependent on worldwide observations. 'The elucidation of [our conclusions]', they

[227]On the contacts established between the Australian Board and the heads of the Watheroo Magnetic Observatory, see Evans (1973, pp. 36, 105, 241, 275–9, 296–7). For more critical insights, see Home (1994b, pp. 156–7).

[228]Hales (1992, pp. 6–7).

[229]Fleming to Berkner, 8 February 1938—quoted in Home (1994b, p. 156).

[230]Compare the following works: L.V. Berkner, 1939. 'Concerning the Nature of Radio Fadeouts'. *Phys. Rev.*, 55, 536–44, and D.F. Martyn, 1939. 'Concerning the Nature of Radio Fadeouts'. *Phys. Rev.*, 55, 983; L.V. Berkner & H.G. Booker, 1938. 'An Ionospheric Investigation 'Concerning the Lorentz Polarization Correction'. *Terr. Mag. & Atm. Elect.*, 43, 427–50, and D.F. Martyn & G.H. Munro, 1938. 'The Lorentz Polarization Correction in the Ionosphere'. *Terr. Mag. & Atm. Elect.*, 44, 1–6; L.V. Berkner & S.L. Seaton, 1940. 'Systematic Ionospheric Changes Associated with Geomagnetic Activity'. *Terr. Mag. & Atm. Elect.*, 45, 419–23, and D.F. Martyn et al., 1940. 'The Association of Meteorological Changes with Variations of Ionization in the F Region of the Ionosphere'. *Proc. Roy. Soc. Lond.*, A 174, 298–309.

concluded, 'has resulted in reconciling the apparent divergence of our views with regard to the experimental observations.'[231]

A similar degree of commitment was injected at the institutional level. In a joint report on the way to secure full and effectual cooperation, Martyn and Berkner recommended the systematic exchange of data, reports and drafts of papers, and even their joint publication when advisable. They counselled, in addition, the interchange of typescripts of criticisms prior to publication. Furthermore, they invited the Australian Board to follow the recommendations of the URSI and IUGG for publishing regularly 'the monthly average hourly values of the penetration frequency and the equivalent heights of the various regions in the ionosphere', as the Carnegie Institution itself did.[232] The *entente cordiale* between the two institutions was the ultimate ideal, something which was worthwhile attempting.

Berkner and Martyn's respectful and conciliatory tone adjusted itself to Madsen's flexible character. As Chairman of the Board, Madsen invited Berkner to expose their recommendations in the next meeting of the RRB, in which Rivett was also in attendance. Following Fleming's instructions, Berkner offered the possibility of the board's officers spending some time in Washington. Brown applauded the 'charming gesture' on the part of the Carnegie authorities, albeit he feared that the work of the small Australian team could 'be seriously affected' if someone left for a period.[233] In the words of Berkner, Rivett 'especially stressed the isolation from other research units and mentioned his feeling that while they had been sending most of their workers to England, he felt it most desirable that some be sent to America to make contact with the work there'.[234] Although much of the Berkner–Martyn spirit of cooperation was dispelled by the threats hanging over Europe, in the organization of a wartime ionospheric prediction service Australia would have a chance for glory.[235]

[231]D.F. Martyn & L.V. Berkner, 'Report on Collaboration between the Radio Research Board and the Department of Terrestrial Magnetism of the Carnegie Institution, Washington', unpublished report. Sydney, 1939. See Evans (1973, p. 277).

[232]Evans (1973, p. 275).

[233]NAA. [File A9873/2], Minutes of the 37th Meeting of the Radio Research Board, 22 February 1939.

[234]Fleming to Berkner, 27 February 1939—quoted in Home (1994b, p. 156).

[235]The pursuit of international prestige through science—especially in the fields related to radio—has been a repetitive issue in Australia. See, for example, Munns (1997).

4

RADIO COMMUNICATIONS, EDUCATION, MANUFACTURING, AND INNOVATION IN THE AUSTRALIAN RADIO INDUSTRY

Introduction

During the interwar years, the strength of the Australian radio industry was perceived as an authentic national achievement. With the outbreak of World War II and Australia's subsequent participation in the secret radar programme, the widespread impression of success and excellence was symbolized by science and industry. Australia stood out as the 'Wireless Centre of the Southern Pacific',[1] and even of the Southern Hemisphere. Especially in the creation of the Radiophysics Laboratory in 1940, in its undisputed leadership in postwar radio astronomy (along with England), and in the boom of the avant-garde sector of radio supply and manufacturing, Australia was placed so frequently on a par with Britain and the USA than one might be tempted to analyse the reasons in terms of education and research.[2]

How was it that a nation still developing, with a relatively weak university structure, and a marked proclivity for agricultural research, rather than for the production of secondary goods, succeeded so extraordinarily in such an abstruse field? The anxiety for a quick response might lead us to a distorted image. It is unquestionable that new population arrivals (an increment of 31% in two decades, from 5.4 to 7.1 million) brought with it new opportunities and avid entrepreneurs for communication and broadcasting ventures. It is also true that patent and tariff policy helped to support radio manufacturing, and

[1]NSWA. 'AWA Records', E.A. Hormer, the works manager at AWA, to H.S. Gullet, Australia's Minister for Trade and Customs, 2nd February 1932.

[2]The contribution of Australians to radio astronomy is admirably examined by Woodruff T. Sullivan, III, 'Early Years of Australian Radio Astronomy', in R. Home, ed. *Australian Science in the Making*. Cambridge: Cambridge University Press, 308–44. For radioastronomy in Britain and the USA, see Sullivan (1984), which also contains F.J. Kerr, 'Early Days in Radio and Radar Astronomy in Australia', pp. 133–45, and E.G. Bowen, 'The Origins of Radio Astronomy in Australia', pp. 85–111. See also Wild (1972).

that the practice of apprenticeship still abounded, which would seem to ensure a permanent source of both manpower and skilled hands to Australia.[3]

In recent studies of the science-based radio industry, however, the thrust of the criticism has been more intense. As Ann Moyal notes, the prism should focus on the abstract, theoretical, intellectualist condition of Australian tertiary education and on the neglect of applied fields. The most prominent universities and engineering schools of Europe and America were training electrical engineers from the last quarter of the nineteenth century, and highly trained staff in radio by the 1920s. Australia, by contrast, was characterized as an area in which practical technological training had been frequently scorned. There were illustrious exceptions, of course, such as the technical colleges in Victoria and New South Wales, especially after World War I in certain fields.[4] But, in general, the insensitivity, not to say contempt, for industry and business affairs became, in the words of Ann Moyal, 'a marked compass error in university education in Australia'.[5] It was not until the mid-twentieth century that the University of Technology (founded in 1946 and eventually becoming the University of New South Wales) offered degrees in electrical engineering that satisfied the demands from the PMG and private radio industry.[6]

Clearly, these perceptions might precipitately lead one to deem industrial success as independent of the educational system. In attempting to analyse this, however, one discovers camouflaged aspects in the interaction between university and radio manufacturing and supply. Here, it is important to emphasize two points. On the one hand, Australia had a vocation to become a nation, an attitude (shared by radio entrepreneurs) favouring an atmosphere of self-sufficiency and autonomous innovation, rather than systematic technological importation. On the other, the new departures in radio research that Madsen and Laby instituted in Sydney and Melbourne (with the RRB's support) in

[3]For an appreciation of the state of electrical communication in Australia, see Australian Academy of Technological Sciences and Engineering (Comp.), 1988. *Technology in Australia, 1788–1988: A Condensed History of Australian Technological Innovation and Adaptation during the First Two Hundred Years*. Melbourne, 529–44.

[4]For an exploration of some theoretical dimensions of engineering history in Australia, see Roy MacLeod, 1995. 'Colonial Engineers and the "Cult of Practicality": Themes and Dimensions in the History of Australian Engineering'. *History and Technology*, 12, 147–62. See also Corbett (1961).

[5]Moyal (1984, p. 129).

[6]Ann Moyal, 1983. 'Telecommunications in Australia: An Historical Perspective, 1854–1930'. *Prometheus*, 1(1), 23–41, p. 36. Moyal argues that 'the universities' failure to grasp the significance of professionally trained manpower...had serious consequences' for the promotion of local telecommunications R&D and for the training of a civil service of engineers.

the 1920s were to burgeon in a terrain propitious to the creation of a seedbed of young researchers, but unfavourable to the training of radio specialists in considerable numbers.[7] Within the university classrooms at Sydney—in its Department of Electrical Engineering—and Melbourne, professional training with radio components was not available, and outside those places the engineering schools established at the universities of Queensland, Adelaide, and Western Australia offered an instruction unsatisfactory to the kind of needs that radio industry demanded.[8]

Accordingly, radio training evolved as a particular mission of a few audacious enthusiasts (above all within AWA and, to a lesser extent, the PMG). It benefited from the expertise and experience of RRB's ex-officers and was impelled by the imperious demands of an industrial sector in vertiginous growth which, because of AWA's overwhelming monopoly in the Southern Pacific, allowed patenting and large-scale investigation in the same vein as the Marconi Company in Britain.[9]

Short wave and the elliptical picture of the Empire

The appearance of radiotelegraphy at the beginning of the new century had generated a competitive struggle against telephone and telegraph systems for the domination of the space of international communications. Whereas wireless technology assisted the Navy and merchant marine, telegraph cables perpetuated postal and telegraph companies, whose presence in financial and stock markets demanded the use of stable systems such as cables. The very scope of high-powered long-wave radio transmitters, along with the security—if higher cost—of submarine cables allowed both technologies to coexist for years. Long-wave systems competed with cables. By 1920, the oligopolies created around each had found their niche in a market sufficiently buoyant to provide mouth-watering profits for both.[10]

[7]This situation was to change with the outbreak of war, when the Department of Electrical Engineering at Sydney began to train all the 'radar boys'. See MacLeod (1999).

[8]For a summary of the early communication work in Australia, including radio, broadcasting, and picture transmission services, see Crawford (1931).

[9]The theme 'Technical Education and Industry between the Wars' was discussed in the seminar held at History House, Sydney, on 24 April 1993. The programme included the following lectures: Norm Neill, 'The relationship between industry and technical education', and Joan Cobb, 'The structure of technical education in NSW before World War II'.

[10]For an excellent study of the growth of cable and radio technologies within a global political context, see D.R. Headrick, 1991. *The Invisible Weapon. Telecommunications and International Politics, 1851–1945.* New York: Oxford University Press.

That, after World War I, Britain was able to hold its central position, its role of functional cerebrum within the nervous system of the Empire, would depend in part on British diplomacy reconciling interests among metropolis and periphery.[11] In Britain, after its successful first wireless message from Caernarvon to Sydney in 1918, Marconi's offered to establish a regular and direct radio service between England and Australia. The proposal was lauded in Australia but not in England. The British Government refused to delegate the service to private enterprises; instead, the British Post Office was vested with the responsibility of furthering its section of radio, acting in conjunction with its counterparts in the Dominions.[12]

Communication turned into a thorny affair. To satisfy the different sensibilities and interests at stake, Britain proposed a chain system for imperial communications (based on low-power relay stations scattered at intervals) which was presented by Sir Henry Norman at the Imperial Conference of London of 1921.[13] Prime Minister Hughes, following the recommendations of his intimate friend Fisk, AWA's managing director, opposed energetically the fact that Australia, being the final link in the chain, remained captive to the reliability of Singapore and other intermediate stations.[14] New Zealand and Canada had been excluded from the scheme because of previously signed agreements, and South Africa did not form part of the plan. With those setbacks, the endorsement of Norman's scheme 'appeared to be in the nature of a Judas kiss'.[15]

Although promises and expectations abounded in the Norman scheme, the Dominions were exasperated by the long delays and the inconsistent policies. In Australia the viability of direct communication was a source of parliamentary debate, but also, and with far-reaching connotations, the morality of an accord between the government and a commercial company like AWA.[16] Despite a certain reticence, Hughes authorized the agreement in March 1922. Under it, the government doubled the capital of the company and became its majority shareholder (acquiring half the shares plus one) while AWA committed

[11]Hugill, (1999, pp. 109–38).

[12]Sturmey (1958, pp. 101–23).

[13]For an account from the AWA's viewpoint with valuable excerpts see Myers (1925, pp. 54–64).

[14]E.T. Fisk, 1923. 'Wireless Service with Australia'. *United Empire. R.C.I. Journal*, 14, 89–98, pp. 92 and 97.

[15]Sturmey (1958, p. 108); J. Wood (1992, pp. 19–21). See also NSWA. [Box 24], 'Fisk, E.T., Correspondence, 1926–38', 'Transoceanic communication' by E.T. Fisk.

[16]See R. Curnow, 1963. 'Communications and Political Power'. In I. Bedford & R. Curnow, *Initiative and Organization*. Melbourne: F.W. Cheshire, esp. pp. 50–2, 80–97.

itself to providing a direct service between England and Australia (at a lower price than that of cable rates). AWA invested in long wave, erecting high-power stations on Australian soil.[17] In this, the agreement followed the same lines as those subscribed to by Marconi's with the Canadian and South African governments. Only New Zealand maintained intact its umbilical cord with the mother country, refraining from deals with private firms, in accordance with London's desires.[18]

Then appeared another contender, short-wave radio. The success of Marconi's short-wave experiments in February 1924 confirmed the superiority of 'beam' radio and rendered completely obsolete the long-wave stations the company was building for Australia.[19] Both from the economic and technical viewpoint, short wave enjoyed notable advantages: it was by far the cheapest long-distance communication system and it was in many aspects superior to cables and long waves. Moreover, it enabled the simultaneous transmission and reception of radio telephony and telegraphy.[20] For a quarter of a century, long-distance radio had been characterized by its trend towards greater cost, size, and complexity, becoming a luxury technology affordable only by giant companies, and prohibitive for the poorest countries of the world. Short wave broke this trend.[21]

The new system had an immediate impact. Marconi's offered to replace the Norman scheme with a chain of short-wave stations which would be 95% cheaper and three times quicker than that of long wave.[22] Against a backdrop of controversy, the dissenting stance of the British Postmaster General, on both the experimental character of short wave and its use restricted only to night, and the reticence of the British Admiralty being preoccupied with the security of transmission, the Imperial Communications Committee adopted a Solomonic position: while it recognized the danger that amateurs would interfere with signals in peacetime, at the same time it admitted the advisability in wartime of having an alternative system to that of cables. Attending to these recommendations, the British government signed a contract with Marconi's

[17]Moyal (1984, p. 132); Sturmey (1958, p. 109).

[18]Hugill (1999, pp. 118–19).

[19]See the classic paper, Hidetsugu Yang, 1928. 'Beam Transmission of Ultra Short Waves'. *Proc. IRE*, 16, 715–41; repr. *Proc. IEEE*, 1984, 72, 635–45, pp. 644–5.

[20]For the technological advantages of short wave, see Hezlet (1975, p. 157), W.J. Baker (1970, pp. 157–9), and Hugill (1999, pp. 125–8).

[21]Daniel R. Headrick, 1994. 'Shortwave Radio and its Impact on International Telecommunications Between the Wars'. *History and Technology*, 11, 21–32, pp. 22–4.

[22]Vyvyan (1974, pp. 75–7).

on 28 July 1924 to erect the chain in 26 weeks. Under these premises, communications with foreign countries were to be conducted by the private company, whereas intra-imperial radio would be operated and managed by the Post Office.[23]

Certainly, short wave caused a political upheaval of the first magnitude in the Empire. Cables constituted the 'nervous system of the British Empire', one of the three pillars which had ensured Britain's worldwide communication during World War I (together with the Royal Navy and the merchant marine).[24] The British government had hearing invested in a submarine cable network and the long-wave station at Rugby, investments on which it believed its military security and its leadership within the Empire depended. Hence, short wave signified much more than an innovation: it was a menace to the communication oligarchs—the Post Office and cable companies. Its triumph at the expense of its competitors was to depend on the political stature of the entrepreneurs and authorities in the Dominions.[25]

For Fisk, short wave would be as paramount as political alliances. As the British government consummated its agreement with Marconi's in 1924, wide-ranging changes seemed inevitable. The new situation transformed Fisk. On the one hand, AWA, together with Marconi's, established a 'beam wireless service' between England and Australia at Fiskville in 1927 (following the high-frequency transmission technique developed by Franklin in England).[26] On the other, a wireless telephone service comprising direct links with England, New Zealand, Java, and New Guinea was inaugurated by AWA, as well as telegraphic communication with most neighbouring Pacific Islands. Fisk observed that short wave was the vehicle to combat the supremacy of cables and, at the same time, the isolation that, in his opinion, Australia was suffering.[27] From these circumstances might emerge a promising geopolitical panorama. Fisk would express picturesquely his dream in a speech broadcast in 1933:

> I like to visualise the British Empire very much as a sort of ellipse—an ellipse is a figure which has two focal points—, one point [being] the old country and the other one Australia...By the use of the Beam Wireless

[23]W.J. Baker (1970, p. 214).

[24]Headrick (1994, p. 26).

[25]For contemporary accounts of the impact of short wave in the imperial context, see Tribolet (1929, pp. 215–18), Denny (1930, pp. 369–402), and Brown (1927).

[26]AWA expanded the service to North America in 1928. See L.A. Hooke, 1938. 'Australian Radio Communication Services'. *AWA Tech. Rev.*, 3, 229–51, pp. 232–4.

[27]E.T. Fisk, 1923. 'The Application and Development of Wireless'. *Proceedings of the 2nd Pan-Pacific Science Congress, Australia*, 1, 619–31.

> Service, a great deal has been done for [overcoming its isolation]. It has served as a link, a bond of friendship between our people here and their friends and relatives in other parts of the Empire.[28]

The paradoxes of the elliptical picture—the nation-building, the apology of the Empire, AWA's predominance (often against Britain's interests)—apparently did not bother Fisk. Quite the contrary, they appealed to him: as he often promulgated in his speeches, the paradoxes arise only in those individuals sensitive to daily reality. The entrepreneurs' sensitivity to intuit the legitimate aspirations of a country and to transcend even such deep-rooted notions as those of nation and Empire supported the hope that Australians could acquire a sense of their own identity, as realizable through politics as through science and industry.

About the same time Fisk propounded a similar nostrum to aid in the digestion of his particular conception of Empire: short wave was for 'his' Empire what cables purported to be for London. 'No scientific discovery', he stated confidently, 'offers such great possibility for binding together the parts of our far-flung Empire, and for developing its social, commercial and defence welfare.'[29] Plainly, the power of short wave was the leitmotiv of Fisk's speeches. The resistance or indifference of most Australian scientists (and also politicians and entrepreneurs) to thinking systematically outside their discipline may help explain why Fisk insisted on defending the validity of his message over and over again.[30]

From the moment of his first arrival in Australia in 1911, this English ship's wireless officer cherished that dream. A 'self-made man' in the old sense of the phrase (and proud of it), Fisk's confidence in himself and his ideas increased in step with his pioneering proofs in wireless communication. In this process of self-education, Fisk identified his own development so completely with Australia's that the realization of its national identity was inseparable from the realization of his personal ideals and professional life. Above all these ideals prevailed that of unity, which in the political sphere connoted the construction of a powerful Empire and in the business sphere connoted faith

[28]NSWA. 'AWA Records', [Box 3], 'Speech broadcasted by Mr. E.T. Fisk at the Scott's Hotel, 9th March 1933', p. 3.

[29]Murray Goot, 'Fisk, Sir Ernest Thomas (1886–1965)' in *ADB*, 508–10, p. 509.

[30]See, among others, E.T. Fisk, 1923. 'Wireless Service with Australia'. *United Empire. R.C.I. Journal*, 14, 89–98; and 'Transoceanic communications', 'The progress of wireless in Australia; speech delivered by E.T. Fisk on 25th February 1937', NSWA. 'AWA Records', [Box 24]; and 'Control and management of National Broadcasting Service, 5th February 1932', 'New Zealand Telegraph, Wireless Telegraph and Cable Service', 17th July 1925, in Fold 'Beam', NSWA. 'E.T. Fisk Correspondence'.

in the interconnectedness of manufacturing, communication, and innovation. Fisk's commitment to the business ideal of the unity of action, his profound national vision (as a member of the National Party), and his pride in the Empire were the pillars on which he erected his business career.[31]

Liberality of thinking, unlimited ambition, deep mistrust of universities, personal magnetism for public relations, and propensity to notoriety—as AWA's two icons prove, the 'Fisk radiola' receivers and 'Fiskville'—were the hallmarks of Fisk's character. Fisk's exquisiteness in dress and speech—indeed sometimes an excess of vanity and narcissism—, his shrewd eye for far-reaching projects, and his long shadow in political circles were frequently remarked upon. Fisk's devotion to communication went beyond imaginable limits. To take but one example, after the death of his son, Thomas Maxwell, in World War II, he fervently embraced spiritualism as a means of communicating with the dead. Fisk appreciated success (of AWA, Australia, and Empire) and celebrity (all his children bore the names of illustrious scientists) as much as he desired to transcend the frontiers of life.[32]

Fisk observed with pleasure the decline of the cable companies. By 1928, in the Pacific region 45 % of the traffic of the Pacific Cable had been diverted to the beam. Indeed, the fact that the British Post Office's Beam Service charged only one sixth of the cable rate proved decisive for it. For Fisk's plans, the ruin of the cable companies was crucial; for Britain, however, it was catastrophic. The British Admiralty expressed its fears that a properly trained cryptographic bureau could 'break down the security of practically any code or cipher'.[33] The British government faced a real dilemma: whether to preserve the cable network at the expense of losing competition as regards the USA, or to allow its collapse at the cost of putting national security at risk in the event of war. That is, security versus business.[34] Yet, for Fisk the dilemma was another: either he was loyal to Britain, thus renouncing his elliptical picture, or was loyal to Australia, opting for business. The duality of short wave and cable gave rise to a far-reaching geopolitical dilemma: centralism versus a two-headed Empire.[35]

[31]For further details on Fisk, see the autobiography of his son, Fred Fisk, an officer in the Australian Signal Corps. E.K. Fisk, 1995. *Hardly Ever a Dull Moment*. Canberra: ANU, 13–15. See also Williams (1965).

[32]See F.W. Larkins, 'The Man Fisk', *Wireless Newspapers*, 1927, August 15; and Murray Goot, 'Fisk, Sir Ernest Thomas (1886–1965)', op. cit., p. 510.

[33]Headrick (1994, p. 26).

[34]On the British dilemma, see Headrick (1991, pp. 211–13).

[35]NSWA. 'E.T. Fisk Papers and Correspondence', 'Overseas Wireless and Submarine Cables (Australia)', by E.T. Fisk, 20th September 1939, pp. 10–19.

The implicit political meaning of communication technology was evident. The 'British dilemma' was solved in July 1928, on the occasion of the Imperial Conference, when the British government promoted the merger of all radio and cable interests into a new company called Imperial and International Communications (I&IC)—renamed Cable and Wireless in 1934.[36] Britain could not afford to obliterate its cable network. The creation of this conglomerate—supposedly private but actually acting as a cartel and a guardian of Britain's global communication system—was approved at the end of 1928 by Canada, India, and South Africa, as well as by the Bruce Conservative government in Australia. But unlike those countries, in which the management of cable and beam services was placed under a local operating company, in Australia it would remain at the mercy of the English enterprise, at least up to the outbreak of war.[37]

Fisk received the news with stupefaction and consternation. 'The best procedure', he stated, 'is undoubtedly to have an Australian-controlled organisation sufficiently experienced.'[38] After long deliberation, Prime Minister Bruce tried to transfer the beam service to the Australian Post Office. But that left AWA in a delicate position. The internal dispute remained in a dormant condition for years. As a result, Australia, which had pioneered short-wave communication, would possess only two permanent overseas beam services in 1939. Ironically, short wave—Fisk's weapon *par excellence*—removed Britain from the centre of imperial communication towards one node in a decentralized network, but Australia was incapable of taking advantage of the situation.[39]

Skilled training for the radio age

As in many other countries, radio in Australia succeeded the age of telegraphy and telephony. Colonial evolvement in establishing communication networks across the continent had been cemented on the basis of willingness and amateurism, rather than professional instruction.[40] Although all the Colonial Departments of Posts and Telegraphs had instituted schools of telegraphy for the training of telephonists and telegraphists, the deficit of mechanics

[36]Sturmey (1958, pp. 117–23); Headrick (1991, pp. 206–10); W.J. Baker (1970, pp. 229–31).

[37]NSWA. [Box 24], 'AWA Records', 'Transoceanic Communications', by E.T. Fisk.

[38]See 'Overseas Wireless and Submarine Cables (Australia)', op. cit., p. 19.

[39]The global communication has been well analysed in Ralph Bown, 1937. 'Transoceanic Radio Telephone Development'. *Proc. IRE*, 24, 1124–35.

[40]Unlike the UK and Canada, there are few studies of the impact of telecommunications on Australian society. Besides the aforementioned work of Moyal (1984), see Harcourt (1987) and Livingstone (1996).

and engineers was more than worrying.[41] In the main towns, the Mechanics Institutes organized occasional lectures and opened their facilities to those who showed enough determination to learn on their own. But generally, telegraph departments employed self-trained technicians lacking a solid theoretical background.[42]

With the advent of broadcasting in the early 1920s, the sense of opportunity that was generated by the sudden appearance of a previously non-existent market altered the lethargic state of technical education. The occasion could not have been more promising to promote changes. By 1923, the plans of H.P. Brown (the newly appointed head of the PMG's Department) and Crawford (another British post office engineer) for a new scheme for cadet training that would be at once practical in orientation and intellectually exigent were beginning to motivate the young officers at the PMG.

The shortage of skilled engineers was an endemic problem at the PMG. The Royal Commission into the Post Office of 1910 had already detected scarcity and shortcomings in the training of technical members. But the efforts made during more than one decade had had a poor outcome. Brown's plan for cadet training, on the contrary, broke this apathy. Several technicians expended half their time in acquiring a science degree at an Australian university and the other half in carrying out routine tasks on the job. Although the percentage was not exactly commendable, their contribution to the PMG's growth in its headquarters and the states more than repaid the time and effort invested in education.[43]

Despite all their deficiencies, the advances at the PMG's Department reveal that there was by no means a divorce between higher education and industry. Throughout the 1920s, however, further advancements in Australia, if not insignificant, were dismally slow and, by the standards of a country aspiring to dominate the region, modest in quantity and mediocre in quality. Radio training, said Ronald R. Mackay, the supervisor of the Melbourne Technical College, was 'perhaps the most backward' among the professions.[44] It is true that at that time radio was viewed as a specialized section of electrical engineering, rather than an autonomous discipline, and that changes in radio occurred

[41]For a general insight into technical education in the colonial era, see Stephen Murray-Smith, 1966. *A History of Technical Education in Australia, with Special Reference to the Period before 1914*, PhD thesis. Melbourne.

[42]Bradley (1934, p. 20).

[43]Moyal (1984, pp. 128–9).

[44]R.R. Mackay, 1938. 'Education Aspects of the Radio Profession in Australia'. In *Proc. Wor. Rad. Con.*, pp. 1–8.

overnight, making textbooks and apparatus rapidly obsolete. But these features were common everywhere.[45] By the early 1930s, the usual practice at the Technical Colleges of Sydney and Melbourne was to provide one full-time officer assisted by part-time specialists. But, because the most desirable teachers were the best men in the profession, it was extremely difficult to obtain their services. Even their own training was symptomatic of the weakness of educational facilities: they were electrical engineers or physicists specialized in radio work either as a hobby or for labour market reasons.[46]

The case of the Technical College of Melbourne is indicative of the exiguousness of existing means. Diploma courses like those that R. Mackay had outlined by 1936 were intended to provide a basic knowledge in radio and communications engineering. Nevertheless, the lack of facilities delayed the project that was originally termed the 'Electrical Trades School' and then was popularized as the 'Radio School'.[47] There could be no pretence that an impartial, overseas visitor, as was F.H. Spencer, a retired English inspector of schools who reviewed the state of technical education in Australia and New Zealand for the Carnegie Corporation, regarded the old buildings as 'grim and forbidding'. Essentially, he judged the diploma work to be 'very satisfactory', but the poor accommodation detracted from the value of effort.[48]

In the 1930s, the relative lack of advanced education from university and technical colleges seems not to have inhibited the flurry of expansion that occurred in the sector of communication. The possibilities of employment in the field of radio were increasingly numerous, and instruction turned out to be almost an obligation for individuals combining working commitments with partly-completed day or evening courses at Technical Colleges or in private enterprise.[49] The disparate character of the process of instruction is illustrated

[45]The curriculum of Physics at the Faculty of Science of Sydney, for instance, did little to counter the sense of insufficiency that distinguished radio teaching at university. A course of about 20 lectures on Electrical Oscillations in the third year was the only subject. See ASU. [G3/158, box 35], O.V. Vonwiller to the Managing Editor of the *Radio Trade Annual of Australia*, 18th February 1936. See also 'Radio Education in Australia', *Radio Trade Annual,* 1936.

[46]Murray-Smith & Dare (1987); and Neill (1991). See also V.A. Edgeloe, 1989. *Engineering Education in The University of Adelaide 1889–1980: A Registrar's Retrospect.* Adelaide: Kinhill Engineers Pty Ltd.

[47]The Radio School was officially opened in November 1938 and expanded in 1942. See Murray-Smith & Dare (1987, pp. 238–9).

[48]E.H. Spencer, 1939. *A Report on Technical Education in Australia and New Zealand.* New York, 37.

[49]For a general valuation of the incidence of post-school training undertaken in Australia in the interwar years and of the importance of 'on the job' apprenticeship, see Hatton & Chapman (1987, pp. 9–13). See also Hyde (1982, pp. 105–40).

by the wide spectrum of the labour market for radio. This demanded specialists for pure and industrial research (mainly in the RRB, AWA, and Technical Colleges), experts in designing devices (for the PMG, Civil Aviation, and State Departments), and technicians and operators for radio stations and plants. The case of the issuing of different certificates (up to eight) by the PMG (such as First Class Commercial Operators for telephony, telegraphy, and broadcasting, Aircraft Operators, etc.) to cover the various types of stations is evocative of the close interaction between technical education and market in Australia.[50]

A longing for expansion also pervaded the specialized press. The somewhat exaggerated perception that Australia was destined to be the pivot of the communication systems of the Pacific and Indian Oceans helped the number of radio magazines mushroom in Australia from the mid-1920s. From 1923 to 1939, a total of 23 journals (see Table 5) propagated the grandeur of radio, and convinced both connoisseurs and the lay public of the commercial lure of electronics.[51] The pace was set by the Wireless Institute of Australia (the world's first national radio society, founded in 1910, and the publisher of the *Radio Experimenter and Broadcaster*, the first specialized journal of its kind), though here also AWA would be the epitome of wireless advance (editing *A.W.A. Radio guide*, *A.W.A. Technical Review*, *Radiogram*, and the popular *Radiotronics*).[52]

No initiative encapsulated the model of education and formation in an industrial context more explicitly than the Marconi School of Wireless within AWA.[53] From its establishment in 1913, following the pattern of its counterpart in Britain, it was devoted essentially to the training of radio operators for positions as wireless officers at the Mercantile Marine, the Navy, Army, and Air Force.[54] The professionalism of the instructors and the applicability of activities attracted military and marine authorities, with whom AWA had signed an agreement to instruct the applicants to fill their vacancies, and from whose fees the school obtained most of income. It was the unanimous

[50]Mackay (1938); and A.S. McDonald, 1938. 1913–38. 'A quarter -century of Radio Engineering in Australia'. *Proc. Wor. Rad. Con.*, 8.

[51]See NSWA. [Box 30], 'AWA Records'. See also 'Australian Radio Journals. Pre 1939', *Radio Waves*, 1985, 14:16; 1986, 15:15; and 'Australian Radio Journals before 1950 (an Extensive Review)', *Radio Waves*, 1994, 49:6.

[52]On the importance of the Wireless Institute of Australia in Australia's radiodiffusion, see J.F. Ross (1978, pp. 28–62).

[53]For a chronologic summary of the history of the School, see William A. Scholes, 'Marconi Radio School's Eighty Years of History', *Nautical Magazine*; and Marconi School of Wireless Communication, *The Marconi School of Wireless Communication* (London, n.d.).

[54]Fisk himself had studied temporarily at the Marconi School of Wireless in England. See Myers (1925, p. 45).

Table 5 List of the Australian radio and electronics journals in the interwar years

Australian Journals	Period
A.W.A. Radio guide	1926–36
A.W.A. Technical Review	1935–
Australian Radio and Electronics (formerly Australian Radio World, Sydney)	1936–
Australian Radio News	July 1932–June 1934
Broadcasting Business Yearbook of Australia, Sydney	1936–39
ERDA Journal (Electrical and Radio Development Association of New South Wales)	1928–
A.R.S.T. and P. Bulletin (Australian Radio Technical Services and Patents, Co. Pty. Ltd., Sydney)	1934–1949
Electronic Australia (formerly Radio TV and hobbies, formerly Wireless Weekly, Sydney)	1939/40–
Institution of Radio and Electronic Engineers, Australia, Proceedings (formerly Institution of Radio Engineers, Australia, formerly incorporated in Radio Review of Australia)	1937/38–
Mingay's Electrical Weekly (formerly Radio and Electrical Retailer, formerly Radio Retailer of Australia, Sydney)	1930–
Radio Broadcast (formerly Experimental Radio and Broadcast News—Wireless Institute of Australia, Melbourne)	1924–1926
Radio and Home, Perth	1933–34
Radio Experimenter and Broadcaster, Melbourne (Official organ of the Wireless Institute of Australia)	1923–24
Radio in Australia and New Zealand, Sydney	1923–1927
Radio Journal of Australia, Sydney	1927–28
Radio Monthly, Sydney	1931/32–1933/34
Radio Review of Australia (Institution of Radio Engineers, Australia, Sydney)	1933–38
Radio Technician, Sydney	1939–41
Radio Trade Annual and Service Manual (formerly Radio Trade Annual and Directory of Australia, Sydney)	1933–37
Radiogram (AWA staff magazine)	1929–36
Radiotronics (A.W. Valve Co. Pty. Ltd.), Sydney	July 1930–Dec. 1950
Telecommunication Journal of Australia (Telecommunication Society of Australia, Melbourne)	1935–37
Tele-Technician, formerly Journal, Melbourne (Postal Telecommunication Technicians' Association, Australia)	1915

recognition and high appreciation of these young officers which allowed succouring the precariousness of the school and the instability of temporary staff at its early stages. Indeed, its earlier activities were confined almost exclusively to the installation of wireless equipment in ships and associated shore stations. The fidelity of military clientele augmented the flexibility and range of courses (the initial personal instruction was redoubled with correspondence courses, in which attendance was compulsory only in the practical exercises).[55]

The fascination of Australian apprentices for learning wireless technology in the interwar years is beyond all doubt.[56] It was mirrored in one of the flagships of the school: the range of five diplomas issued by 'the prestigious AWA, the second largest wireless organisation in the British Empire'.[57] Thanks to this wide assortment and to, its competence, the Marconi School of Wireless was a notable success. From its inauguration, up to 700 of its graduates had obtained employment in the radio industry, 70% of whom continued in AWA. To be sure, the school carved for itself a pre-eminent role as the most prolific 'manufacturer' of radio-skilled personnel. The corollary of this policy of motivation and education was a high presence of accolades. Of the 4,096 certificates awarded by the Commonwealth to operators and technicians from 1913 to 1937, 2,374 (almost 60%) were issued by the school.[58] Meanwhile, other institutions—the Technical Colleges of Melbourne and Sydney, the PMG's Department, and the engineering schools of Adelaide, Queensland, and Western Australia—dispensed an instruction more finely tuned to electrical engineers with the daily demands of the world of telecommunications.[59]

The repute of the Marconi School was strongly shaped by the standards of the five courses of formation, namely 'radio engineering', 'radio technician', 'radio operator', 'talking picture operator', and 'radio mechanic'. Their main function was to guarantee a high technical knowledge in the whole field of wireless. The courses were of different length and were conceived as providing a range of specialized formation, subject to the availability and ability of students. The commitment to both motivation and improvement in quality of

[55]See NSWA. [Box 46], 'AWA Records', 'Marconi School of Wireless'.

[56]An illustrative example is A.W.A., *Marconi School of Wireless: Radio, the Industry of the Future* (Sydney: AWA, 1931).

[57]NSWA. [Box 46], 'AWA Records', *Radio. The Industry of Opportunity. Marconi School of Wireless* (AWA, 1937).

[58]Ibid., 'Marconi School of Wireless'.

[59]From 1920 to 1940 electrical engineering rose from 24% to 28%. For the professionalization of engineering during that period, see B.E. Lloyd, 1988. *In Search of Identity: Engineering in Australia.* University of Melbourne, PhD. thesis, esp. Chapter 6.

AWA's staff was reflected in the practice of reducing the fees of its employees (by up to 33%).[60]

The course of 'radio engineering' led to a 'Diploma of Radio Engineer' issued by AWA. Unlike the other courses, its main advantage was that, after a first three-year cycle of theoretical training and practice in telegraphy, most students obtained the First Class Certificate of Proficiency issued by the PMG, a guarantee for the labour market for telecommunications.[61] The knowledge of notions of general electricity and physics, indeed, formed the basis of radio engineering. The theory of electronic valves, transmitters, aerials, wave reception, amplification, direction finding, provided an attractive lure, as did the technology itself. Here, in contrast to the engineering schools, the practical nature of the teaching was defended with pride and honour, and in most cases the feeling was justified. For the instruction combined workshop exercises (such as the construction of short-wave sets) with an aggressively practical initiation into wireless techniques (three months at a radio-transmitting station), laboratory procedures (six months at the AWA laboratory at Ashfield), broadcasting (three months at a commercial receiving station), and visits to plants and factories.[62]

The progress of the Marconi School is a gauge of the demand for skilled technical education in radio. Like the students of engineering schools, apprentices and employees of their own company saw radio as rich in opportunities, or (in some cases) as a way of progressing within the company. Here, the paradigms to emulate were those graduates who had achieved posts of responsibility in military institutions and private industry, as well as minor positions as technicians at broadcasting stations, radio salesmen, and service mechanics. The figure of Adrian George Brown, an expert in radiotelephone circuits, is a good example, as are John Godkin Downes and C.E. Bardwell (the School's manager), two radio engineers who carved out a laudable professional trajectory for themselves.[63]

The Marconi School quickly became a symbol of the commitment of radio employers to advanced instruction, and a response to governmental indifference to the practical teaching of engineering. By the outbreak of World War II, it came to be regarded as pivotal for the machinery of Australia's defence. From

[60]NSWA. [Box 1], 'AWA Records', 'Management Committee—Minutes of Meetings, 1925–1936', Minutes of the School Committee Meeting, 2 July 1935.

[61]See *Radio. The Industry of Opportunity. Marconi School of Wireless* (AWA, 1937).

[62]NSWA. [Box 46], 'AWA Records', AWA Marconi School of Wireless, Syllabuses, 1918–1968.

[63]Scholes, *Marconi Radio School*, op. cit.

1939 to 1945, 1438 wireless operators for the Royal Australian Air Force, 1607 signallers for the Army, 300 operators for ships of the Mercantile Marine, and 560 telegraphists, were trained by AWA's staff. The school's ability to maintain high standards of proficiency in comparison with other nations, which were forced to accept much lower qualifications and issue provisional licences to semi-skilled operators, did not pass unnoticed by the allies. Their recognition might be perceived as the prize for the effort of radio entrepreneurs to demonstrate the usefulness of education to the new industrial order that emerged with the age of radio.[64]

Towards a self-contained radio industry

The most palpable sign of the evolution of radio manufacturing in Australia from a state of dependence to self-sufficiency lies in the production of individual components, assembly, and finishing by local firms. At the turn of the century, a common pattern was for the early sophisticated equipment to be produced and supplied from Britain and America, but for the installation and maintenance of communication stations to be undertaken by Australian mechanics. The gradual emergence of Australian firms specialized in sub-assembly—that is, in constructing a large and complex product from a number of component parts—marks a turning point in this pattern. The final product of one sector became only a component part for another sector, as was the case for valves for radio sets.[65]

The Australian office of the Western Electric, London (set up in 1895 and the founder of the Standard Telephone and Cables, STC, in 1914), for example, came to prominence through the installation of complex devices imported from Britain, the manufacture of which could not be supported economically by the limited Australian market.[66] By contrast, the Australian Wireless Co. and the Australian branch office of Marconi's Wireless Co., absorbed in 1913 into AWA, immediately appropriated all Australasian marine radio telegraph services, and, gradually, built up a comprehensive wireless organization covering all sectors, including manufacturing.[67]

[64]See NSWA. 'AWA Records', *AWA and the War*. Amalgamated Wireless (Australasia) Limited; 'AWA in World War II', *Radio Waves*, 2003, 85:11; and McIlwaine, (2004).

[65]One of the best syntheses on Australia's electrical industry in the 1920s is Colin Forster, 1964. *Industrial Development in Australia, 1920–1930*. Canberra: ANU, esp. Chapter V, 'Innovations in Power: Electrical Manufactures', pp. 103–23.

[66]The Company's history of manufacturing is chronicled in *Makers of Australian Telephone Cables* (Austral Standard Cables Pty. Ltd., 1958). See also Young (1983).

[67]Jones (1972, p. 34). See also 'The Coastal Radio Service', *Radio Waves*, 1999, 68, 19.

The expansion of radio manufacturing in Australia received considerable assistance through the tariff.[68] Before the protective policy of the Tariff Board in 1925, American manufacturers had used Australian markets as an outlet to dispose of surplus stocks at the end of their radio season. This operation incidentally coincided with the beginning of the Australian season. For American firms, it was far more profitable to establish formally constituted subsidiaries in Australia and then to sell at cost (or less than cost). In so doing, they avoided any carry-over into American new season productions. The consequences were apparatuses of inferior quality for Australia.[69] The underlying reason for this fact was, without doubt, the weakness of the Australian competitive position itself. It was no secret that its cost structure was at a much higher level than the USA's and Britain's, the main sources of radio imports.[70]

With these precedents, one might imagine the uncomfortable position of Australian investors at a time of aggressive world trading, such as the 1920s, in which the international market's sentiment was practically insensitive to national endeavours to industrialize. The emergence of numerous Australian suppliers assembling sets from imported parts probably alleviated their unease. In fact, by 1925, American, British and Australian suppliers shared about equally the Australian market.[71]

The introduction of protective tariffs eliminated the importation of inferior goods. Simultaneously, it endowed Australia with an infant radio industry beginning to be praised not only for economic reasons but also for its aspirations to become a self-contained industry. By the end of 1928, local manufacturers held up to 60% of the markets in sets, parts, and accessories (with the exception of cabinets and valves). It was a time in which employees and shareholders began to feel a profound satisfaction with profits and dividends.

The climate of excitement became more pronounced with the *politicization* of the radio market. From the foundation of AWA in 1913, Fisk had conceived local manufacture as a deterrent to foreign intrusion, and had varnished it with

[68]See *Tariff Board Report on Wireless Receiving Sets*, 1929.

[69]For the impact of the American intromission, see NSWA. 'AWA Records', 'Notes on the Effect of the Protective Tariff on Radio Manufacture in Australia' (probably 1932).

[70]Some valuable secondary sources on the history of the radio industry and electronics in Australia are N.S. Smith (1948, esp. pp. 6–7); 'Daily Telegraph' special report 'The Post Office, Partner in Industry', 10th February 1964, pp. 1–64; and H.R. Wood, 'The Electronics Industry in Australia', paper delivered to A.N.Z.A.A.S. Jubilee Congress at the University of Sydney, 24 August 1962, pp. 1–6 [also in *Telecommunication Journal of Australia*, 1963, 14(2)].

[71]L.D. Batson, 'Electrical Development and Guide to Marketing of Electrical Equipment in Australia', United States Department of Commerce, *Trade Information Bulletin*, 487, p. 34.

a strong dose of imperial sentiment. A consistent policy of self-reliance and a sense of Australian-ness had prevailed throughout AWA's history. During World War I, for instance, the company had achieved the complete manufacture of wireless telegraph equipment for use on ships.[72] The purchase of half AWA's shares plus one by the Commonwealth government in 1922 reduced the weight of English shareholders, and appealed to Australian prospective investors. But, above all, AWA would endeavour to respond to the foreign invasion with the irruption of radio broadcasting, especially through the installation of stations across the Pacific. Broadcasting stations were designed, equipped, and erected in Sydney, Melbourne, and Perth in 1924, and 'exported' to New Zealand, Tonga, and Fiji.[73]

It was the first of a series of decisions that marked AWA's industrial evolution. Manufacture and marketing of radio receivers would be the next strategy.[74] Crystal sets, two-valve sets with loudspeakers, the luxurious Sheraton cabinet type, direction-finders, and lifeboat equipment, were all products through which the company aimed to slow the tide of technological subservience and to counteract the overwhelming foreign presence in the rest of the Australian electrical industrial sector.[75] Plainly, AWA's ambitious dreams about self-contained industry came to fruition with the manufacturing of *Radiola*, the Australian radio receiver par excellence and the emblem of its new identity. The output of *Radiola* sales was relatively moderate at the beginning but spectacular in the late 1930s (from 3,000 receivers and 177 employers in 1927 to 40,000 units and more than 1,700 workers in 1937).[76]

AWA's superiority over other smaller firms lies, among other reasons, in certain privileges. Its vast control of patents was at once a cause and consequence

[72]The Marine Division was of paramount importance for AWA's consolidation. See Y. Scott, *History of Marine Department, 1913–62* (Sydney: AWA, 1962), pp. 1–6.

[73]See *Wireless Progress in Australia*. Sydney: AWA, 1930, chapters 'Network of Australian Controlled Stations in the Pacific Islands' and 'Coastal Radio Service'.

[74]Among the numerous reports and accounts of the history of AWA, see Philip Geeves, 1993. *The Dawn of Australia's Radio Broadcasting*. N.S.W.: Federal Publishing (probably the most complete, if unfinished work); AWA, 1932. *Facts Regarding the Wireless Industry in Australia*. Sydney; Myers (1925); and NSWA. 'AWA Records', 67 boxes.

[75]For a contemporary account of the company, see NSWA. [Box 30], 'AWA Records', 'AWA and its History, Notes'.

[76]For marketing details, see Amalgamated Wireless, *Radiola Broadcast Receivers: The Supreme Achievement in Broadcast Reception* (Melbourne: AWA, 193?). For a technical account, Peter Hughes, *AWA Radiolettes, 1932–1949* (Victoria, 1998), 2nd ed.; and 'AWA Radiolas 1925–1932', *Radio Waves*, 1983, 6, 8; 'AWA Radiolas 1932–1935', *Radio Waves*, 1984, 7, 6 ; 'AWA Radiolas 1936–1938', *Radio Waves*, 1984, 8, 4.

of its specific status. Before the agreement with the government in 1922, AWA owned extensive patent rights on equipment from its two paternal firms, the Marconi and Telefunken Companies, polarizing practically the whole market. But this domination was perceived almost as an institutional *casus belli*, inasmuch as AWA needed a licence from the government to broadcast, whereas the government resisted leaving patent possession in private hands. The semi-nationalization allayed the fears existing on both sides: it guaranteed a form of effective state control and endowed AWA with the potential to compete, *a priori*, in advantaged conditions with the rest.[77] What had begun as a dispute concerning the tenure and concession of AWA's patents had acquired, thanks to Fisk's diplomatic dexterity, the character of symbiotic operation, from which both AWA and the national industry obtained fruitful benefits.[78]

Australian economic fortunes reached their nadir during the Great Depression. However, its aftermath was felt in disparate ways. Recovery in manufacturing proceeded more rapidly than the primary sector, too vulnerable to international market mechanisms. The high tariffs and import controls, introduced during the Scullin government 1929–1931, had certainly to do with the creation of an environment favourable to, if not the growth, at least the maintenance of radio industry, despite the Depression.[79] The crisis had negative consequences in employment. For the first time the number of workers in radio manufacturing declined—unofficial estimates produced figures of 1030 in New South Wales for 1928 and only 908 in the whole of Australia for 1930–31.[80]

The growth in size and importance of the radio manufacturing sector for the rest of the decade shows how well Australian enterprises responded to the expectations and demands of their consumers. Thus, as the pace of licensed listeners trebled (from 350,000 in 1932 to more than 1 million in 1938), the

[77]In this context, it is instructive to observe that AWA possessed exclusive rights to 33 Australian radio patents in 1913, 258 in 1922, 1,008 in 1932, and more than 2,000 in 1939. In the late 1930s, 1,200 new inventions were received each year. See NSWA. 'AWA Records', 'Patents', 4th March 1939.

[78]For an historical account, see Ibid., 'Wireless Patent Rights', by Fisk, 1932. For the list of AWA patents, see Ibid., 'AWA Patents—Patent Specifications, 1920–1937'.

[79]The high tariff protection provoked hostile responses in Britain –Australia's major trading partner. For the British attitude towards the expansion of Australia's secondary industry, see A.T. Ross (1990, p. 187).

[80]According to the classical viewpoint in the historiography of Australia in the interwar years, while the primary sector turned to be the weak point of a relatively stagnant economy, the manufacturing sector appeared to be the main factor in the recovery and prosperity of the country. See Schedvin (1970, pp. 291–302); Copland & Janes (1937). For a dissenting opinion, see Thomas (1985, pp. 1–3).

response of radio companies, in the form of reinforcement of efficiency and precision, reached even greater proportions. By 1938, for instance, the number of employees in the whole of the radio industry was 30,000—8,000 in manufacturing (nine times more than in 1930), 15,000 in distribution and selling, and the remainder in communication services and broadcasting. The driving force behind this economic expansion lay in product, rather than in service or installation.[81] Unlike the latter, manufacturing sector was subject to market law, and hence was stimulated (and sometimes bounded) by immediate, rather than long-term, or even medium-term, priorities. That feature left its mark, for if it satisfied the most desired material demands, it also transmitted some of the limitations that impeded the progress of long-distance communications.

It is necessary, in this context, to resume, once again, the progress of the one enterprise enjoying the status of an exception, AWA. Indeed, this case reveals the originality by which Fisk's ambitious projects, and hence his ideals about business entities, were characterized. The expansion of AWA, following the economic revival after the Depression, adjusted itself to the parameters that distinguished the boom of the sector. Prospect of success was more than hopeful. But it was an expansion erected on the basis not only of manufacturing but also of the installation of stations for communication and broadcasting purposes. Certainly, the production of valves and radio receivers was indispensable to the company, but the emphasis of AWA's tactic was commercial.

Basically, Fisk's strategy consisted in organizing an intra-imperial network of communications systems (through the foundation of the Beam Wireless Service in 1927), as well as in seizing commercial broadcasting stations.[82] According to his conception of the radio industry, the organization of manufacturing and research was as essential as operating services. In the absence of the former, operating companies became merely radio transmission, as occurred in New Zealand. On the contrary, the American companies amalgamated communication services with manufacturing. He aspired precisely to that.[83]

[81]Of £5,525,000 from capital directly employed in the Australian radio industry in 1938, over 63% was earmarked for manufacturing and distribution, 13% for communication services, 12% for national broadcasting, and 10% for commercial broadcasting. See NSWA. 'AWA Records', AWA Ltd. Miscellaneous Papers, 1926–1944.

[82]Fisk's plans of a system of inter-communication between all parts of the Empire included high-speed telegraph, interlinking telephone, wireless picture and facsimile, and interconnected and world-range broadcasting. See NSWA. 'AWA Records', 'British Empire Communications', by Fisk, 19th January 1944.

[83]The main aspects of Fisk's ideas are described in E.T. Fisk, 'Ideals in Modern Business', *Business Lectures for Business Men*, 1933, 6; and his manuscripts 'British Empire Communications', 'The Progress of Wireless in Australia' (25th February 1937), 'Transoceanic Communications',

Fisk soon learned how profitable it was to trade with the USA without renouncing his imperial ideals, especially if it was about ensuring the future of the company. Britain's conservative attitude—too predisposed to restricting wireless services, in order to safeguard its revenue derived from submarine cables—left no room for hope. The inauguration of the Amalgamated Wireless Valve Co. in 1933, as a result of an agreement between AWA and the giant American RCA, IGE, and Westinghouse (AWA's shareholdings being 55%), fired the sales of electronic valves and tubes in the enterprise.[84] The main objective was to be as self-contained as possible: preparation of filaments and cathodes, assembly, ageing, testing, all for the sale of Australian-designed valves (although following American-type models).[85] While the electrical nature of radio manufacturing turned progressively into electronics, the nature of business relations was Americanized.[86]

The contribution of the rest of radio manufacturing industry in Australia was by no means insignificant.[87] Such companies (over 40 in 1932) could export to the countries where their parent firms had production and distribution networks or benefit, in the case of Australian firms proper, from the climate of accumulation of patents that was generated. In the second half of the 1920s, many manufacturers saw either in the production of components on a large scale or in the specialization as assemblers of radio sets an incentive to invest substantial capital.[88] Electricity Meter Manufacturing Co. (registered in 1912), Electrolux (1925), Hoover (1927), Stromberg-Carlson (as agent of its American parent Company, 1928), Radio Corporation (1929), Mechanical Products (later Kelvinator Australia, 1932), Email Ltd (1934), to mention the major firms,

'Communications and Business. How Wireless Has Developed', 'Wireless in Australia. Work of National Company', in NSWA. 'AWA Records'.

[84]By 1938, the production reached a level of 1 million radiotron valves a year, and the numbers of the staff at the new A.W. Valve works at Ashfield had been extended from 30 to 300 employees. See NSWA. 'AWA Records', 'Amalgamated Wireless Valve Company Pty. Ltd. Some Notes on its History', unsigned manuscript, 15th January 1963.

[85]E.G. Bailey & R.H. Healey, 1938. 'Some Aspects of Valve Manufacture in Australia'. *Proc. Wor. Rad. Con.*, 1–7.

[86]It is practically impossible to see a line of demarcation between radio and wireless technology and electronics at this time. For an example of this metamorphosis, see C.W. Vaughan, 'A Brief Survey of Wet Electrolytic Condensers and their Manufacture in Australia' and P.S. Parker, 'Radio Coil Development in Australia', both papers in *Proc. Wor. Rad. Con.*, 1938.

[87]For details of individual firms and their production, see *The Electrical Manufacturing Industry and its Importance to Australia* (Publication of the Electrical Manufacturers' Association of N.S.W., 1930), pp. 40–1; C. Forster (1964, pp. 111–12).

[88]Pratt (1977, p. 358).

made their reputation mostly through the manufacture of 'Australian-made radio receivers'.[89]

It may be that there were some reservations, not to say reluctance, among entrepreneurs when observing that AWA's demeanour tended to create a monopoly, as occurred in the Empire with Marconi's or in Germany with Telefunken. But the truth is that most of companies increased their investment of capital in accordance with the character of their activities. To be sure, one might believe that the imposition of tariffs and AWA's semi-nationalization would have eliminated the competition and would have restrained the appetites of investors. Quite the contrary. It was precisely the bravery and risk assumed by Fisk and AWA, particularly in the difficult years of the Depression, which galvanized the market and infused confidence among shareholders more than satisfied with competitive prizes and the performance of Australian radio receivers.[90]

The sense of Australian pre-eminence in radio in the late 1930s was increasingly tangible. It was reflected by the World Radio Convention, held in Sydney in 1938, under the auspices of the Australian Institution of Radio Engineers, on the occasion of the celebration of the 150th anniversary of the European settlement of the country. The convention, which gathered several of the world's leaders in radio development, reinforced the status of Australia as the most pioneering Dominion in radio.[91] The 51 papers read by the representatives of national institutions on the progress of radio research, manufacturing, and communication, amply fulfilled the objectives of its organizers, Fisk and O.F. Mingay.[92] Fisk aspired to promote radio as integral to Australia's nation-building, a task needing international impact and ostentation. The event cemented the association between the modernity of self-contained radio industry and that of a buoyant country pursuing, as a prime objective, its metamorphosis from Dominion to nation, all within the framework of Empire.

[89]For further data, see the questionnaires forwarded for completion to the largest companies in the Australian radio industry, in NSWA. [K61970/14c], 'AWA Records', Radio Industry Report, 24th May 1932.

[90]The Australian prices were on a par with those of overseas; for example, the retail price of a three-valve all-electric table model was £15.15 (Philips model), £23.20 (H.M.V.), £29.80 (Ferranti), and £16.17 (AWA) in 1932. See NSWA. 'AWA Records', 'Notes on the Effect of the Protective Tariff on Radio Manufacture in Australia' (probably 1932).

[91]'The 1938 World Radio Convention—60th Anniversary', *Radio Waves*, 1998, 64, 19.

[92]Institute of Radio Engineers (Aust.) eds., *Complete Proceedings of the World Radio Convention* (Sydney, April 4–14 1938). One of these rare copies is deposited in NSWA. [Box 54], 'AWA Records', 'World Radio Convention'.

AWA demonstrated how the combination of commercial success and manufacturing competitiveness could be achieved in an electrical sector of technological dependence that perceived investment in research and development as not only a risky venture but also precipitate. Most electronics manufacturers, representatives of local subsidiaries of European and American conglomerates, preferred to manufacture or assemble imported components, rather than to do research. But the situation was not catastrophic at any rate. Radio industry in Australia had a wide range of resources, comparable to the most advanced countries, including experienced technicians and electronics supplies. This was especially so during World War II. The event had the crucial effect of stimulating the vague readiness for becoming seriously involved in self-content technology that seems to have characterized hitherto only Fisk and his co-workers at AWA.[93]

Research for a conception of business

In the circumstances described, what might be seen as the paradigm of a thriving innovating enterprise can also be interpreted as a mark of failure in the Australian radio industry. Let us see why. In the 1920s, even the most bitter critics of the dependent character of Australian technology rarely conceived of laboratories as vehicles for innovation. The emphasis, in fact, was on adaptation and emulation rather than on invention and originality. The example of AWA's Technical and Research Department, with a well-equipped laboratory and skilled staff since the early 1920s, is revealing. The objective was to experiment with improved methods and to test and adapt the very latest inventions to Australian conditions.[94] On the eve of World War II, apart from AWA, few companies financed the laboratories from which a handful of patents and technological improvements had emerged.[95] In most Australian subsidiaries of the giant international corporations, where innovation was no priority, the emergence of laboratories was associated with the needs for routine testing and quality control that manufacturing processes required.

By the mid-1930s, the AWA Research Laboratory at Ashfield competed with the laboratories of Marconi's, GEC, and Bell in the race of innovation in the region. It serves little purpose to insist on the unequal market conditions

[93]Mellor (1958, pp. 481–511); 'Wartime Valve Production by AWA', *Radio Waves*, 2003, 84, 26;

[94]Myers (1925, p. 46).

[95]The fact that few workers of private companies cooperated with the RRB is indicative of that neglect: R.N. Morse from the Westinghouse Rosebery (1936–37), R.H. Healey from the A.W. Valve Co (1939), and five researchers from the AWA.

of contenders, but recent works on radio communication in Australia inexorably provoke reflections on a comparison and on the degree of originality of the Australian initiative.[96] Might the AWA Research Laboratory be seen as the Southern Hemisphere's counterpart of Marconi's or GEC's? Did ionospheric research in AWA have the fundamental character and commercial role which W.J. Baker ascribed to it in the case of Marconi?[97] And to what extent was the research component in AWA the corollary of an original conception of business entity?

AWA's involvement was, in part, a response to the economic realities of the sector that made the progress of companies depend on the deliberate concentration of patents and on independent innovation. As Fisk observed, it was essential to devote investment and manpower to laboratory projects that might lead not only to the development of apparatus invented and patented by foreign firms but also (and most importantly of all) to inventions.[98] The profusion of overseas patented devices and their immediate promotion in a market avid for new products did little for the advancement of local industry. This is reflected particularly aptly in the overwhelming foreign invasion of applications in the Australian Patent Office (see, for instance, the case of 1937 in Table 6). Here, Australia's contribution scarcely managed a paltry 1.7% being related exclusively to radio communication.[99] In these circumstances, AWA strengthened the dominance on patent rights in Australian soil with the control of 2,000 radio patents. Moreover, over the next two years AWA trebled, as a result of research effort, the number of patents involved in the construction of new models.[100]

The attention towards research as a source of stimulus for competitive innovation had its repercussions at university and the RRB. The first consequence was an increment of competitiveness. The officers of the RRB involved in atmospherics and ionospheric investigation at Sydney and Melbourne soon were tempted by an economic stability that the RRB could not guarantee. As Green stated after joining AWA in 1935, 'my resignation was prompted solely

[96]M. Keentok, 2004. AWA in Ashfield. *Journal of the Ashfield and District Historical Society*, 15, 22–41; and J.F. Ross (1998).

[97]W.J. Baker (1970, p. 289). See also Vyvyan (1933, p. 83).

[98]See NSWA. [Box 25], 'AWA Records', 'British Empire Communication', by Fisk, 19th January 1944.

[99]S.A. Mathews et al., 1938. 'A Review of Radio Development as Reflected in Applications in the Australian Patent Office during 1936–37'. *Proc. Wor. Rad. Con.*, 7.

[100]See NSWA. [Box 4], 'AWA Records', 'Patents', 4 March 1939, and 'Wireless Patent Rights', by Fisk, 1932.

Table 6 Number of applications in the fields of radio and electronics in the Australian Patent Office during the year 1937

	Printing telegraph transmission	Radio communication	Thermoionic devices vacuum tubes	Total
Great Britain	51	217	132	400
Germany	64	50	66	80
USA	31	59	52	142
Holland	5	17	53	75
Australia	–	5	–	15
Miscellaneous	7	18	14	39

by the need for a greater security of tenure'.[101] In so far as the AWA Research Laboratory at Ashfield might guarantee laboratory-based careers, the most immediate result would be the incorporation of young graduates, an unusual event in the industrial world.[102] Certainly, salary increases and extended contracts of employment were alluring incentives, but also the access to imaginative and unconventional activities. In this respect, the recruitment of Baker in 1931, Builder and Benson in 1934, Pulley in 1935, and Healey in 1939, served as a legitimation of the orientation adopted by AWA. The attractive remunerations, together with exciting expectations of innovative works with a high dose of creativity, offered an opportunity difficult to reject.

The modest but sustained flow of researchers from university to private enterprise brought with it a variety of means of collaboration. These tended to be personal and intermittent relationships, rather than formal agreements between institutions.[103] The nature of physical research started to be a scientific venture in which commercial applications were obtained as by-products.[104] Indeed, it was precisely because industrial laboratories

[101]Evans (1973, p. 389).

[102]E.T. Fisk, 1935. Foreword. *AWA Tech. Rev.*, 1, 2.

[103]On the occasion of the Sesqui Centenary of Science at the University of Sydney, Matti Keentok gave a talk on such relations. See the document 'Synergy and Symbiosis between University and Industry: Radio Research in the University of Sydney, the Radio Research Board and AWA', 2002, deposited at the SUA.

[104]In the opening speech of the IVth Conference of Australian Physicists, Kerr Grant expressed his fears of the new commercial turn acquired by physical science. See AAAS, [Box MS 86/6/38], Kerr Grant, 'The Place and Value of Physical Science in the Modern State', *Proceedings and abstracts of the IVth Conference of Australian Physicists, Melbourne, 15 to 18 August 1933.*

ensured an approach to market and competition that collaboration was so highly esteemed. It was reflected at the Fifth Conference of Physicists and Astronomers, held in Sydney in May 1936 under the auspices of the Australian Branch of the Institute of Physics.[105] The event, which was organized by the Departments of Physics and Electrical Engineering at Sydney together with AWA, allowed delegates to inspect the laboratories and valve works at Ashfield. But the participation of radio industry went beyond a courteous, albeit perhaps opportunist invitation. That, in fact, a quarter of the total number of papers presented came from employees of the AWA and other firms was symptomatic of change.[106]

The tendency towards innovation clearly coloured the nature of laboratory work at AWA. The laboratories provided a considerable flow of skilled employees. Their function, however, extended to the fields of both electronics and atmospheric physics. It is easy to understand why laboratory specialists engaged in invention projects attached especial importance to communication conditions. In Fisk's conception of the business entity, the manufacturing of apparatus was intimately associated with its operation in service. Hence, in terms of investment evaluation, the uncertainty of a buoyant but infant market was as pernicious as the unpredictable behaviour of the waves in the upper atmosphere. Persuaded as Fisk was that communication was an investment security, atmospheric research could provide the immunity to those short-wave services being apparently much more vulnerable to ionospheric meteorology than to market volatility.

No initiative embodied Fisk's aspiration of applied research within an industrial context more patently that the journal *AWA Technical Review*. Founded in March 1935, it was published quarterly by AWA at Sydney, Melbourne, Wellington, and London. This initiative would not have succeeded without the acuity of its first editor, W.G. Baker, who, despite the scarcity of resources, endeavoured to emulate the seriousness of the *Proceedings of the Institute of Radio Engineers*—the doyen of radio engineering journals. His stay at General Electric's radio laboratories in Schenectedy, New York, allowed him to keep close ties with its editing house, the American Institute of Radio Engineers.[107] In pursuing the elucidation of techniques and principles which might be exploited by

[105]The Proceedings and abstracts of the 1st, 2nd, 4th and 5th Conferences of Australian Physicists (1928, 1929, 1933, and 1936, respectively) are deposited at the AAAS.

[106]Of the 29 papers read out at the conference, 16 belonged to the field of radio and atmospheric physics—five of which were read out by the AWA's staff.

[107]Gillmor (1991, p. 191).

the company, the journal symbolized the public rostrum that AWA's researchers used to spread their works.[108]

As in other overseas laboratories, research in AWA was a desperate attempt to turn art into pure science. Its unreliability for commercial use notwithstanding, short wave was the backbone of Fisk's scheme of intra-imperial communication. Combating their seemingly random behaviour, however, was at once a scientific challenge and a risky investment. Fading, unpredictability, loss of signal at certain times of the day and night, were inopportune characteristics requiring rigour and imagination.[109] The priority was to understand the causes in order to remedy their effects. But the elucidation of the former involved rigorous theoretical standards that few employees of AWA were able to attain. This is why researchers examined effects rather than seeking their causes in the structure of the upper atmosphere.[110]

Short waves had already proved to be effective, albeit capriciously for England–Australia communication purposes. On the occasion of the establishment of the 'Beam wireless' service in 1927, Fisk had decided to solve these difficulties using twin aerials, erected back to back and directed in opposite directions around the globe. Thus, when fading commenced on one aerial, the other one was immediately activated, emitting the signals in the opposite direction.[111]

By the 1930s, the policy of the AWA Research Laboratories was guided by the quest for effectiveness and profitability through the adoption of a series of practical measures. Usually, in contrast with other exclusively electronic firms, this signified an emphasis on the selection of frequencies and the improvement of transmitting and receiving antennas. The spectacular growth of broadcasting was perceived as an opportunity to develop higher-gain antennas and to design new prototypes. The aim was ambitious but feasible, for it was regarded as perfectly viable to combine the more refined laboratory procedures with the most sophisticated theories in order to improve the picturegram, telephonic, and telegraphic services.[112]

[108]One of the rare copies of the journal can be found at the State Library, Melbourne.

[109]A valuable work of research analysing the results of measurements made in Europe, the USA, and Australia is A.L. Green, 1936. 'Non-Fading Noise-Free Broadcasting Stations'. *Journal of the Institution of Engineers, Australia*, 8, 203–15.

[110]See the contrasting cases of Baker and Healey, in W.G. Baker, 1938. Studies in the propagation of radio waves in an isotropic ionosphere. *AWA Tech. Rev.*, 3(6), 297–320; and R.H. Healey, 1938. The effect of a thunderstorm on the upper atmosphere. *AWA Tech. Rev.*, 3(4), 215–24.

[111]Sturmey (1987, p. 113); Moyal (1984, pp. 140, 132–3).

[112]See, for example, Builder & Benson (1936), Maynard (1936), and A.L. Green (1937).

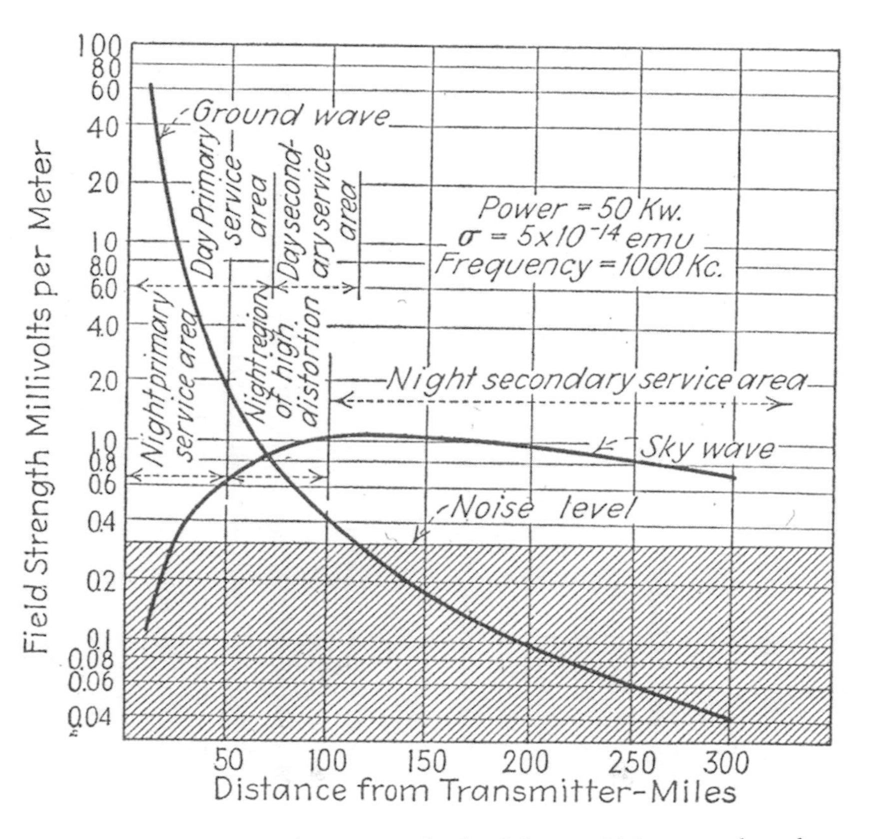

Fig. 22: Diagram illustrating different types of coverage obtained from a high-power broadcast station during the day and night periods. The top of the shaded area represents the lowest field strength that completely overrides the noise level. From Terman (1938), p. 346.

Priorities in the field of electronics were even more practical and imperious. Most of them were associated with testing and control, which were then used by staff for improving the quality of product. Part of the time was devoted to precise measurement and to maintaining physical sub-standards in the valve manufacturing. Here, practice, often far removed from theoretical considerations, fomented a conception of research very close to the contingent exigencies prevailing in market.[113] The personnel of the laboratory, evidently immersed in the reality of local demand, tried to be informed about the latest novelties. The design of the first automobile radio receivers, televisions, radiotelephone circuits, loudspeakers, reflexed amplifiers, and cathode ray oscillographs illustrate the disparate nature of the process of electronic innovation. The combination of novelty and reliability would be the pragmatic response to the needs derived from commercial rivalry.[114]

It would be rash to conclude that the radio industry in Australia was the paradigm of what the interaction between manufacturing, research, and the educational system should be in quest of a self-sufficient state. Radio companies certainly remained bound, as has been seen, to the inertia of erstwhile practices dazzled by the purchase of foreign products and too receptive to the setting-up of tempting but economically disputable subsidiaries. Here, plainly, there was a breeding ground apt to financially satisfy entrepreneurs and leave a trail of prejudice towards the ideals of self-reliance in Empire circles. The attachment to import-replacement manufacturing and the reticence towards exportation were facets characterizing industry as a whole, and in particular that of radio.[115] But all this did not prevent industrial research and technical education being fomented at a pace roughly commensurate with immediate needs.

Evidently, the true agent of this new reality was an almost exceptional case, namely AWA, which emerged as a windfall capable of animating supply after two decades of fruitless attempts. But for all this to come to fruition, an industrial radio fabric was necessary that was capable of generating an environment for competition, despite the state educational system not rising to the occasion. Hence, if it is possible to talk about success, it is so not thanks to the tariff policy or the ups and downs of economic circumstances but thanks

[113]See G. Builder, 'Report on A.W.A. Ltd. Research Laboratories', *Proceedings and Abstracts of the Vth Conference of Australian Physicists and Astronomers, Sydney, 25 to 28 May 1936.*

[114]See, among others, Builder & Benson (1938), Rudd (1938), and Schröter (1938).

[115]The huge imbalance between imports and exports characterized the balance of trade even in the 1940s. For comparative statistics in 1938 and 1949 in the Australian electrical and communication engineering industries, see *The Structure and Capacity of Australian Manufacturing Industries* (Commonwealth of Australia, 1947), pp. 267–72, p. 275.

to Fisk and his company's perspicacity, reflected in their zeal to unite manufacturing and research with communication and to conquer national pride. For, as has been described, it was the strength of AWA's subtle conception of combining self-contained industry and intra-imperial communication services, sprinkled with nationalist connotations, that marked one of the most fundamental differences between the company and the rest of the Australian firms.

AWA's Research Laboratory and Marconi's School of Wireless, each in its own way, contributed decisively to remedy what in retrospect might appear a failure of the electrical sector, and the Australian industry as a whole, for example by promoting the notion of the radio engineer as the innovator of devices. In the ways indicated, the potentiality of short wave for long-distance communications fomented the research of the upper atmosphere. AWA was curbing a tendency towards scientific and technological national dependence, and at the same time turning Australia into the centre of communications in the Southern Pacific. Like the RRB, AWA was preparing the ground on which innovation and self-sufficiency were programmed to yield rich dividends in the near future, as would be demonstrated by the Radiophysics Laboratory and radar.

Conclusion

In the last two chapters it has not been our intention to examine in detail the theoretical developments of ionospheric physics in Australia. This would lead us astray from our designated objectives. Rather, we have suggested that certain structural, not incidental, factors actively fomented the emergence and continuance of a radio research tradition, to the extent of strengthening it. Indeed, if it is possible to condense a panoramic valuation of the Australian contribution, this must certainly be as much concerned with scientific as with geopolitical and economic issues. Let us delve more deeply into this area with two long reflections.

Firstly, the multiple connotations of what we have labelled as Australian ionospheric physics and radio R&D reflect the multiple connections of intertwined interests, in both the educational and commercial sectors. In retrospect, it seems clear that major advances in ionospheric physics were closely related to innovation and development in radio technology; that radio research was an essential integrand in the pursuit of atmospheric knowledge; that the practice of radio physicists was influenced by the vicissitudes of the radio industry. The allure of a burgeoning sector, incentives to patency and to innovation, vested interests in private firms, all had a direct impact upon the research activities in the RRB. In retrospect also, it seems clear that the industrialization of the country was matched by rapid growth in professional engineering.[116] In this field, the particular relationships between universities and technical colleges varied enormously—compare, for instance, the function of both as *producers of engineers* during the course of the century—, as did the corresponding accounts of academic modernization and adaptation to industrial demands.[117]

However, in stark contrast with this substantial increment in engineering specializations, there is abundant evidence of the shortage of courses of radio engineering in technical colleges, and, even worse, of their 'non-existence' in universities. Here, the precise instructive dimension of the latter in this discipline must be sought in the electrical engineering programmes at the Universities of Sydney and Melbourne. In this respect, it is more than probable that university graduations in Madsen's department were more significant than the college evidence seems to suggest. But it is clear that in its most general sense, technical training became an ineluctable, essential *task* of the radio industry itself, or, more exactly, an integral component of what the firm might produce. The ethos, importance and methods of technical preparation were all weighed up by AWA's entrepreneurs in their strategies of expansion, of 'Australianization', of self-sufficiency. The Marconi School of Wireless, the PMG's Department, apprenticeship systems, remedied the lacunae of those state educational institutions unwilling to accept advanced syllabuses, promoting the sense of an 'electrical training' for the radio profession. But the effort of this training never proved to be in vain. We have only to recall the alacrity of the private firms to recruit the young researchers of the RRB, and the numerous postgraduates of the electrical engineering departments who, in facing specialization and work retraining, drew upon the openings and prospects which the radio industry would bring.

And, secondly, 'short wave and the elliptic picture of the Empire' is a political *formula* with economic and cultural connotations. Indubitably, there were vital constituents of commercial expansion integrated in the art of radio communications and in the quest for atmospheric knowledge. But here what is remarkable is that such constituents pertained to specific strategies which imperialist radio entrepreneurs shared. The aforesaid strategy, pursued by Fisk and AWA's co-workers, conflating research, manufacture, and services for

[116]Butlin (1970, p. 290).

[117]The number of newly qualified university engineers grew 30.1% from the 1920s to the 1930s, and 77.8 % from the 1930s to the 1940s, as opposed to 273.3% and 200.3% from technical colleges over the same decades. Moreover, whereas in the late nineteenth century university was the main source of engineering degrees, in the 1930s Sydney Technical College produced two thirds of the higher rate of engineers. See Edelstein (1987, p. 18). See also Corbett, (1961).

communication and broadcasting, was in essence that initiated by Marconi's and the American giants GEC and Bell.[118] It was also an adaptation to the specificities and peculiarities of autochthonous development, accentuating nationalist sentiments for instance, or aspiring to an intra-imperial network of broadcasting stations (an issue overlooked by Marconi's). Ironically, a former subsidiary of Marconi's drew upon its commercial leverage and seized its business strategy of self-reliance in order to dominate the region in which Marconi's itself still aimed to wield authority, the Southern Hemisphere.

In sum, we have seen how a multiangular vision of the atmospheric sciences provides a wider outlook upon the Australian experience. There is no historical circumstantiality or fortuitous confluence of personalities concurring simultaneously at a given time and space; there are multiple structural factors which have resonant repercussions. We have proved that the idea of a central body (as was the RRB) galvanizing radio R&D exclusively is insufficient. There are in addition unquestionable 'agents'—an *amalgam* of technical education, radio industry, and consciousness of nation-building, transcending, and overcoming the imperial frontiers. In hindsight, it was the predisposition of industrialists and radio scientists to inject originality and imaginativeness into their initiatives, to pursue leadership and self-sufficiency within both the radio industry and the scientific community without renouncing their imperial ideals, to integrate qualified training (in the case of entrepreneurs) within their business strategies without sacrificing their utilitarian values. On a certain occasion historian Michael Sanderson exhorted us to examine, *contra* a pervasive common practice, the educational and industrial fabrics which, to a greater or lesser degree, used to sustain R&D. When the chapters on New Zealand and Canada are read, the reader may well realize how influential, yet how inconspicuous, those factors could be.

[118] On the major areas of research and activity in the American two giants, see Leonard S. Reich, 1985. *The Making of American Industrial Research. Science and Business at GE and Bell, 1876–1926*. Cambridge: CUP, 86–8, 222–24.

5

ORGANIZING RADIO RESEARCH IN NEW ZEALAND

Government, university, and the radio industry

When Sir Frank Health, the head of the British DSIR, visited New Zealand in 1926 in order to advise on the transplanting of the British model, he found a Dominion in which government authorities viewed not only research but the concept of a central research organization with suspicion.[1] His recommendations had been debated and refuted in Australasia through 'over twenty-five years of thought, discussions, submissions and experimentation'.[2] New Zealand, like Australia, had articulated an organizational structure deficient with respect to the interaction between government, education, and scientific activity. A handful of disseminated and barely interrelated institutions (such as the Dominion Laboratory, Geological Survey, Meteorological Office, Hector Observatory, Magnetic Survey, and the Samoan Observatory) shared resources with the national university network, composed of colleges at Auckland, Wellington, Dunedin, and Christchurch. Their role had been focused on education and services rather than on research investigations.[3]

The desideratum of contributing to the national prosperity and modernity—notions associated with the progressivism of that time—dominated the intellectual landscape of the 1920s.[4] Although the Dominion lacked a tradition of systematic experimentation, opportunities and particular circumstances in the booming field of broadcasting permitted sporadic forays into wireless research.[5] Most of these proceeded from what we might call 'transplanted Britons'. Robert Jack, a very talented Scottish physicist, is an illustrative example. Educated at the Universities of Glasgow, Paris, and Göttingen, Jack researched the effect of magnetic fields on the spectra of atoms (the Zeeman effect) before moving

[1]Frank Heath, 1938. 'Report on the Organization of Scientific and Industrial Research in the Dominion of New Zealand'. *App. J. Ho. Rep. of N.Z.*, II, H-27, 12 March.

[2]Michael E. Hoare, 1976. 'The Relationship between Government and Science in Australia and New Zealand'. *Journal of the Royal Society of New Zealand*, 6(3), 381–94, p. 392.

[3]J.T. Atkinson, 1976. *DSIR's First Fifty Years*. Wellington, pp. 9–18.

[4]Important works with general insights are R.M. Chapman & Malone (1969); Burdon (1965).

[5]Ian Dougherty, 1993. 'Marconi a Phoney? Change and Expansion in New Zealand Telecommunications in the Interwar Years'. *Stout Centre Review*, 9–16 March.

into the public service as professor of physics at the University of Otago in Dunedin.[6] In 1921, he examined *in situ* the naval radio communication system in the United Kingdom; then, importing high-voltage generators and two Edison valves, he assembled the first radio device for transmission of voice and music. His initiative, drawing particularly on the British experience, laid a framework for the promotion of public interest in broadcasting. Two years later, new regulations enacted by the government would accord licenses to amateur experimenters to make transmissions.[7]

Plainly, research as priority and regular practice hardly appealed to government and university authorities, even in its less academic forms.[8] The very few ventures undertaken sporadically as ideas occurred were minor in nature. The physics departments (closely linked to the chair of chemistry) suffered from the great difficulty of assimilating the rapid and tremendous development of physics.[9] Reminiscing about those first years, Auckland professor P.W. Burbidge lamented: 'The university system was terribly cumbrous. The teaching staff was under considerable strain both to acquire the established knowledge and to impart it. There was a considerable tendency to remain highly conservative.'[10] In this, the situation did not differ from their Australian counterparts. But in respect of technological provision, it was even more precarious.[11]

The department of physics at Auckland University College provided the archetypæ research effort carried out in precarious conditions and under the sponsorship of a variety of sources.[12] Aided by funds supplied by the Research Grant Committee, Burbidge constructed apparatus receiving long-wave signals emitted from Lyon and Bordeaux;[13] E. Green measured the intensity of waves over the Pacific Ocean with the support of the P&T; and the skilful experimenter G. Munro, who later worked with Laby in Melbourne, studied the sunrise and sunset reception effects on the signals from Wellington thanks

[6]Jim Sullivan, 1998. Jack, Robert 1877–1957. *DNZB*, 4.

[7]A.C. Wilson, 1994. *Wire and Wireless: A History of Telecommunications in New Zealand, 1890–1987*. Palmerston North: Dunmore Press, 115.

[8]For a contemporary account, see Sidney H. Jenkinson, 1940. *New Zealanders and Science*. Wellington.

[9]AUA. 'P.W. Burbidge Papers', [MSS Archives B930], *New Zealand Science in WW II. The Story of the War Activities of the Department of Science and Industrial Research, New Zealand, 1939–45*, pp. 11–13.

[10]Ibid., 'Memoirs of Early Physics in New Zealand, 1909–1950', p. 4.

[11]For a valuable review of research in New Zealand in the early 1920s, see M. Thomson, 1925. 'Scientific Research in New Zealand'. *N. Z. Jour. Sc. Tech.*, 8(1), 41–61, p. 41.

[12]Beaglehole (1949).

[13]Percy William Burbidge, 'The Intensity of European Radio Signals'.

to a National Research Scholarship.[14] The department reflected what Burbidge described as an enthusiastic spirit subject to multiple vicissitudes, which prodigiously converted a grave shortage of resources into a miraculous line of investigation.[15]

In theory, with the creation of a twin DSIR would come organization, and with organization, research. The search for planning and order—applauded by visiting experts like Rutherford and Heath, rather than by cabinet members—would lead to the application of methods and structures enjoying a very high degree of esteem in Britain.[16] But the paternalistic advice in favour of organization hid deep contradictions. The transplantation involved an aura of vagueness, overtly equivocal in matters of manufacturing and standardization. British authorities' guidance was sought and appreciated; the British DSIR remained the model to emulate and transplant; but local manufacturing industry was not suited to a scheme of imperial science in which New Zealand would be one of its main actors.[17] As a result, the picture emerging in the early 1930s was a triptych in which the centre panel was occupied by government applied science, with agriculture the main recipient of benefits; on one side of the panel appeared the life sciences, still chiefly concerned with native fauna and flora; and, on the other, physics and chemistry, whose development was progressively acquiring overtones of defence policy rather than of fundamental research.[18]

Symbolically, the New Zealand DSIR embraced radio research. Ernest Marsden, the architect of its gestation, played a leading role in bringing physical sciences to the DSIR through several of its existing institutions.[19] The appointment of an experienced physicist, as was M.A.F. Barnett (forged and tempered at Cavendish along with Appleton), with a view to organizing a committee inspired by the principles of the British and Australian RRB, might in principle be seen as a good decision. But radio research was low on the list of priorities. Indeed, Marsden assigned him a cascade of operations and tasks

[14]G.H. Munro, 'The Origin of Errors in Wireless Direction-Finding'.

[15]Other works on wireless research include L.W. Harrison (at Victoria University College), 'Experiments on Short-Wave Length in Wireless Telegraphy'; Robert R. Nimmo (senior scholar in physics at Otago University), 'Measurements of High-Frequency Resistance: A New Substitution Method', *N.Z. Jour. Sc. Tech.*, 1926, 8, 110–17.

[16]Eisemon (1984, pp. 49–50).

[17]Ross Galbreath, 1998a. *DSRI: Making Science Work for New Zealand: Themes from theHistory of the Department of Scientific and Industrial Research, 1926–1992.* Wellington: Victoria University, 111–12.

[18]AUA. 'P.W. Burbidge Papers', [MSS Archives B930], *New Zealand Science in WW II,* op. cit., p. 15.

[19]Galbreath (1998b); C.A. Fleming (1971).

in such motley fields as seismology, refrigeration, and building regulations.[20] These circumstances condemned radio to remain in the background, at least until the potential of radar was grasped. The Geological Survey, by contrast, was a division of DSIR in which searching for oil and gold deposits justified geophysical investigations and in which political leaders did not hesitate to view its creation as consistent with, or pertinent to, the true interests of the economy.[21]

Between 1926 and 1937, as imperial science tried to become national, New Zealand's radio science failed in its attempt at following in the footprints of neighbouring Australia. In 1928, Marsden had highlighted the importance of conjoining the scant human and material resources in radio development.[22] Under his guidance, representatives of the P&T Department (A. Gibbs), the Radio Broadcasting Company or RBC (J.M. Bingham), universities (P.W. Burbidge), and the Research Council (M.A.F. Barnett) formalized the constitution of the Radio Research Committee (RRC) in February 1929.[23] The four bodies comprised the compass of radio application. From 1921 to 1987 the Post Office would be the authority in radio licenses and in the allocation of frequencies for both broadcasting and communications purposes.[24] Furthermore, the RBC, created in 1925 after an agreement with the government, was perceived as the future national broadcasting system (an operation that would crystallize in 1936). Its relay stations network only extended to metropolis and provincial towns, while rural areas were wholly dependent on the poorly financed and private 'B' stations. Within these confines, reviews of short-wave radio technology, regular telephone services, radio broadcasting programmes, and the use of radio for guiding aircrafts became hallmarks of the era.[25]

While it would be correct to speak of these four bodies as lobby groups (above all P&T and RBC) in both style and behaviour, their strategies for research, advancement, and diffusion of radio knowledge could not hide a

[20]See *Sir Ernest Marsden—80th Birthday Book*. Wellington: Reed, 1969, p. 60; Atkinson (1976, pp. 19–32).

[21]Burton (1965).

[22]NZNA. [AAOQ 72/123/1], 'Marsden, 20 November 1928'.

[23]NZNA. [AAOQ 72/123/3], 'Minutes of the First Meeting of the RRC, 19th February 1929'.

[24]NSWA. 'Sir Ernest Fisk Papers', [Fold Beam], 'New Zealand Telegraph, Wireless Telegraph and Cable Service', by E.T. Fisk, 17 July 1925. The early history of the P&T is chronicled in Robinson (1964).

[25]For a general history of broadcasting in New Zealand, see P. Day (1994), Hall (1980), and Harcourt & Downes (1976).

locally ingrained element such as the radio amateur.[26] Barnett attempted to place radio amateurism on side with science, a model explored by the RRB counterparts to unite research and amateur cooperation to the mutual benefit of scientific method.[27] The destiny of New Zealand's radio research would be, therefore, in the hands of both physicists and radio amateurs. This would therefore counteract the lack of human resources. The starting point was the logging of the strength of signals from the Byrd Expedition in the Antarctic in 1929.[28] The New Zealand Association of Radio Transmitters embraced Barnett's plan. Indeed, four standard receivers were constructed and systematic measurements were recorded. But the 'irksome and routine nature' of work was not sufficiently alluring to maintain a continuous and sustained service.[29] Despite the affectionate messages of cooperation from official circles, amateurs preferred sporadic collaborations to lasting and largely unrewarding engagements.

One might be tempted to conclude that this trend to research in New Zealand was the expression of a premeditated and mature change on the part of government, industry, or university. Quite the contrary. Radio research in its New Zealand form was an amalgamation of desires and hopes subject to improvization and enthusiasm, in which figures like Barnett and Gibbs showed neither the stature nor the talent of their Australian counterparts (Madsen and Brown).[30] It was a declaration of intent, rather than of realities or *faits accomplis*, in which the very few physicists involved could barely achieve isolated successes, whereas state administrators, for various reasons, transmitted illusory promises.[31]

Nor can it be precisely stated that manufacturing policy, even in the glory days, contributed to encourage what scientific promoters projected. New Zealand's radio manufacturing sector imported excessively.[32] The country, with

[26]Ian Dougherty, 1977. *Ham Shacks, Brass Pounders and Rag Chewers: A History of Amateur Radio in New Zealand.* Upper Hut.

[27]See, among others, NZNA. [AAOQ 72/123/3], Barnett to the Secr. of N.Z.A.R.T., 27 February 1929; [SIR 1, 26/5], Barnett to the Association of Amateurs of Wireless Transmitters.

[28]Ibid. 'R.R.C.: Report of Progress to Date, from 1929 to 1930'.

[29]Ibid. Secretary of N.Z.A.R.T. to the RRC, 20 January 1930.

[30]Referring to Barnett, F.W.G. White confessed: 'When I went to New Zealand in 1937 I remember being appalled at the fact that he had not begun to research in his subject to any great extent'. See AAAS. [111/4/5], 'F.W.G. White's Biographical Notes, vol. 2', in 'Meteorological Physics', p. 1.

[31]For the expectations created, see, for instance, NZNA. [SIR 1, 26/10/2], Barnett to Laby, 28 March 1928; Barnett to the Australian RRB, 10th March 1930. See also Munro's report on radio research in New Zealand (1930) in Evans (1973, pp. 140–1).

[32]In 1924, the bulk of radio instruments and material used in New Zealand came from the USA, Canada, and England. See 'Radio in New Zealand', *EW & WE*, 1924, November, 107–8.

a weak manufacturing base but a heavy demand for radio receivers, resorted to foreign firms and to local subsidiaries, whose workmanship, paradoxically, was often superior to that of any imported receiver.[33] High distribution costs and added wage expenses gravely penalized the two main local manufacturers, Radio Ltd. of Auckland and Radio Corporation. Radio interests, in the Depression years, concentrated on increasing production and on eluding crisis, whereas research was consigned to oblivion.[34] Unlike Australia and AWA Co., New Zealand companies, because of the divorce existing between manufacturing and communication, remained bound to purist visions and practices which put too much emphasis on the importance of production and broadcasting as independent entities. This eliminated the nodal points that might have arisen from the intersections between both sectors, such as research and education.

To be sure, the economic policy of the Labour government (elected in 1935), while making attempts to diversify and insulate the industry, also generated tensions with important ramifications for the radio sector.[35] Both in content and in form, these tensions would undermine and frustrate entrepreneurs' desires in favour of self-sufficiency. The first source that threatened a self-contained radio industry derived from the weak government position towards patents. The majority of the applications received by the Patent Office stemmed from overseas interests. The government assumed their control, but unlike in England or in the USA, for example, did not conduct any exhaustive examination into the merits of these applications. As a result, their granting was a simple and nearly automatic procedure benefiting vested interests from overseas and prejudicing national efforts.[36]

A second frustration grew from the delicate position of the local firms. Owing to their limited size, these could not house a scientific staff of the necessary calibre for undertaking tests and research. These difficulties notwithstanding, the government abandoned the small companies to the tender mercy of a limited and exigent market. Neither the Government nor the young DSIR injected scientific or engineering methods into their manufacturing processes; and a weary repetition of failed projects was left to haunt the modest

[33]Of 9,557 sets in cabinet imported in 1935, 65% came from the USA, 11% from Britain, and 10% from Australia. See NZNA. [SIR 1, 26/–, General], 'Radio Industry, by W.J. Lyon'.

[34]NSWA. 'AWA Records' [K61970], 'Particulars of Prices and Sets on New Zealand Market, 25 August 1932'. See also, ibid., [Box 21], 'Wellington (New Zealand)—Correspondence, Notes, 1925–55'.

[35]Tom Brooking, 'Economic Transformation', in Oliver & Williams (1981, 226–49, p. 248).

[36]NZNA. [SIR 1, 26/7], 'Radio Patents', A.R. Harris to D.G. Sullivan, Minister of Industry and Commerce, 17 July 1936.

entrepreneur until the end of the war.[37] Attitudes of mind that accepted technological dependence and enshrined individualism underestimated, without a doubt, the capacity of local radio industry to compete in a changing and challenging world, and not to be resigned to mere survival.[38]

Australianizing New Zealand's radio science

Several years after the constitution of the RRC, only a handful of researchers devoted themselves to radio science in New Zealand.[39] Financial constraints and the lack of commitment on the part of the P&T and broadcasting companies, largely occupied with practical routine tasks, gave rise to a near suspension of the projects devised along the lines of the British and Australian RRB.[40] The most important research hub was at Auckland University; here, C.G. Rudd, under the direction of Burbidge, developed a directional recorder plotting the position and movement of the sources of atmospherics over the Tasman Sea.[41] The Victoria College, in Wellington, hosted another of the rare cases of government-supported university investigations: a young F.W.G. White, then 24, was constructing a stable high-frequency amplifier that would enable taking fading curves from broadcasting stations.[42] These and a handful of sporadic initiatives enjoyed a somewhat precarious existence.[43] In most cases investigations were intended as no more than temporary activities to facilitate the realization of masters' theses.[44] In the early stages of the RRC, management was more important than performance or solvency; this management included

[37]NZNA. [SIR 1, 26-], 'Postwar Research for Radio Industry, 27th September 1945'. See also 'Annual Report of the DSIR', *App. J. Ho. Rep. of N.Z.,* 1930, III, H.34, pp. 55–6.

[38]Ross Galbreath, 'New Zealand Scientists in Action: The Radio Development Laboratory and the Pacific War', in MacLeod (1999, pp. 211–27).

[39]Barnett (NZNA. [SIR 1, 26/10/2], Barnett to Cook, August 1934) noted the shortage of human and material resources that the RRC had after the Depression. In 1934, only two research students worked on ionospheric investigations: J.A. Peddie of Victoria University College, Wellington, on the structure of a G region, and a German student and Jewish refugee, Kurt Kreielsheimer of Auckland University, on layer height measurements.

[40]These institutions had made numerous measurements of the field strength of stations, but the data collected had not been examined theoretically. See NZNA. [SIR 1, 26/5], 'Field Strength Survey of New Zealand'.

[41]NZNA. [SIR 1, 26/5], 'Observations on Atmospherics in the Tasman Sea'; and R.W. Boswell, 1936. Sources of Atmospherics over the Tasman Sea. *RRB Rep.*, 10, 9–17.

[42]White (1930).

[43]The most valuable work since 1920 is the calculation of the height of the reflecting layer done by Munro (1928) at Auckland.

[44]AAAS. [111/4/5], 'F.W.G. White's Biographical Notes, vol. 1', p. 42.

exploration of sources, expense evaluation, and a palpable if understandable air of inexperience.[45]

The total solar eclipse of October 1930 was a test to calibrate the possibilities and capabilities of New Zealand radiophysicists. The URSI had recommended Barnett to arrange for records of the strength of signals transmitted from Honolulu, Samoa, and Fiji on both long and short waves.[46] Barnett had witnessed an experiment of similar type in England in 1926, even if, then, the effect of eclipse on the propagation of waves could not be clearly concluded. Perhaps by its uniqueness, the scientific programme seems to have impressed New Zealand radio amateurs, who came to see collaborating in reception tasks as a route to official recognition and prestige.[47] For the first time a team of radio operators was successfully organized to record the effects of changing ionosphere conditions.[48]

Commitment to fundamental research was not only a question of willingness and vocation but also an echo of ideas from overseas. Efforts to strengthen the infant scientific structures were often encouraged from Britain. For the metropolis, atmospheric sciences constituted an especially propitious terrain for the condensation of volatile desires regarding the hypothetical organic unity of Empire and science, which had marked the agenda of imperial relations since before the World War I. The interwar years' debate about British centralism, leadership, and loyalty surfaced in New Zealand in a variety of ways. Soon after the creation of the RRC, the Chief Engineer of the British Post Office, Colonel Lee, earnestly encouraged investigations into the reception conditions for the English beam stations in New Zealand, because, in his opinion, this would benefit the Post Office and BBC's services in the Dominions.[49] Besides radio comunication, the British meteorological authorities demanded meteorological information in connection with the imperial air route and with flights in the Pacific.[50] Edward Kidson, the head of the New Zealand Meteorological Bureau—administrated by the DSIR—, amended the lack of barometric stations with a greater attention to the possibilities of the

[45]Accounts of the New Zealand radio research programme have been given by White (1975, pp. 43–4); Evans (1973, pp. 140–56); and AAAS. [111/4/5], 'F.W.G. White's Biographical Notes, vol. 1', pp. 42–8.

[46]NZNA. [SIR 1, 26/10/2], General Secretary of the URSI to the New Zealand RRC, 10 March 1930.

[47]NZNA. [SIR 1, 26/5], Barnett to the Association of Amateurs of Wireless Transmitters.

[48]M.A.F. Barnett (1930).

[49]NZNA. [SIR 1, 26/5], 'The Reception of Beam Signals in New Zealand'.

[50]Cook to the Australian RRB, March 1936, in Evans (1973, p. 148).

radio methods. The international conference of 1937, convened by Kidson to plan for an organization in aviation meteorology in the South-West Pacific, reflects the magnitude of imperial demands.[51]

To an extent impelled by need rather than by belief, Kidson offered a model of scientific cooperation, translated into the language of metropolis and periphery. From his personal interests in Antarctic meteorology, and the application in the Southern Hemisphere of the Norwegian frontal methods of analysis of forecasting, essential for the trans-oceanic air services, came a response that would satisfy the expectations of Britons, both at home and overseas.[52] At the heart of the programmes of cooperation lay a readiness for complaisance, and a commitment to progress along the line of mutual profit and common consent. In aviation or weather forecasting, meteorology conducted along scientific bases would bring prosperity and welfare to the Empire as a whole.

This history can be contrasted with that of the RRC, which, once instituted, became integrated into the scheme of imperial relations just as the Meteorological Service had been but with a degree of interrelation that was much less than Kidson's. Thereafter, the sense of British cooperation became purely testimonial, a product of the contradictions that propriety often generates. The Australian model, with its controversial vision of an imperial focal point in the Southern Pacific, would prevail not through any preference but thanks to indifference on Britain's part.[53]

Just as New Zealanders, in the interwar years, seemed to regard Australians as neighbours and close friends, rather than part of the same family, sharing separate but parallel lives and histories, so it is possible to perceive a generalized apprehension of being absorbed by Australia's tentacles in science and politics.[54] Belief in the pre-eminence of the imperial partnership provided most New Zealanders with a collective identity, a broad definition of themselves and their role in the world; and thereby they reserved all their loyalty and devotion for the Empire. Belief in the utility of the Australasian federation, by contrast, had provided them with a political framework to satisfy their national

[51]M.A.F. Barnett (1939–40); and Kidson (1941, pp. 130–1).

[52]The value of New Zealander meteorologists' frontal weather analysis was acknowledged by Higgs, Munro, & Webster (1936, pp. 59–62).

[53]New Zealand–Australian relations have been considered in a wide range of studies. See A. & R. Burnett, 1980. *The Australian and New Zealand Nexus Documents.* Canberra; Keith Sinclair, 1972. 'Fruit Fly, Fireblight and Powdery Scab: Australia–New Zealand Trade Relations, 1919–39'. *The Journal of Imperial and Commonwealth History*, I (1), 27–48.

[54]Mary Boyd, 'Australian–New Zealand Relations', in Livingston & Louis (1979, pp. 49–51).

self-interests but also with a sense of 'distrustful independence'.[55] From such perspectives New Zealanders could feel they had a special destiny in the vanguard of British civilization and obligations and duties to share with their neighbouring Australia.[56]

There is, it is necessary to say, a certain tragedy in the *Australianization* of New Zealand. The deliberate pursuit of cooperative programmes was a rational response to the penury, stagnation, and financial constraints of the RRC. The Depression had literally stifled financial sources for radio research in New Zealand whereas broadcasting and post office authorities still offered their unconditional support in Australia. The imperative needs arising from the RRC's predicament—its need for information exchange and for the synchronization of observations in a joint programme of ionospheric pulse sounding—were cruelly exposed in 1934 by Barnett and Marsden in several letters dispatched to the Australian RRB.[57] There, the CSIR authorities, like Rivett and Laby, sounded out the possibilities of cooperation in atmospherics for broadcasting and weather forecasting purposes.[58] In view of Britain's neglect at a time when the RRC was suffering from economic hardship, the approach to Australia might be regarded as the most logical strategy.

The most tangible evidence of the 'Australianization' of New Zealand's radio science lies in the minutes of the influential meeting of the ANZAAS of 1937.[59] Held in Auckland, with the attendance of Madsen, Martyn, and Munro, among other Australian leaders, the event became a rallying point for the perspectives of radio research within New Zealand. In his presidential address, Madsen indicated that the formation of a new body would be advantageous not only for New Zealand but also for Australia.[60] Concerned about the inactivity and apathy of the RRC, Madsen diverted his energies into persuading the New Zealand government of the necessity for establishing a new organization,

[55]André Siegfried, 1914. *Democracy in New Zealand.* London: G. Bell & Sons, p. 358—quoted in Boyd (1979, p. 49).

[56]For an excellent discussion of the extent of British cultural influence on New Zealand's society in the interwar years, see P.J. Gibbons, 'The Climate of Opinion', in Oliver & Williams (1981, p. 303).

[57]NZNA. [SIR 1, 26/10/2], Barnett to Cook, August 1934; Barnett to Rivet, 15 September 1933.

[58]Rivett to Marsden, 25 August 1933; Minutes of the 20 Meeting of the Australian RRB, March 1934. See also 'Co-operation with Ionospheric Research in New Zealand by Builder', August 1934, in Evans (1973, pp. 143–5).

[59]*Report of the 23rd Meeting of the Australian and New Zealand Association for the Advancement of Science.* Auckland, January 1937.

[60]Madsen (1937, pp. 14–18).

a highly diligent committee, not a passive and lethargic one, which would soon become the kindred spirit of the successful Australian RRB.[61] Indeed, judging by Madsen's words transcribed by the Auckland Herald, the polychromatic plan of the new board offered such immediate and manifold benefits that the proposed expenditure would be more than justified.[62]

The URSI authorities had talked of the internationalization and the globalization of radio research, with its additional endorsement of ionospheric prediction, social dimension, and the application of comparative analysis to both Hemispheres.[63] While digesting these precepts, Madsen perceived similar needs and values emerging on both sides of the Tasmanian Sea, pointing out simultaneous coordinated observations as the driving force of the new liaison.[64] Radio telegraphy, broadcasting, aviation, defence, and meteorology formed the five cardinal points of the 'Australianization' plan for New Zealand's radio science, and Madsen was to spare no effort in the mission.

More subtly, perhaps, the gathering storm clouds hanging over Europe may also have appealed to the profound, persistent desire of Madsen and his colleagues not only to export their model but also, and most importantly of all, to build up a protective shield for both countries. External and internal communications, meteorology, and aviation, as Madsen recalled, were vital from a defence point of view during abnormal times. 'The more knowledge we can obtain of such services the greater is our protection', he declared.[65] The moment of battle was approaching, and the tempo of defence policy would be set by the requirements of cooperation, radio science, and its methods.

Aspects associated with defence, such as those associated with bilateral relations, provided an unquestionable allure for the government, and provoked a change of attitude. Once again this change was induced from outside rather than inside the country. On this occasion, however, the venture had more of a military character, as can be seen from the presence of a representative of the Defence Department in the new RRC.[66] Ironically, for radiophysicists, the notion of public interests having underlying military

[61]For strong evidence of this point, see White's judgement on Madsen's role in Evans (1973, p. 149).

[62]*Auckland Herald*, 20 January 1937, ibid., p. 150.

[63]NZNA. [AAOQ, 72/123/1], Memorandum: Proposal of Radio Research Board, DSIR, Wellington, 11 February 1937. See also Evans (1973, pp. 270 and 273).

[64]NZNA. [SIR 1, 26/10/2], *New Zealand Herald*, 13 January 1937.

[65]NZNA. [SIR 1, 26/10/2], *New Zealand Herald*, 13th January 1937.

[66]NZNA. [AAOQ, 72/123/3], Minutes of the preliminary meeting of the RRC, 23 March 1937.

connotations proved beneficial for, thereafter, an annual amount of £2,500 would be earmarked for radio research (about 1% of the broadcasting licence fees), the New Zealand Broadcasting Service contributing 70%, the P&T 15%, and the DSIR 15%.[67] Fundamental university research could no longer justify itself as a neutral and purist activity, a bastion of pure *savoir*, whose purpose was to enlighten the paths leading to scientific knowledge. Above all, it was a government venture invested with a public and military interest; a venture that would gain its credibility from the defence service it would provide in the war, in return for the public money it was granted (a total of £10,658 from 1937 to 1942).[68]

For the re-activation of radio science in New Zealand, the keyword was 'opportunism', an attitude that was exploited by both Australia and Britain, seasoned with discourses of cooperation and equally of opportunity, yet an attitude from which New Zealand knew how to obtain worthwhile benefits. After its creation on the 28 April 1937, the new RRC accepted the Australian research programme, and adapted it to the intrinsic peculiarities of the country.[69] Almost simultaneously, the British RRB invited its two counterparts to participate in a short-wave cooperative project, a project that would have the radar programme as its extension, and that was received cautiously by Australians, if enthusiastically by New Zealanders.[70] This set of projects had on paper a strong local component, mainly in radio broadcasting. Their implementation, however, could harbour other messages. Palpably, the apparent necessity of ionosphere information could be easily translated into profit for Canberra; the supposed weakness before the Japanese threat, into profit for London. Looking at this conjuncture from a distance, one might perceive that these movements were associated with a sagacious, geopolitical opportunism, which appeared just when the conflict was imminent, and would tailor wartime policies in which radio would have a fundamental role to play. At the outbreak of war, New Zealand's ionospheric research would prove to be essential to the Australian RRB in predicting best frequencies for transmission.

[67]NZNA. [SIR 1, 26/4], D.G. Sullivan, Minister of Industry and Commerce, 13 July 1938.

[68]A good summary on New Zealand's radio research, including prewar and wartime activities, is NZNA. [SIR 1, 26/], 'Post War Radio Research, by E. Marsden, 20 November 1945'.

[69] 'Annual Report of the DSIR for 1937/38', *App. J. Ho. Rep. of N.Z.*, 1938, III, H.34, p. 12. See also Evans (1973, pp. 151–2).

[70]The signs of a positive reception for the cooperation in short and ultra-short wave are abundant. See NZNA. [SIR 1, 26–], V.F. Brown, Secretary of the British RRB, to Barnett, 24 December 1937; Barnett to V.F. Brown, 24 February 1938. For the Australian disenchantment, see Evans (1973, p. 273).

In this context, New Zealand's ionosphere stations would become as vital for Australia as radar would be for Britain.[71]

Between 1937 and 1939, the somewhat generalized perception that New Zealand's radio science had been neglected in the last decade had become less strong as a result of the formation of small groups of researchers, notably in the fields of the ionosphere (in Christchurch and Wellington), overseas communication (Auckland and Dunedin), and field strength survey (Wellington and Dunedin).[72] The appearance of these teams coincided with, and certainly contributed to, an unprecedented level of radio research.[73] Whereas until then there had been sporadic efforts and interrupted programmes hardly comparable to those in Australia, to mention a parallel case, the introduction of automatic recorders in Christchurch and Wellington and the exchange of data in 1937–39 provided experience and expertise and, in response, a vote of confidence on the part of Allies in the radar programme and the ionospheric prediction service. Six years later, this feverish prewar activity would lead DSIR's director, E. Marsden, to be proud of New Zealand's defensive potential—six ionospheric stations out of a world total of 50.[74]

One case of cooperation, which very clearly illustrates the revitalizing effect resulting in the re-activation of radio for other research areas which were not to the same extent beneficiaries of financial help, concerned the determination of the maximum auroral frequency. Attempts to accomplish this task had been gestating since the eighteenth century with the legendary expeditions to the Antarctic continent, and later by the Aurorae Section of the New Zealand Astronomical Society.[75] Even from as early as 1921, the regular presence of aurorae australis, in accordance with a work performed by A. Gibbs for the P&T Department, was perceived as a perturbation in the ideals of reliability

[71]The point is made very strongly in AUA. 'P.W. Burbidge Papers', [MSS Archives B930], *New Zealand Science in WW II,* op. cit., pp. 2–19.

[72]In Christchurch, White supervised the work of J.C. Banwell and T.W. Straker; in Auckland, E. Collins and K. Kreielsheimer worked under Burbidge's guidance. NZNA. [AAOQ, 72/123/3], Minutes of the meeting of the RRC, 25 May 1939.

[73]The organization and high level of participation in the Meeting of the Australian and New Zealand Association for the Advancement of Science, held in Auckland in January 1937, illustrate this new turn.

[74]At Lincoln, Campbell Islands, Kermadec Islands, Pitcairn Islands, Rarotonga, and Suva Fiji. See NZNA. [SIR 1, 26/], 'Postwar Radio Research, by E. Marsden, 20 November 1945'. See also Galbreath (1998a, pp. 109–39).

[75]As part of the programme of the RRC, White collected early reports of aurora australis from 1773 up to 1898 to form a valuable auroral log. See White (1939).

and audibility of wireless communication.[76] But it was only after the creation of the new RRC and New Zealand's total integration into the schemes of the URSI, when the existence of aurorae and magnetic storms was recognized as a major factor in the ionization of the lower atmosphere, that New Zealand radiophysicists began to collaborate with the magnetic observatories with a view to establishing the relative positions of ionospheric stations. Military considerations also contributed to engage the government's interest, for they dictated a route which led to the formation of the Allied Ionospheric Prediction Service for the South-West Pacific Area that, if it were not for the climate of war, would probably never have given rise to this sensitivity. At last, the outcome was unquestionable. A higher number of papers on auroral geophysical research were published between 1937 and 1939 than in the two preceding decades.[77] In 1938, White, in collaboration with M. Geddes of the Carter Observatory in Wellington, conclusively proved that an extraordinary activity of radio fade-outs preceded by about 30 hours the appearance of aurorae and magnetic storms.[78] And, in the next year, they plotted the approximate position of the zone of maximum auroral frequency.[79] It is hard to imagine that in other circumstances a phenomenon of ostensibly academic interest would have generated such excitement.

The outbreak of war precipitated events and accelerated radio research. As a deployment of forces from the metropolis and the periphery, the upper atmosphere had the appearances of a theatre of operations in which agents with different stakes placed their bets. The ionosphere encouraged fundamental research, and, through it, prediction, in which Australia had a special interest; while the reception of distant signals was especially gratifying to Britain. State agencies surfaced—the Broadcasting Service to study the attenuation of local signals, and P&T to allocate frequencies for both broadcasting and communications—where their application was of public interest. But, as in other spheres, Australia weaved networks of influence as subtle and with as fine as mesh as Britain's imperial networks of dominance.

[76] A. Gibbs, 1921. 'Effects of the Recent Aurora on Telegraph-Lines, Telephone-Lines, and Wireless Stations'. *N.Z. Jour. Sc. Tech.*, 4, 183–8.

[77] Cooperative works supported by the RRC include, among others, Geddes (1939) and White (1939).

[78] White, Geddes, & Skey (1938).

[79] White & Geddes (1939).

6

GOVERNMENT, UNIVERSITY, RESEARCH, AND RADIO INDUSTRY IN CANADA

Introduction

The interwar period witnessed a growing interest in the use of radio for the betterment of imperial communications. In Britain, this was reflected by a special preoccupation with relations with the Dominions and by an intensification of the 'spirit' of cooperation, with Appleton as a functional cerebrum and the London–Slough–Cambridge triangle as the nervous system. In Canada, one might study the ways in which the organization of radio R&D differs from the imperial precedents set by countries such as Australia and New Zealand. It should mirror the greater weakness of her imperial bonds in the process. The situation is even more complicated by the double polarization of the Anglophone and Francophone communities. Prevailing perceptions on the role of radio in national identity and in international relations often assumed a diametrically opposed perspective when seen from Montreal or from Ottawa. In this respect, this chapter is an invitation to reflect on the nature of the relations with the Empire and with the United States. It is also a reflection on the wider structural factors which gave Canadian radio experience, in a context of approximation to Washington and of gradual detachment from London, the status of an exceptional and fascinating case.[1]

In Canada, the pursuit of the knowledge of the upper atmosphere also became a relative priority, as much in connection with government-funded military applications as with economic development.[2] This pursuit was so

[1]The proliferation of voices in the 1920s craving for a more vigorous expression of nationhood is symptomatic of the social debate that such relations aroused. See, for instance, R.S. Sommerville, 1926. 'Is Canada Becoming Americanized?'. *Empire Review*, 43, 537–40. 'Canadians are gradually losing their inherited British reticence and dignity... [But] Love and admiration for the Motherland are inbred deeply and are not to be cast out lightly' (p. 540); J.C. Hopkins, 1926. 'Relations with the Empire'. *Canadian Annual Review*, 52—3. See also the introduction in Lyon (1976).

[2]The sense of priority and of importance was much more marked after World War II. Exhaustive and valuable studies on the history of postwar ionospheric physics in Canada include: Jones-Imhotep (2000, 2001).

intertwined with radio development that the cultivation or procrastination of ionospheric physics depended upon the industrial, academic, and geopolitical conditions in which the latter matured. In our present quest for structural factors, the elucidation of such binomials as 'ionosphere—radio industry', 'wireless—Empire', 'broadcasting—manufacture', and 'communication—national-building' becomes critical. Canada's experience will permit us to identify the complex association of motives underlying the gestation of upper atmospheric sciences that has usually passed unnoticed in the history of interwar physics. This has been so because radio has been traditionally regarded as apolitical, non-aligned, achromatic, its association with the ionosphere, merely through the technological umbilical cord. Radio, it was often said, is little more than an application of the electromagnetic waves theory. Ionospheric physicists map the upper atmosphere and radio facilitates the means for sounding, but this inflicts no onus upon radio, no connotation beyond the compass of engineering. The distinctive, cognitive features of the new picture of the atmosphere, such as its stratified nature, appeared thereby deracinated from their cultural and socio-economic habitat. However, if the thesis that structural reasons intrinsic to radio industry, education, and to the geopolitical reality of the time underlie ionospheric physics is accepted, then Canadian radio also should have a fundamental role to play in its demonstration.

Radio research, government, and universities

In Canada, as in other Dominions, the interwar years constitute one of the most significant and intriguing periods in which to study the development of government participation in support of research. In the late 1920s, the National Research Council (NRC), which had been founded in 1916, introduced new procedures and modes of attending to the exigencies of science.[3] One may recognize important advances in the organizational framework within which the Division of Physics (one of the four laboratory divisions created in 1929) came to operate, and in the administration of research funds mainly orientated towards national defence.[4] This period witnessed the development of

[3]A valuable guide on NRC is Donald J.C. Phillipson, 1991. 'The National Research Council of Canada: its Historiography, its Chronology, its Bibliography'. *Scientia Canadensis*, 15, 177–200. For an indispensable book on the early history of the NRC with abundant primary sources, see Thistle (1966). See also Eggleston (1978).

[4]W.E.K. Middleton, 1979. *Physics at the National Research Council of Canada, 1929–1952*. Waterloo: Wilfrid Laurier University Press, 3–16.

government policy towards the cooperation with universities, which had hitherto been almost the sole foci of physical research.[5]

The institutionalization of radio science came late and haltingly to Canada. When, in March 1931, representatives of the NRC, defence, marine, universities, and industrial firms established the directives of what came to be known as the Associate Committee on Radio Research (ACRR)—the counterpart of the British and Australian RRB—, the sense of Canadian backwardness in radio organization was palpable. This was both disclosed and highlighted by the minutes of its first meeting.[6] In the tortuous process of delimiting what criteria should govern state investment in radio research and which fields should benefit from it, ACRR's responses were mainly guided by a series of imperatives common to the Departments of National Defence and of Marine and Fisheries. Firstly, the lack of knowledge of broadcasting conditions was cited as a significant shortcoming. The development of a high-precision frequency standard that could serve as a primary standard was regarded as essential for self-government in radio communications but also for the fulfilment of international prerogatives in that connection.[7] The disparate nature of the state of autonomy in radio regulation and standardization is exemplified by the contrasting cases of the USA, whose government took active part in securing radio bands, on the basis of a defence of her existing wavelengths jurisdiction, and the Canadian government, which experienced difficulty in maintaining her international rights in the radio field. Between these extremes in the spectrum of interests between standardization and broadcasting practicability, there was little space for fundamental research on the ionosphere. The case of the

[5]On the growth and diversification of research in universities, see Yves Gingras, 1991. *Physics and the Rise of Scientific Research in Canada.* Montreal: McGill-Queen's University Press, 58–82. See also R.A. Jarrell, 1988. *The Cold Light of Dawn: the History of Canadian Astronomy.* Toronto: Toronto University Press.

[6]NAC [MG 30 B157, vol. 7/16], 'Proceedings of the First Meeting of the Associate Committee on Radio Research, Montreal January 3rd, 1931'. The list of ACRR's members includes important individuals and institutions associated with radio in Canada: A.S. Eve (McGill University, Montreal); R.W. Boyle (National Research Laboratory, Ottawa); C.P. Edwards (Dept. Marine); A. Frigon (École Polytechnique, Montreal); H.J. MacLeod (University of Alberta, Edmonton); J.C. MacLennan (University of Toronto); A.S. Runciman (Shawinigan Water & Power Co., Montreal); V.G. Smith (University of Toronto); W.A. Steel (National Defence, Ottawa); H.J. Vennes (Northern Electric Co.); H. Vickers (University of British Columbia, Vancouver); and B.G. Ballard (National Research Laboratory, Ottawa).

[7]Such concerns as 'the use of a foreign standard would place us at the mercy of other countries' and 'Canadian role before the future International Radio Conferences to be held at Madrid and Stockholm' were vigorously expressed. See NAC [MG 30 B157, vol. 7/15], 'Report of the ACRR for the Period March 31, 1930 to March 31, 1931'.

exploration of the ionospheric layers, for instance, attracted ACRR's attention, not so much for altruistic reasons of knowledge as for pragmatic objectives such as the betterment of local radio transmission.[8]

In the early stages of the new venture, the aura of excitement that was exceeded by joint research initiatives in a sector particularly lacking in institutional collaborations was as intense in academia as it was in the manufacturing and broadcasting industries. The occasion coincided with a natural event of special magnitude. In August 1932, on the occasion of the total eclipse of the sun, the plans contrived by Colonel W.A. Steel and Professor A.S. Eve—along with Appleton's inestimable help[9]—for a hitherto unknown effort of cooperation that would at once have scientific objectives and incentives to commercial firms were to come to fruition.[10] It was around this phenomenon, in a project aiming to do in Canada what was usually exclusive of the promoters of metropolitan science in Britain and the US, that a joint programme of solar observation was developed. Thus, while McGill University and the NRC Laboratory performed measurements of the Heaviside layer,[11] Canadian Marconi and Northern Electric companies recorded radio signal fading.[12] The result was certainly promising. For besides the fact that a total of three papers were published in the *Canadian Journal of Research*—NRC's channel of diffusion—,[13] the experience served to acquaint inexpert Canadian observers with the pulse technique for ionosphere sounding and to establish close contacts with eminent figures such as Appleton and Dellinger.[14]

[8]The work done at McGill University by W. Bruce Ross and H.R. Smyth in 1933 in connection with the finding of 'two well-defined reflecting regions' is perhaps the most significant. See NAC [MG 30 B157, vol. 3/9], 'Notes on Some Recent Work on the Ionosphere in Canada', by T.T. Henderson, 23 April 1934.

[9]E.V. Appleton, 1933. 'Co-operative Radio Research in the Empire'. *World-Radio*, May, 19, 669.

[10]The preparations for the event are well described in NAC [MG 30 B157, vol. 7/15], 'Annual Report of the Associate Committee on Radio Research by B.G. Ballard, 28th April 1933'.

[11]The incentive for private companies is clearly encapsulated in Henderson to Boyle, 26 October , 1933: 'Exact knowledge [on the upper ionized layers] is essential for the complete theory of wireless propagation and in turn the need for such a working theory is being increasingly felt by commercial, communication and broadcasting companies', in NAC [MG 30 B157, vol. 21/9].

[12]NAC [MG 30 B157, vol. 8/1], 'Canadian Marconi Co.: Report on Signal Strength Observations during the Solar Eclipse, August 31st 1932'; and 'Report on Eclipse Measurements by the Northern Electric Co., Sept. 21st 1932'.

[13]Henderson (1933); Rose (1933); Henderson & Rose (1933).

[14]The contacts mentioned emerge from Steel's visit to the Bureau of Standards at Washington in the months previous to the eclipse. See NAC [MG 30 B157, vol. 9/6], 'Report 3: Heaviside

Despite all its limitations, the intermarriage between universities and private firms in an issue on paper as abstruse to industry as that of eclipse proves that in the organization of joint ventures Canada was in all respects as mature as any other developed country. She could arrange as many human and material resources as Australia in the exaltation of radio research that took place in the early 1930s. But this achievement would prove to be adventitious. During almost all the rest of the decade, significant advances in radio and ionospheric research, albeit not totally non-existent, were intermittent in assiduity, and modest in magnitude and scale. It was so not only by the standards of nations in parallel situations such as Australia but also by the proficiency shown by Canada herself on the occasion of the eclipse.[15] It is true that this phenomenon was relatively rare, and that its enterprise was encouraged by international personalities and entities such as the URSI, but the flurry of the engagements generated quickly evanesced. Between 1934 and 1937, ACRR's activity focused in great part on laboratory testing and precise calibration. It was Steel, now in 1934 the chief of the Canadian Radio Broadcasting Commission, who secured a contract with the NRC to test the sensitivity, selectivity, fidelity, and distortion of commercial radio receivers, with a stipend of $3,000 a year.[16] This income certainly contributed to alleviate ACRR's distressing financial situation, but at the expense of supplanting creativity by routine work. In the conditions prevailing in other industrialized countries such a task would have been assumed by industrial laboratories.[17] The insufficiencies of the Canadian radio industry have, therefore, their echoes in state-funded research.[18]

Layer Measurements by W.A. Steel, February 1932'. For the correspondence with Appleton, see NAC [MG 30 B157, vol. 8/1].

[15]The lack of staff, insufficiency of funds, and intermittency in ionospheric research are highlighted over and over again by Henderson in his correspondence with foreign colleagues. See, for instance, NAC [MG 30 B157, vol. 3/16], 'Henderson to Dellinger, November 28th, 1933; October 1st, 1935; February 25th, 1937'.

[16]NAC [MG 30 B157, vol. 1/8], 'R.W. Boyle to Henderson, April 19th, 1933'. See also Middleton (1979, pp. 54–5).

[17]Henderson bewailed the fact that laboratory testing 'occupied a considerable portion of our time...We are supposed to test a representative sample of each different model receiver made in Canada [125 different types]'. In NAC [MG 30 B157, vol. 2/31], 'Henderson to Appleton, March 20, 1936'.

[18]The case of the radio industry was no exception. According to a questionnaire prepared by the council, of some 2,400 leading Canadian firms engaged in manufacturing, only 37 had laboratories. See *Annual Report of the National Research Council Containing the Report of the President and Financial Statement for the Fiscal Year1936–37.* Ottawa: NRC of Canada, 1937, 13.

Table 7 Financial statement of expenditure at the National Research Council of Canada and the Associate Committee on Radio Research, 1931–1940

Fiscal year	Total expenditure ($)	Radio Committee	Radio expenditure (%)	Aeronautical Committee	Grain Research Committee
1931-32	556,832	0	0	—	—
1932-33	453,347	0	0	—	—
1933-34	444,614	0	0	—	—
1934-35	522,449	267	0.05	753	24,772
1935-36	658,326	2,868	0.4	1,167	21,790
1936-37	708,233	5,387	0.76	5,696	—
1938-39	1,041,431	22,199	2.1	14,450	35,736
1939-40	1,226,981	27,421	2.2	17,949	37,424

Source: *Annual Report of the National Research Council Containing the Report of the President and Financial Statement* (Ottawa: NRC of Canada), for the period 1930–41, vols. 14–23.

The austerity in radio research at its earliest stages was powerfully conditioned by the established priorities of the NRC in such fields as agriculture and mining (compare ACRR's expenditure with that of the most subsidized, the Grain Research Committee, in Table 7). Accordingly, ACRR's non-existent or exiguous budgets had to be supported by additional revenue and supplementary contributions which had been obtained from the broadcasting industry. The marginality of government-funded research projects in the matter of radio was reflected in the meagre granting of scholarships to the ACRR (of 239 bestowed by 1938, only three pertained to the issue). It was not until that year that C.W. McLeish, a young graduate of the University of British Columbia, was awarded a special scholarship to construct an apparatus for measurements of the ionosphere.[19] Essentially, thereafter, the ACRR was endowed with sufficient funds for the technological development of cathode ray direction-finding (CRDF) and, thereby, was able to anticipate Allies' needs significantly.

It was precisely the somewhat generalized perception that Canada was pioneering the use of CRDF technology which facilitated research on atmospherics performed in the same vein as that of the most advanced European centres to achieve a prominent position in the ACRR in the 1930s. The path had been initiated by General A.G.L. McNaughton (NRC's president from 1935) and Steel, who had jointly patented a prototype direction-finder in Canada

[19]McLeish (1940).

circa 1923.[20] By the time of the Imperial Economic Conference in 1930, the British RRB had confided its knowledge of CRDF to Canadian colleagues. As a result of a cooperation agreement with the NRC in the study of atmospherics, a set of equipment was constructed at their Slough workshops for the Ottawa Laboratory.[21] Even the new and first director of the Radio Section of the NRC, John Tasker Henderson, had a marked profile in the field. He had been trained under Professor Appleton at King's College, London, where he had received his PhD in 1932 (on the transients induced in radio receivers by lighting flashes),[22] and then at Slough. There he had become familiar with the technicalities in direction-finding and with the recording of atmospherics.[23]

The influence of the British school in sustaining the programme on atmospherics in Canada is beyond any doubt.[24] In the reports and dispatches that proliferated in the ARCC between 1934 and 1936 (the year in which the second set was set up in Forrest, Manitoba), the correlation between atmospherics and meteorology in the direction indicated by Watson-Watt, was described as solid and precise enough to justify the contracting of an engineer and two technicians for the systematic observations that were being effected.[25] However, in the desire to link atmospherics with weather forecasting, and, thereby, to gratify the Meteorological Service and the department of aeronautics, there were inevitable risks.[26] One menace resided in the disparity of meteorological conditions. While England's insular position led to frequently disturbed weather (and, thence, the tendency to associate atmospherics with the numerous existing lighting flashes), Canada's propitiated large air masses (and, thence, the

[20]Here also these are clear signs of a peripheral science relegated . Apparently, the British had been unaware of their work. For more details, see NAC [MG 30 B157, vol. 11/8], 'The Cathode Ray Direction Finder History, 1924–64'.

[21]NAC [MG 30 B157, vol. 9/6], 'The Study of Atmospherics', by W.A. Steel.

[22]NAC [MG 30 B157, vol. 1/19], 'On the Sudden Changes in the Earth's Electric Field', by J.T. Henderson, 1932, Ph Diss, University of London.

[23]Autobiographical notes can be found in NAC [MG 30 B157, vol. 1/8], 'J.T. Henderson to H.M. Tory, President of the National Research Laboratories, October 7th, 1932'. See also NAC [MG 30 B157, vol. 1/2], 'J.T. Henderson's Obituaries, including that from *The Proceedings of the Royal Society of Canada*, 1983, 21(4)'.

[24]Henderson (1935).

[25]See, for instance, NAC [MG 30 B157, vol. 7/15], 'Reports of the Associate Committee of Radio Research to the National Research Council, April 1934; 7th January 1935; and 1st June 1935'.

[26]Here it is worthwhile remembering the fact that while British observers concluded atmospherics were caused by lighting flashes, a school of thought headed by M.R. Bureau in France pointed to a local origin. See NAC [MG 30 B157, vol. 1/37], 'Direction Finding of Atmospherics'; and Bureau (1926).

tendency to associate atmospherics with cold fronts requiring no full-scale thunderstorms).[27] It is hard to appraise the repercussion of this contrast on the tone of the ethos and methods with which the phenomenon was pursued in Canada. But the truth is that by 1937 Henderson felt a greater proclivity towards placing the emphasis on hurricane disturbances, which had the advantage of having been exhaustively studied in the neighbouring country, notably at the Universities of Puerto Rico (by G.W. Kenrick) and of Florida (J. Weil and S.P. Sashoff).[28]

The sustained exchange of correspondence between Henderson and the British and American leaders in the field shows how engaged the NRC was with the pursuit of precision in forecasting.[29] Thus, as the authority and exigencies of the committee of aeronautics augmented from the late 1930s (see Table 7 figures), the might of official prediction, based upon the reliability and accuracy of an alleged correlation, assumed even greater prominence. But the hopes and expectations of the council's members did not come to fruition.[30] Atmospherics appeared to be of little help for prediction purposes. This miscarriage was doubly significant, for if it evinced the scant usefulness of what came to be accepted as the paradigm of the successful radio application (especially in virtually uninhabited areas as Canada's), it also confirmed the serious reservations which most meteorologists, such as John Patterson, the Director of the Meteorological Service, had expressed.[31] Not only was such correlation not found but also and perhaps more decisively the faint intimations of recurrent patterns in the evolvement of hurricanes, associated with atmospherics, made the almost exclusive expenditure in the programme on observations hard

[27]The question is expressed very clearly in NAC [MG 30 B157, vol. 4/1], 'Henderson to J. Patterson, 17th March 1936'.

[28]NAC [RG 77, vol. 172/ File 45-2-2], 'Meeting of the I.R.E. (New York) and visit to Washington for Demonstration of C.R.D.F. for Aircraft, by J.T. Henderson, 8th June 1937'. See also J. Weil & W. Mason, 1936. The Locating of Tropical Storms by Means of Associated Static. *University of Florida Engineering Experiment Station, Bulletin,* 3.

[29]The exchange of missives includes Watson-Watt, Appleton, Lutkin, and the abovementioned, American professors but also Australian personalities such as Laby and Lightfoot. See NAC [MG 30 B157, vol. 4/1], 'Correspondence on Atmospherics'. For the ample correspondence with Americans, NAC [RG 77, vol. 172/ File 45-2-5], 'C.R.D.F. for atmospherics'.

[30]The sense of disappointment is patent in such statements as those of 'J.T. Henderson to R.L. Smith-Rose, 5th January 1939', and 'J.T. Henderson to J. Patterson, 13th February 1939', in NAC [MG 30 B157, vol. 4/1].

[31]For Canadian meteorologists, radio potentiality was a dilemma, not so much in connection with its value to broadcast forecasts, as with its service for weather mapping. See Thomas (1996, pp. 64–65, 153–8).

to defend or afford. If the notion of atmospherics as indicators or predictors of lighting flashes had once been regarded as alluring to the forecaster's practice, that property had became unsubstantial by mid-1938.[32] In Canada at least, this allowed defence systems for aircraft (rather than for forecast) to progress on the eve of the war.

The disbandment of the ACRR in 1938 and the ensuing formation of two subcommittees (on 'cathode ray compass for use in aircraft' and on 'marine direction-finder') was a gauge of the preoccupation for safety in those fields. In both subcommittees, the new research projects embraced defence and commercial applications of CRDF based upon the experience and expertise that had been harvested in the abortive atmospherics schedule. The most illustrative exponent of this precedent was the memorandum on the use of CRDF in aircraft that Henderson submitted in 1936 to McNaughton and Boyle.[33] In perceiving the advantages that an adaptation of the erstwhile cathode ray compass would offer for the safety of air and marine navigation, newly appointed President McNaughton showed himself to be far more open-minded and far-sighted that his predecessors. He soon sensed that NRC's regeneration lay in the introduction of 'wartime specialities' such as aviation medicine.[34] In the event, the funds for CRDF research far surpassed the total of all investment hitherto made in radio (from $267 in 1934–35 to over $27,000 in 1939, becoming the second-best funded committee). By the outbreak of World War II, an active core of four engineers and seven technicians, headed by Henderson, had been firmly consolidated for research work,[35] and CRDF development subsumed not only long wave essential to marine but also shortwave and its applicability in aircraft.[36]

[32]For an instructive contemporary account of the Canadian experience in atmospherics, see NAC [MG 30 B157, vol. 1/21], 'Brief Notes on the Historical Development of the C.R.D.F. in Canada, by J.T. Henderson'; and NAC [MG 30 B157, vol. 1/53], 'Draft of Paper Giving the History of the Atmospherics Programme Conducted from 1937–39', 1939.

[33]NAC [RG 77, vol. 172/45-2-2], 'Memorandum to Dr. R.W. Boyle re C.R.D.F. in Aircraft, September 14, 1936'.

[34]Phillipson (1991, p. 179); Swettenham (1968, pp. 319–43).

[35]The magnitude of the venture should not be overestimated: most staff (all except two) were working under temporary contracts, and their salaries, because of NRC's insolvency, had to be covered by funds from the Army and the Department of Transport. See NAC [MG 30 B157, vol. 12/24] 'Radio History. Part One: Draft of the War History of the Radio Branch; ed. by J.F. Breese, n.d.', pp. 1–2.

[36]NAC [MG 30 B157, vol. 7/15], 'Report of Progress on the Work of the Radio Research Committee, 11th December 1939'; 'Annual Report of the Radio Research Committee for the Year 1940–41, by J.T. Henderson, 2nd May 1941'.

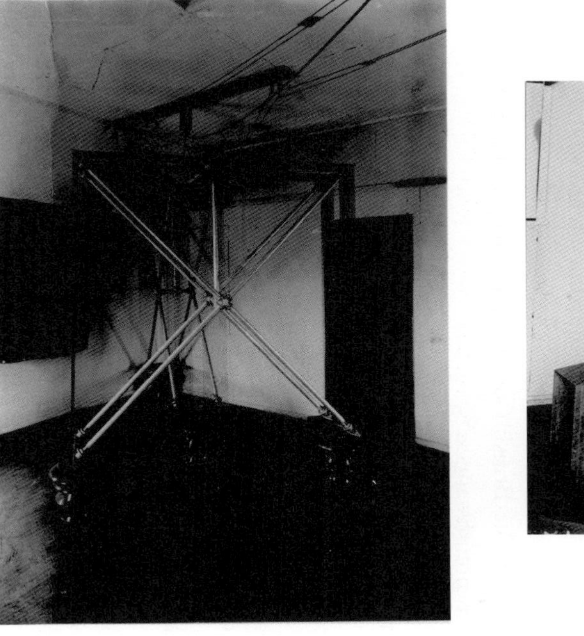

(a)

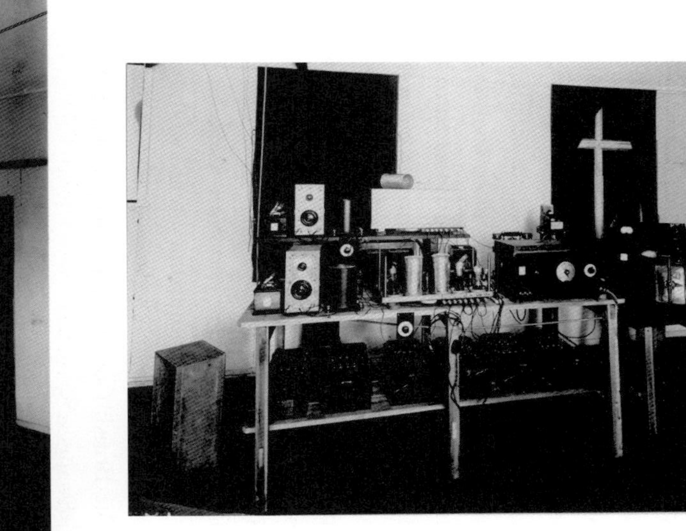

(b)

Fig. 23: (a) 'All Metal' Frame Aerials for Cathode-Ray Direction Finder
(b) Cathode-Ray Direction Finder operating on 10 Kc/s. Courtesy of the Science & Technology Facilities Council, Rutherford Appleton Laboratory.

Although the success of the restructuring owed much to NRC's privileged position in the field of CRDF (for thanks to McNaughton's and Steel's 1923 patent there was no fear of rights infringement),[37] it is equally significant to note the anxiety for systems for detection that had taken hold of the committee of aeronautics from the mid-1930s.[38] The confidence of public administrators in CRDF's potentiality was generalized, and such confidence, which was often confounded with confidentiality, was reasonable, not to say justified. On the brink of war, Henderson boasted of having an available regiment of 300 qualified men who 'would be capable of undertaking R.D.F. work at once'.[39] Clearly, a distinctive ambiance of optimism and self-confidence imbued in the circles of the council.

The successful assignment of radar development to Canada, and the subsequent delegation of components manufacturing, owe much to that aura.[40] In a remarkable memorandum, which discloses the exceptional role that Canada might exercise in allied defence by reason of her proximity to the US, safety from enemy bombing, and accessibility to US technical help, the secretary of the Royal Society of London, A.V. Hill, exhorted allies earnestly to trust in the Canadian capability in a grandiosely laudatory tone: 'Everything I saw and heard [in Canada] convinced me that there had been a grave lack of imagination and foresight on our part in failing to make full use of the excellent facilities and personnel available in Canada.'[41] This was in May 1940. Three months later, Sir Henry Tizard, the British Commissioner for liaison with Canada, was beguiled by similar circumstances.[42]

By 1943, the Radio Branch of the NRC would become the symbol of government commitment to allied defence.[43] Moreover, there are well-founded reasons to think that Canadian science, in such fields as radio and ionospheric

[37]NAC [RG 77, vol. 172/45-2-2], 'Meeting re C.R.D.F. for Aircraft, November 7, 1936'.

[38]The increasing need for the location of aircraft in flight in this department is expressed strongly in 'A.G.L. McNaughton to J.W. Wilson (Controller of Civil Aviation, Department of Transport), 12th Nov. 1936', NAC [RG 77, vol. 172/45-2-2].

[39]NAUK, [AVIA 22/2286], 'R.D.F. in Canada and the United States, by A.V. Hill'—also quoted in Middleton (1981, p. 20).

[40]On radar work in Canada, a valuable source for the information it contains is Middleton (1981, esp. Chapters 4 and 5). See also Zimmerman (1986, 1988).

[41]NAUK, [AVIA 22/2286], 'Research and Development for War Purposes in Canada, by A.V. Hill'—also quoted in Middleton (1981, p. 20).

[42]R.W. Clark, 1965. *Tizard.* Cambridge, Mass.: MIT Press, 259–64.

[43]The intensity of a real sense of commitment to this branch is reflected in such data as those of staff size (over 200 employees in 1943) and budget ($59,800 in 1940–41). To explain this explosion, see Middleton (1981, pp. 28–30).

physics, came of age during the war. In the light of these achievements, the pre-war CRDF experience in Canada seems to show how small size and limited resources were no impediment to an endeavour of independent innovation that concentrated on one specific topic to prove to be determinant for success.[44] On a defence and government level at least, the risk that this research strategy involved seems not to have deterred public administrators or the NRC's authorities. In this respect, they were fully conscious of the advantages of acting as a purveyor of first-rate technology to allied governments on one specific issue while they in turn obtained confidential technical knowledge from those countries in diverse fields.[45] This was particularly evident during the implementation of the CRDF enterprise between 1938 and 1940 which, despite all its shortcomings and its modest dimension, had the *special* virtue of being credible and solid enough to persuade British envoys, through a balanced organizational exercise of human and logistic resources in three critical agents as were universities, NRC laboratories, and manufacturing firms.

Manufacture and technological dependence

As outlined in the foregoing chapters, radio industry comprises three interrelated applications: navigation, point-to-point communication, and broadcasting. These involve three generally interrelated processes: assembling, manufacturing, and innovation. The burgeoning interdependence between these applications and the processes evolved determine the disparate patterns in the emergence and sustenance of radio R&D in the Dominions in the interwar years. Nourishing this interrelationship away from self-sufficiency and self-reliance had been a bidirectional process with manifold beneficiaries in countries such as Australia. While the private firms endeavoured to cultivate a terrain as technologically self-contained as possible, the radio engineers and physicists drew upon the *impetus* of a blossoming industry to secure technical and financial resources for research.[46]

[44]See, for example, J.T. Henderson, H.R. Smyth, & J.W. Bell, 'A Cathode Ray Direction Finder for Long Waves', presented to the URSI Meeting, Washington, 29 April, 1939.

[45]Henderson's visit to the Bureau of Standards and RCA in Washington, and to the Bell Telephone Labs and Airplane and Marine Direction Finding Corporation in New York, met those requirements. See NAC [RG 77, vol. 172/45-2-2], 'Meeting of I.R.E. and Visit to Washington for Demonstration of CRDF for Aircraft', by J.T. Henderson, 8th June 1937.

[46]For an excellent and admirably documented study of the development and magnitude of the early broadcasting industry in Canada, see Mary Vipond, 1992. *Listening In: The First Decade of Canadian Broadcasting, 1922–1932*. Montreal: McGill-Queen's University Press, esp. Chapter 2.

In Canada, however, this relationship evolved from a technological dependence. The clearest exponent of this circumstance lies in the provenance of the technology that was used in early broadcasting and coastal stations, and in long-distance communications. After World War I, the dominant pattern was for the radio facilities to marine navigation and to communications over land to be controlled and operated by the Department of Marine and Fisheries, while for the equipment to be purveyed and manufactured by foreign companies.[47] The 47 coastal stations belonging to the government in 1919, for example, achieved their modernization through the installation of radio beacons supplied by the British Marconi's.[48] Likewise, the six stations established in the Northwest Territories and Northern Saskatchewan were operated by the Department of the Interior (via the Royal Canadian Corps of Signals), while their technical material was provided by subsidiary organizations.[49] But even the outstanding community of radio amateurs assembled most of their apparatus from household items and bits and pieces from the local hardware firms that had close relations with the American electrical giants.[50]

The penetration of foreign investors into Canadian radio manufacturing was characterized by the diversity of forms and by the strong influence of the US. Before the creation of the RCA in 1918 at the request of the US Navy, American Marconi's took pride in its dominance in the communications field throughout the continent.[51] But, thereafter, with GE as the main manufacturer and RCA as Marconi's strongest rival, it became ineluctable for the big companies to redefine the relations of their subsidiaries in third countries. Thus, Marconi's and Canadian GE went on in 1921 to found Canadian Marconi Co.

[47]For postwar radio aids to marine navigation, see Department of the Naval Service, 1920. *Annual Report for 1918–19.* Ottawa: King's Printer, 48; Department of Marine and Fisheries, 1924. *Annual Report for 1922–23.* Ottawa: King's Printer, 140.

[48]Babaian (1992, p. 34). For an official list of radio stations in Canada, see *Wireless World and Radio Review*, 1922–23, 11, 279.

[49]Royal Canadian Corps of Signals, 1960. *A Short History of the Northwest Territories and Yukon Radio System, RC Signs.* Edmonton, 2–16.

[50]On the crucial role played by Canadian amateurs in connecting isolated regions through a reliable long- distance communication system, see Saskatoon Amateur Radio Club, 1968. *From Spark to Space. The Story of Amateur Radio in Canada.* Saskatoon, 15–25; C.B. DeSoto (1936, pp. 29–30). See also NAC [RG 97/123, file 4003-4-1], 'Canadian Amateurs Make their Views Known Prior to International Conference 1927'.

[51]This is well analysed in Alain Canuel, 1986. 'La présence de l'Impérialisme dans les Débuts de la Radiophonie au Canada, 1900–1928. *Jour. Can. St.*, 20(4), 45–76, pp. 46–49. See also W. Hopkins, *History of Canadian Marconi Company.* Montreal: CMC, n.d.

(CMC).[52] Simultaneously, there arose negotiations to reproduce in Canada the larger pool already extant in the US. These precedents prove that it was much more economical (and therefore advantageous) for them to set up manufacturing factories on Canadian soil, often adopting cross-licensing agreements and sharing some local capital, through a varied gamut of ownership arrangements.[53] Strikingly, these seem to have arisen a special allure among Canadian shareholders to exploit not only the patents but also the illustrious names of the parent firms. This is the case, for instance, with Westinghouse of Canada (subsidiary of the US firm), CGE (controlled by GE), Bell Canada (AT&T owned 23% of its stock), Northern Electric (43% by Western Electric), CMC and International Western Electric, all of which subscribed to the Canadian Radio Licence Agreement in 1923, whereby they shared out radio patents and fields of interest.[54] Similarly, it was a well-known fact that British Marconi's and CGE founded in 1922 a jointly-owned company to produce radio tubes in Canada, the Radio Valve Company RVC, in spite of being pompously heralded as a national enterprise;[55] or that the sumptuous inauguration of beam stations at Drummondville and Yamachiche by CMC (for communication with similar stations erected in England),[56] was in reality, despite all the patriotic connotations of the event, a prolongation of Marconi's commercial influence on Canadian territory.[57]

The allocation, with the liberal government's connivance, of manufacturing rights and patent control to a select group of firms endowed the country with a radio industry that was competitive, but technologically derivative and organizationally oligopolistic. It gave rise moreover to a disconcerting situation in which the patent holders' satisfaction due to proceeds and dividends intermingled with the fears of small manufacturers of a possible litigation and prosecution. In this situation, the former scarcely helped to allay any intimidation that the latter may have felt. Neither did the formation of Canadian

[52]NAC [MG28 III 72/86, file 19-6], 'Patent Agreements'. See also Vipond (1992, p. 27) and Aitken (1985, pp. 392–400).

[53]Wilkins (1974, p. 71).

[54]NAC [MG28 III 72/86, file 19-1], 'Canadian Radio Licence Agreement, 12 March 1923'.

[55]NAC [MG28 III 72/6, file 16], 'The C.M.C. Patent Situation, Montreal, October 24, 1956', p. 18; NAC [MG28 III 72/151], 'History of Radio Valve Company, 1922–5'.

[56]See, for example, the editorial 'The Inauguration of the Canadian Beam Service', *EW & WE*, 1926, 3, 715–16; 'The Canadian Beam Service', *The Electrician*, 1926, 97, 474, and 498; and 'Wireless Development in Canada', *Engineer*, 1930, 150, 561–63.

[57]The policy of Marconi's is expressed clearly in the CMC annual reports: NAC [MG28 III 72/6, file 19], 'CMC Annual Reports', 20 Sept 1927; 12 June 1928; 15 July 1929; 17 June 1930. See also Babaian (1992, pp. 65–6).

Radio Patents Ltd (CRPL) in 1926 contribute to it. This was constituted by the four major members of the foregoing pool along with Rogers-Majestic (the only Canadian-owned). CRPL issued manufacturing licences to other radio firms, provided that these utilized the patents controlled by the signatories. The sense of monopolistic domination is clearly manifest in the reports of auditors on the size and production of the giant manufacturers.[58] The five biggest companies included in the non-royalty group (CMC, CGE, Canadian Westinghouse, Northern Electric, Rogers-Majestic, along with RCA Victor) manufactured between 1928 and 1932 'only' a 43% share of the domestic market, while holding a practically unbeatable position before any possible patent litigation.[59] The small firms, by contrast, would pay royalties at the rate of 10% of the selling price, besides diverse extra remittances. Fixed minimum payments and additional quantities 'in settlement of past infringements' were all measures with which the agency CRPL felt it could face up to the high manufacturing offer, but *at the cost of perpetuating* technological subservience. The fact that most of the 20 firms licensed by CRPL by mid-1931 were American subsidiaries would confirm that pattern. Likewise, the high revenues procured in 1932 in current royalties (over $2m), precisely at a time in which the licensed firms could hardly survive the Depression, would seem to corroborate the profitability of the venture.

Certainly, the venture was profitable but not without stricture. By the early 1930s, the displays of uneasiness among small entrepreneurship did become more frequent. On account of some public outcries against the parent firms' policy in Canada, Professor K.W. Taylor of McMaster University remitted to the minister of labour a report in which he impeached the oligopolistic consortium for having operated 'against the interest of the public'.[60] It was effectively with a good dose of nationalistic sentiment that his accusations on the abuse of patent control to lessen competition and on the levy of 'onerous' royalties on small firms were cast. The submission to high political spheres of Taylor's report was the tip of an iceberg of resentment which was to be accentuated in the next years with the publication of numerous public protests against CRPL's alleged strategy to 'restrict production and to regulate prices'. Indeed, it was only as the crisis of small manufacturers intensified (of 32 companies operating between 1927 and 1936, only 12 remained in 1937) and the sense of

[58]NAC [RG 79/46, file 104], Report of the Auditors, Exhibit 13, 'Patents: Sales of Radio Sets by Years 1928–1933 Showing Division between Royalty Paying and Non-Paying Licenses'.

[59]Vipond (1992, p. 29).

[60]NAC [MG 26k, M-987], 'Taylor Memorandum', 32029. See also Vipond (1992, p. 31).

American invasion was exacerbated (all the patent holders excepting Marconi's and Rogers-Majestic were American)[61] that occasional remonstrations ended in pervasive discontentment.[62] Significantly, what had begun at the turn of the century as a concession before British interests involving Marconi's monopoly had assumed, over three decades of British enfeeblement, the character of a strong oligopoly subordinate to a great extent to the American 'radio trust'. Ironically, what then was viewed as British political dominance was now viewed as the oblique, non-political, dominance of American economic power.[63]

The technological dependence—and therefore the weakness—of Canadian radio manufacturing industry was a consequence of the strength of the oligopoly (see Tables 8 and 9). Their members sought risk neither in aspects of development and instrumentation, which had already been to varying degrees carried out by their parent companies, nor in engineering processes going beyond those occasionally required by the Canadian specificness. Or, in other words, there was no necessity to alter the policy of capital investment focused exclusively on assembly and production and the strategic control of patents elicited through cross-licensing agreements upon which the strength of its hegemonic position rested.[64] It is likely that the stability provided by the monopolistic condition injected a high dose of contentedness and continence among shareholders, dissuading them from independent innovation and impelling them towards moderate positions in conformity with the conservative nature of their activities. That in 1937 the radio sector expended on engineering and development a mere 2.1% of the value of radio sets in terms of sales or 2.5% of the total capital invested may be a clear indicator that such policy scarcely proved to be alluring to investors.[65]

[61]The dimension of the investment and management of American capital in the Canadian radio manufacturing industry emerges clearly from the following data. In 1932, 60.1% of the funds invested were American, 33.8% Canadian, and 4.8 per cent British. See: Vipond (1992, p. 33).

[62]For an exponent of that indignation, see NAC [MG28 III72/155, File 9], 'Radio Patents. Extract from Hansard, January 31, 1939', by W.K. Esling. For a strong defence of the current patent system, see A.E. MacRae, 'Some Aspects of Patent Protection', *The Queen's Review*, November 1938. For other testimonies, see NAC [MG28 III72/155], File 'Canadian Radio-Patents Limited, 1938. Philco Products Limited'.

[63]Canuel (1986, p. 54).

[64]CRPL's strategy of patent control is expressed clearly in the licence agreements and assignments of patents deposited in NAC [MG 28 III 72/152 and 153], 'Canadian Radio Patent Limited, CRPL'.

[65]NAC [MG 30 B157/21, File 13], Radio Manufacturers Association of Canada, 'Brief of the Radio Manufacturers Association of Canada, for Presentation to the Tariff Board, 1937', on p. 19.

Table 8 Number of sets sold in Canada, by composition, 1929 and 1930

	1929		1930	
	No. Sold	% of Total	No. Sold	% of Total
Wholly made in Canada	74,312	38	83,755	37
Partly made in Canada, partly assembled from foreign parts	20,286	11	25,140	11
Wholly assembled from foreign parts,chassis imported complete	54,551	28	36,798	16
	44,957	23	79,688	36
Total	194,086	100	225,399	100

Source: M. Vipond, *Listening* In, op. cit., p. 35

Table 9 Estimated value of materials or parts purchased by Canadian radio firms during 1937

	Set manufacture ($)	Tube manufacture ($)	Parts manufacture ($)	Total ($)	% of Total
Purchased and manufactured in Canada	3,163,315	107,873	158,841	3,429,729	65.7
Manufactured in the USA	1,272,019	276,805	186,828	1,735,562	33.2
Manufactured in other countries	47,804		6,505	54,309	1.1

Source: Radio Manufacturers Association of Canada, op. cit., p. 10.

At this stage, it is necessary to examine in detail a characteristic that seems, in principle, to have endowed the country with the imprint of a singularity, the fabric of small industry.[66] In reality, this feature defines the constraints to

[66]Most of the production was concentrated in non-Canadian- owned plants set up in Montreal, Toronto, and Hamilton. In 1937, radio set division companies included CGE, Toronto; CMC, Montreal; C. Westinghouse, Hamilton Ontario; Northern Electric, Montreal; RCA Victor, Montreal; Rogers-Majestic Corp., Toronto; Dominion Electrohome Industries Ltd., Kitchener, Ontario; Sparton of Canada, London Ontario; Stewart-Warner-Alemite Corp., Belleville Ontario; and Stromberg-Carlson Telephone, Toronto. Again, the companies specializing in radio parts (cabinet and accessory division) were Aerovox Canada, Hamilton Ontario; Canada Wire & Cable, Toronto; CGE; CMC; C. National Carbon, Toronto; C. Transformer, Kitchener Ontario; C. Westinghouse; Continental Carbon, Toronto; Diamond State Fibre,

which Canadian radio firms (and, in consequence, their policies and pursuits) were subject, but at the same time explains the relative success of the sector. Until 1927, the Canadian radio sets market was primarily dominated by imports. But this pattern was drastically altered during the next five years (61% in 1927, 36% in 1930, 11% in 1931, and 0.4% in 1932).[67] The revival of the national radio industry (nourished substantially by American capital) in the late 1930s, after the sharp crisis of the Depression, was comparable with that of the leading countries in the field (Table 10). In 1937, Canada ranked fourth in the world list of number of persons per set, and fifth in that of radio sets in use.[68] However, this was a revival sustained despite, rather than by, the strength of the oligopoly. In fact, it was effected by and because of an industrial fabric that specialized in the assembly of receivers and transmitters (to a significant degree, of imported parts), rather than in the manufacture of components. The injection of foreign capital and the production in the big companies were all-important to the expansion. But the impelling force lay in small industry and in the rich fabric of purveyance, distribution, and retail trade upon which it drew. Radio subsidiary strategy was predicated basically upon procuring a share of market, materials (both raw and partially fabricated) from various other national industries, rather than from their parent companies abroad,[69] and commercial bonds dealing in provinces in which they had been set up to do business.[70] Once secured a segment of market spectrum, subsidiaries would

Toronto; Erie Resistor, Toronto; Hammond Manuf., Guelph Ontario; Int. Resistance, Toronto; Andrew Malcolm Furniture, Kincardine Ontario; Natural Fibre, Toronto; Radio Condenser, Toronto; RVC, Toronto; RCA Victor, Montreal; Rogers-Majestic, Toronto; M. Slater, Hamilton; Stark Tube, Toronto; United-Carr Fastener, Hamilton; and White Radio, Hamilton. See Radio Manufacturers Association of Canada, op. cit., p. 1.

[67]These figures are deceptive, for a considerable part of the domestic manufacture merely consisted of assembling American-made parts. See Vipond (1992, p. 32).

[68]When the number of persons per set in Canada was estimated at six in 1937, the average was comparable with the corresponding ones for the USA (4), Denmark (5), and UK (5.3). Moreover, there were only four countries in that year which outstripped the figures for the number of radio sets in use in Canada (1.9 m units): the USA (33 m), Germany (8.5 m), UK (8.2 m), and France (3.9 m). See Radio Manufacturers Association of Canada, op. cit., p. 16.

[69]It was estimated that some 270 Canadian manufacturers other than those directly connected with the radio set makers supplied materials in 1937, and that 186 distributors and over 7,000 retailers were involved in the sector of radio. See Radio Manufacturers Association of Canada, op. cit., pp. 6–7.

[70]These bonds are mentioned clearly in the Radio Manufacturers Association of Canada, op. cit., Appendices A and B, as well as in the archival material of radio companies deposited in the NAC. See, for example, NAC [MG 28 III72/155], 'CMC—Philco Products Limited, November 24th 1938'.

Table 10 Production and sales of radio sets in Canada, 1925–1937

Year	Quantity	Selling value ($)	Growth (%)
1925	48,531	2,278,292	−1.1
1926	42,430	2,253,098	66.3
1927	47,500	3,748,622	50
1928	81,032	7,486,127	47.9
1929	150,050	15,604,145	81.2
1930	170,082	19,196,936	−3.4
1931	291,711	18,555,710	
1932	121,468	6,808,877	
1933	112,273	4,401,313	
1934	188,710	8,196,248	
1935	191,293	9,493,399	
1936	253,896	11,388,173	
1937	293,729	12,878,591	

Source: Radio Manufacturers Association of Canada, op. cit., p. 10.

Table 11 Sales of radio sets by provinces in Canada, 1937

Province	Number	% of Total
Maritimes	19,046	7.2
Quebec	49,306	18.5
Ontario	115,234	43.4
Manitoba	28,191	10.6
Saskatchewan	13,125	4.8
Alberta	20,738	7.9
British Columbia	20,215	7.6
Total	265,855	100

Source: 'Production and Sales of Radio Receiving Sets. Fourth Quarter, 1937, Dominion Bureau of Statistics'—quoted in ibid., p. 14.

impose, often with little preoccupation for patents, price, or innovation, the technology that they would bring with them to the assembly and manufacture of receivers. In most cases this technology had already been developed and commercialized in the neighbouring country.

The foregoing data show how high productivity and market saturation could triumph in a sector contingent upon foreign technology that made a firm gamble on independent innovation that seemed not only expendable but also counter-productive. Part of the explanation lies in the subsidiary nature of the sector. For it was the automatic absorption of imported technological knowledge, especially with the installation of national plants, that marked it with the character of small entrepreneurs more than content with moderate but exigent profits, which they regarded as the reward of industrious assembly and manufacture. This was particularly perceivable during the recovery following the Depression between 1932 and 1939 that had the determinant but, in the light of other Dominions' experience, pernicious effect of anaesthetizing the modest innovator spirit and thereby of augmenting the disparities in innovation which seem to have distanced the Canadian radio companies so conspicuously from the Australian AWA—an exemplar of venture in the pursuance of technological independence.[71]

Broadcasting and Canadian idiosyncrasy

In a mode that is not exclusive to the Canadian experience, the circumstances in which radio manufacturing burgeoned are not reduced to those of a mere response to the demand of receiver apparatus. In analysing the economic expansion of the radio industry, other aspects also must be taken into consideration. The development of broadcasting stations, the existing public radio configuration, and size, coverage, and population issues, all conditioned the potentialities of market.[72] The specific nature of the interrelationship between manufacture and broadcasting in Canada makes it all the more comprehensible (perhaps justifiable?) that investments in R&D were as limited as they were. As will be seen, these limitations are in great part attributable to the powerful influence of American telecommunications giants, who knew how to exploit the weaknesses in the Canadian market. But this was not the sole factor.[73]

As the postwar rash of enthusiasm materialized through the installation of early broadcasting stations, any American capital expansion into the Canadian

[71]On Canadian technological dependence, see,for example, John Vardalas, 2001. *The Computer Revolution in Canada: Building National Technological Competence.* Cambridge, Mass.: MIT Press. See also its review by E.C. Jones-Imhotep in *Technology and Culture*, 2004, 45, 659–61.

[72]Normally, increasing demand for radio receivers, closely linked to increasing investment in broadcasting, would lead to an increasing rate of development and innovation. See the case of radio tuning: Harrison (1979).

[73]There is abundant literature on early broadcasting history in Canada. For an emphasis on public broadcasting and regulation, see Peers (1969, pp. 1–191), Weir (1965, pp. 1–204),Prang (1965), Vipond (1992, pp. 43–53), and Canuel (1986). For perspectives on communication issues, Lorimer & Wilson (1988), which includes Robert E. Babe, 'Emergence and Development of Canadian Communication: Dispelling the Myths', 58–79, and Jean McNulty, 'Technology and Nation-Building in Canadian Broadcasting', 175–98. For the role of private broadcasters, see Nolan (1989).

market was perceived as an increasingly grave menace to cultural sovereignty. In fact, even before early radio programmes had evinced the imminent threats to indigenous mass culture, fears of invasion had been expressed.[74] Radio became a *sine qua non* in Canadian–American relations, as much in connection with industry as with culture.[75] One can see this duality in the three types of organization which promoted Canadian broadcasting in its earliest stage: on the one hand, newspapers (the most engaged in the early 1920s); and on the other, retailers (mainly small electrical stores), and big electrical companies (those who previously controlled the communications field).[76]

The early success and subsequent languishment of newspapers, probably the most vehement defenders of the cause of fomenting autochthonous culture in broadcasting, is clearly illustrative of what can be described as a mark of American domination. Since the introduction of broadcasting licenses in April 1922, even the businessmen least enthusiastic about the revolutionary character of broadcasting regarded radio as a natural extension of, and additional component for, their expansion in the world of communications. Their participation, in fact, owed much to self-promotion and publicity, not importation or subservience.[77] The 14 stations licensed by the Department of the Naval Service, among them the Toronto *Star* (CFCA), the Vancouver *Daily Province* (CKCD), and the Winnipeg *Tribune* (CJNC) promised to offer the public a service combining culture with entertainment and news. For the first years, they became the 'most active and best-financed'.[78] But, by 1932, only two of the eight publishers still owning stations, the Montreal *La Presse* and London *Free Press*, would show enough will to expend a great part of their modest profits upon the installation of powerful but costly transmitters. In this respect, the advent of advertising-financed broadcasting, sharply rising costs, competitiveness from American stations, and listener preference for American frequencies, all made a sustained substantial investment vital for competition to appear inevitable, but unluckily also incompatible with their solvency.

Retailing incursion, the second broadcaster of our troika, was a practical response to the novelty of radio. Electrical retailers produced programmes for

[74]The fears of American imperialism are conveyed in an increasing number of nationalist discourses by figures such as Henri Bourassa, Augustin Frigon, and Sir Henry Thornton. See, for example, Bourassa (1926) and Frigon (1929).

[75]Spry (1935, p. 107).

[76]A.L. Neal, 1941. 'Development of Radio Communication in Canada'. *Canadian Geographical Journal*, 22, 164–91.

[77]Jome (1925, pp. 240–41).

[78]Vipond (1992, p. 44).

the owners of small shops and department stores in order to allure potential purchasers of receivers. While the profusion of insufficiently equipped stations—up to 16 in the 1922 license list—transmitting only during opening hours, and often locally, contributed to the advancement of broadcasting in the earliest stages, their role was nonetheless marginal and geographically limited. As broadcasting augmented in size, power, and capital invested, retailers' engagement diminished in intensity and conviction. By the early 1930s, their presence was almost negligible. Yet, it permitted a number of small and isolated centres—such as Saskatoon, Port Arthur, and Fredericton—to enjoy services which otherwise would not have come to fruition.[79]

In the circumstances described, it is reasonable to think that telecommunications companies engaged in radio manufacture from its earliest times could have played a crucial role in broadcasting expansion. In fact, such firms as the five which figure in the 1922 licence list,[80] had entered the field either to expand the radio set market in which they manufactured or to safeguard other interests, as in the case of telephone companies.[81] The British-owned Marconi's, merging manufacture and point-to-point communication with broadcasting (through its pioneering station CFCF), is a good example of an integred business.[82] The domestically-owned Rogers-Majestic and its station CRFB in Toronto is another example, even if it did not incorporate long-distance communication. However, by the early 1930s, only these two companies maintained stations from which they obtained a modest flow of revenue and trade bonuses. Significantly, in the Canadian subsidiaries of the more diversified, powerful American giants, in which neither technology nor capitalization was an insurmountable obstacle, the average life of stations rarely went beyond a short period of probationary transmission. This fact was in stark contrast with the modus operandi of the parent firms themselves in the USA. Here, leading corporations such as GE (via the station WGY Schenectady) and

[79]Vipond (1992, p. 45–46).

[80]Although the August 1922 list includes 15 licences conferred on Canadian Westinghouse, Bell Telephone, Northern Electric, Eastern Telephone and Telegraph, Canadian Independent Telephone, and Marconi's, only five were in reality operating in 1922 (CFCF Montreal, CFCE Halifax, CFCB Vancouver, and CKCE and CHCB Toronto).

[81]On the Bell Telephone stations, see Babe (1990, pp. 202–03).

[82]The first experimental broadcasting station in Canada, Marconi's XWA, was set up in the William Street factory (Montreal) in September 1918. In addition to the manufacture of broadcast receivers, the Company specialized in marine direction- finders and in the installation of vacuum tube equipment in coastal stations and ships. See NAC [MG 28 III72/ 7, File 7], 'Marconi, the Company' by W.J. O'Brien; and Godfrey (1982).

Westinghouse (KDKA Pittsburg) countenanced the risk implicit in the capital injection into broadcasting that, ironically, was denied to their neighbours.[83]

It would be absurd to dismiss these actions as inconsequential or anecdotal. Clearly, there was no necessity or incentive to promote the conjunction of local programme production, substantial capital, and a policy directed to the erection of powerful stations upon which the success of broadcasting industry rested. The fact that in the late 1920s over 80% of the programmes listened to were American,[84] that 93% of high school students preferred to tune in American rather than Canadian frequencies,[85] that in 1925 a good number of the over 600 American stations (as opposed to barely 40 Canadians) were easily received in the south of the country,[86] and so on and so forth, all these facts point to a direct, steadfast engagement appearing, from the big companies' point of view, unnecessary, unjustified, and gratuitous.[87]

There are two reflections to make at this juncture. Firstly, as noted in the preceding section, it was the secondariness of the development of manufacture based on self-sufficiency and on innovation in an electrical sector that had been by far a branch-plant industry since the nineteenth century that significantly distinguished the rise of radio industry in Canada from that in countries such as the US and Britain. The Canadian subsidiaries of the American communications giants, albeit dominant and relatively buoyant, did practically nothing to preclude what with hindsight, and certainly from the Canadian contemporary standpoint, might seem an industrial and media invasion, for instance by financing programme production to further the *figura* of broadcaster as the facilitator of indigenous culture.[88] In this respect, however, it would be unfair to regard them as recanting the overt but still unconsummated calls for sovereignty.[89] Quite simply, they were exploiting the weaknesses of a

[83]L. White (1947, pp. 10–13); Barnouw (1978, pp. 14–20); Peers (1969, p. 8).

[84]John Herd Thompson & Stephen J. Randall, 1997. *Canada and the United States. Ambivalent Allies.* 2nd ed. London: University of Georgia Press, 121.

[85]F.H. Soward, 1938. 'What Radio Programs School Children Hear'. In H.F. Angus, *Canada and her Great Neighbor: Sociological Surveys of Opinions and Attitudes in Canada concerning the United States.* New Haven, 369.

[86]Edelgard E. Mahant & Graeme S. Mount, 1984. *An Introduction to Canadian-American Relations.* Ontario: Methuen Publishers, 138.

[87]This point is well analysed by Vipond (1992, p. 47).

[88]For a vivid account of the debate about the American influence on Canadian broadcasting, see S.M. Crean, 1976. *Who's Afraid of Canadian Culture?* Don Mills, Ont.: General Publishers.

[89]Such demands (and concerns) for cultural sovereignty proliferated in parallel with, and largely in consequence of, the process of legislation of the Canadian Radio Broadcasting Commission in 1932 and the ensuing creation of the Canadian Broadcasting Corporation in 1936. See, for

state of technological subservience that made Canada a fruitful market for her more powerful neighbours. It was a commercial market in which the security and oligopoly of manufacture were planned to produce more month-watering profits than the speculation and politicization of broadcasting. It was, moreover, a cultural market in which the north/south hierarchical structure of the radio industry tended to exacerbate, not to counterbalance, the 'north/south pull of continentalism'—*contra* the alleged historical centrality of communication media in nation-building.[90] And, finally, not only did the subsidiaries do little to avert, in the way indicated, such an invasion but the parent firms, with their strategic accumulation of powerful stations almost on border soil, helped to nurture and perpetuate in Canada a small or at best mid-sized broadcasting entrepreneurship.[91]

In the light of these reflections, it is even more necessary to examine in detail the confluence of factors that endowed Canada with a truly idiosyncratic character, the interconnection between population, geography, and coverage (Tables 12 and 13). For this concurrence seems to disclose that the smallness of most parts of the broadcasting business in Canada was not only a consequence of its dependent condition. The consolidation of broadcasting facilities in the late 1930s, in parallel with the revival in manufacturing radio after the Depression, was as laudable as that of the leading countries. In fact, there were 83 stations in 1937 vs. 77 in 1932, more than the total of Germany, the UK, and France put together. Estimates of radio sets in use were also reasonably sanguine (fifth in the world ranking in 1937). Likewise, the degree of market saturation was more than notable (six persons per set; 138,554 persons per station; 23,192 sets per station). Yet, it was a consolidation resting upon an uneven geographical distribution of stations and upon imbalanced statistics (44,000 sq. miles per station!). The concentration of transmitter stations along a southern strip being highly exposed to American wave invasion reflected

example, the nationalist tone with which the speeches of one of the founders of the Canadian Radio League are imbued, in G. Spry, 1929. 'One Nation, Two Cultures'. *Canadian Nation*, 1, 21; and G. Spry, 1931. 'The Case for Nationalized Broadcasting'. *Queen's Quarterly*, 38, 151–69.

[90]According to the standard viewpoint, telecommunication systems, services, and industries had been essential for counterbalancing the strong north/south pull of continentalism existing in Canada in the twentieth century, and thereby instrumental in forging Canadian nationhood. For an analysis against this myth, see Babe (1988), who states that 'Canada persists despite, not because of, media of communication' (p. 59).

[91]The American influence on Canadian broadcasting industry is analysed in Canuel (1988, pp. 345–6). See also F.W. Peers, 1966. 'The Nationalist Dilemma in Canadian Broadcasting'. In P. Russell ed. *Nationalism in Canada*. Toronto: McGraw Hill, 252–71.

Table 12 Estimated figures of population, geographical size, and broadcasting facilities in 1937

Country	Population	Area in sq. miles	Pop. density person/sq. miles	Sets in use	Broadcasting stations
USA	130,000,000	3,737,421	34.7	33,000,000	735
Germany	63,000,000	180,970	348.1	8,500,000	38
UK	44,174,000	94,180	469	8,269,000	18
France	40,746,000	212,659	191.6	3,915,000	26
Canada	11,500,000	3,684,700	3.1	1,925,000	83

Source: Radio Manufacturers Association of Canada, op. cit., p. 16.

Table 13 Statistics derived from population, geographical size, and broadcasting data in 1937

Country	Persons per set	Persons per station	Sets per station	Sq. miles per station	Total power in kw
USA	3.9	176870	44897	5,084	2,309
Germany	7.4	1657894	223684	4,700	
UK	5.3	2454111	459388	5,200	
France	10.4	1567153	150576	8,200	
Canada	5.9	138554	23192	44,000	80*

* This figure corresponds to the total power of 71 Canadian stations in November 1937.

a marked north/south asymmetry.[92] More significant still, this consolidation reposed upon a technically very fragile toehold. For the provision of high-power stations (over 5 kw), which as a rule marked the degree of coverage adequacy in a country, was marginal in Canada (only 80 kw in 1937 vs. 2,309 kw in the US!). Her common denominator was short-range antennae (in 1930, for example, Chicago stations alone had five times the total power of all the Canadianones!).[93] The characteristic pattern in Canadian broadcasting consisted essentially in installing a low-power service in the major metropolitan centres, mainly by local entities (notably churches, universities, and grain

[92]Ontario concentrated 51% of stations and about 33% of the total population. The disparity in accessibility to radio broadcasts, as well as the location of the most powerful stations, are described in Vipond (1992, pp. 48–9).

[93]The sense of inadequacy in coverage was vigorously voiced in the 1932 Report of the House of Commons Special Committee on Radio Broadcasting (Report 1, 10th March 1932), p. 14.

companies) but also by public corporations (the Canadian National Railways[94] and the Manitoba Telephone System[95]) whose prime motivation was publicity and market protection. Thus, from a position of purely municipal interest, a myriad of small and mid-sized entrepreneurs would vie for an unduly limited domestic market, often with exiguous means to finance the high costs of super stations.

The previous figures also serve to illustrate how factors such as population (barely 11.5 m, 80 % of whom were living within 60 miles of the US border) and a vast territory (3.6 m sq. miles) could mould not only the nature of dependent broadcasting but also the few incursions into R&D. Peculiar broadcasting and reception conditions would require radio sets of a higher sensitivity and selectivity than those commercialized in other countries.[96] Here, however, entrepreneurs were to divert their modest readiness for risk-taking in innovation to lines of product development, rather than long-term applied research. It was especially so in those manufacturers having no foreign affiliations. These lighted upon the imperious but insidious situation of having to expend energy and resources upon 'long, arduous and expensive development works' which seem not to have affected the Canadian subsidiaries that benefited from and were aided by technological knowledge from their American parent companies.[97]

Ironically, while Canada may reasonably be regarded as a victim of what has come here to be depicted as a state of technological dependence, it also can sanguinely be seen as a beneficiary of knowledge transference. Indeed, even the most captious contemporary critics of the dependent character of the Canadian radio industry often emphasized the case for subsidiaries as an agency for technological transference.[98] For the staunchest defenders, advantages lay in technical assistance and diffusion, rather than in self-sufficiency. Canada's vital participation in the production of radar components during the war speaks eloquently of that irony of fate. All began in the summer of 1940, when *Tizard* mission's members urged 'to move as much research as

[94]Among the numerous studies on CNR's special and singular involvement in broadcasting development, see Collins (1977, pp. 214–17), Weir (1965), and Vipond (1992, pp. 50–1).

[95]The fullest treatment is in Vipond (1986).

[96]While American radio sets were relatively simple, the average receiver in Canada was a seven-tube set in 1931. See Radio Manufacturers Association of Canada, op. cit., p. 27.

[97]The point is strongly emphasized by Radio Manufacturers Association, op. cit., p. 28.

[98]At the other extreme, there were administrators and investors who had no doubt that the present degree of industrialization and standard of living owed much to American and British expertise and technical competence.

possible to Canada', particularly when 'it may be followed up by production of such equipment in Canada or the US'.[99] Almost simultaneously, Henderson strongly appealed to C.J. Mackenzie, the controversial new President of the NRC, for the commencement of 'RDF equipment' manufacture, for 'it could readily be constructed in Canada'.[100] As a result of these pressures, Research Enterprises Ltd. (REL), a Crown corporation recently created to manufacture optical glass and precision instruments, assumed responsibility for radar production.[101] By the conclusion of the war in 1946, it would come to be regarded as a fundamental ingredient in the radar supply to naval and air forces worldwide. In this honours list, the radio manufacturing fabric had much to commend it. The capability of the Canadian radio subsidiaries was so good by the standards required (certainly comparable to that of their parent firms in the US) that the Canadian government had resolved that the major portion of works be constructed by private firms, and the most secret parts, assembly, and calibration of apparatus by REL.[102] The maturity of that fabric proved to be, therefore, crucial.

Teaching and training for the mastery of radio waves

The first presentation of radio broadcasting in Canada was as glamorous and exciting as elsewhere. On 20 May 1920, Professor A.S. Eve of McGill's physics department arranged, as part of a demonstration of wartime inventions, a special concert for the fellows of the eminent Royal Society of Canada in Ottawa. The guest of honour was the vocalist Dorothy Lutton, who was singing in a room from Marconi's plant over one hundred miles away in Montreal. Despite the probationary nature of the event, the sentiment of excitement and receptiveness soon became manifest and pervasive. It was reflected not only

[99] H. Tizard, 'Future Liaison with the National Research Council of Canada', Memorandum—quoted in Middleton (1981, p. 22).

[100] Quoted in Zimmerman (1988, p. 124). For an accusation on REL's mismanagement on Mackenzie's part, see Zimmerman (1986, pp. 95–103).

[101] Indispensable contemporary sources on Research Enterprises are NAC [RG 28A, vol. 17], 'The Development of Research Enterprises, Ltd.', by W.E. Phillips, n.d.; and Kennedy (1950). See also Middleton (1981, pp. 41–5); and Zimmerman (1988).

[102] The government drew upon private firms' experience not only as competent manufacturers but also as a source of qualified personnel. In this respect, the participation of entrepreneurs and engineers from the radio industry in managing Research Enterprises was particularly notable. R.A. Hackbusch (vice-president of Stromberg Carlson Ltd.) as the manager of the Radio Division, K.A. MacKinnon (from the Canadian Broadcasting Corporation), and V.V.E. Ross and D.L. West (with a great industrial experience) are good examples. See NAC [MG30 B157, vol. 1/ File 56], 'Research Enterprises Limited, 6th Sep. 1940'.

by the numerous displays that followed the popular phenomenon but also by the proliferation of radio magazines, such as *Canadian Wireless* and *Radio News of Canada* (both Marconi's publications), *Radio, Wireless and Aviation News*, and *The Radio Bug*, and by the furore over radio manuals, such as those by A.H. Verrill, *The Home Radio: How to Make and Use It*, and A.C. Lescarboura, *Radio for Everybody*, both best-sellers in 1922.[103] Certainly, this spurt would be hardly understandable without the presence of a hitherto-unparalleled number of skilled radio amateurs. They had been released from military obligations after the war, and now would assume the role of technicians, popularizers, and apologists of the new art.

Under these circumstances, the necessity for technical education soon was to be regarded as imperative. It is true that the opportunities for employment in the sector were not too numerous yet, but demand would ascend in an arithmetical progression to the number of stations and radio sets sold. Training had been hitherto predominantly the function of Marconi's, while had inaugurated a Wireless School in Montreal in 1916 for the preparation of radio operators as a result of wartime requirements.[104] This had inculcated wireless principles into the whole generation of radio officials specialized in marine navigation and licensing, and had cemented the association between the prodigy of radio communication and that of a victorious country open to modernity.[105] Even better, it had helped to consecrate Montreal as the pre-eminent *ville des télécommunications*.[106] At the opposite extreme in the balance of the interface between technical education and employment, there lay the typical cases of young innovators proceeding to establish their own company, often with scant solid preparation in sciences or with no industrial experience. Perhaps the most remarkable example is E.S. (Ted) Rogers, the son of the founder of Standard Radio Manufacturing and a dexterous handyman in circuitry, who abandoned his engineering studies at the University of Toronto before graduating to found the most prominent Canadian-owned Company.[107] By 1925, he had elaborated his commercial version of the vacuum tubes whose

[103]See Vipond (1992, p. 22).

[104]The main architect of the establishment of the School was D.R.P. Coats, a graduate of the British School of Telegraphy in London. See C.M.C., 1952. *This is Canadian Marconi Co.: 50 Years of Progress*. Montreal; and W. Hopkins, *History of Canadian Marconi Company*, op. cit.

[105]M.K. MacLeod (1992).

[106]Alain Canuel, 1992. 'Les Télécommunications à Montréal entre 1846 et 1946'. *Scientia Canadensis*, 16, 5–24, p. 8.

[107]NAC [RG 79, vol. 50/File 104-11], 'Rogers-Majestic Corporation Ltd. Statement Under Reference 104 to the Tariff Board'. See McDougall (1995).

patent had been purchased from Westinghouse's in Philadelphia, and his Company successfully launched its flagship 'batteryless radio'.[108]

These early cases in radio training seem to show that, while some private firms grasped the atmosphere of opening that was occasioned by the emergence of a previously non-existent sector, universities experienced difficulties in following in their footsteps. Throughout the 1920s, in fact, further advances in technical education, albeit not imperceptible, were notably limited, and, due to the increasing demand in such an exigent and incipient industry, somewhat insufficient. It is true that in these years many broadcast exhibitions were arranged on academic campuses and that a good number of student Radio Clubs, such as that of Toronto in 1926, were formed. But enthusiasm for these ventures emanated from a small cadre of professors rather than from the institutions themselves. It was precisely at the University of Toronto, whose physics department was to dominate, together with McGill's, physical research in the interwar years, that one of the first regular courses in radiotelegraphy was offered from the 1924–25 academic year. The initial results were without doubt modest. For, although electrical and engineering physics students could enjoy a programme of two courses on the general foundations of radio communication, they received neither laboratory teaching nor workshop practice that was necessary if they were to aspire to posts of responsibility in the radio industry on an equal footing with those employees trained within the firm eager to be promoted. It was not until about 1930 that a small 'radio laboratory' equipped with a basic instrumentation permitted forth-year pupils to test measurements at radio frequencies and audio-frequency circuits. By the mid-1930s, in two courses of theory and laboratory, the principles and applications of radiotelegraphy and of thermionic tubes were taught to electrical and engineering physics students, respectively.[109] Its artificer was a former instructor of radiotelegraphy, Benjamin De Forest Bayly, who was very highly regarded and much sought after as a public lecturer on this subject.[110]

[108]Ted Rogers developed the world's first alternating current (A/C) tube which allowed radios to operate from regular household electricity rather than batteries. See Anthony (2000) and Vipond (1992, p. 29).

[109]For the contents and characteristics of the courses, see *Faculty of Applied Science and Engineering Calendar, 1936–37* (Toronto: University of Toronto), courses 147–48 (Radiotelegraphy); and 150–1 (Thermionic Tubes).

[110]In a university magazine, Bayly was described as an 'Instructor of Radio Telegraphy in the Radio Laboratory of the University'. See *Toronto Globe*, 11 October 1928, in UTA [Department of Graduate Records, A1973-0026/022(53)].

Conclusion

Let us conclude with a series of reflections in this quest for structural factors which actively fomented the emergence and maintenance of radio research. This book has proposed to examine the complex effects of empire, industry, and education on the development of radio and ionospheric physics in the interwar years. In the light of the evidence accumulated, it is clear that the process evolved in the Dominions diverges considerably from the course of that in Britain. Research organizations such as the British DSIR do not reproduce in the Dominions as they germinate in the mother country, but gradually acquire certain of their own lineaments and tones to engender a mature, distinctive model—yet a model containing residues of centralism. We might add, moreover, that the emergence of an enthusiastic endeavour in radio research runs practically parallel in these countries [*circa* 1930]. The three Dominions—Australia, New Zealand, and Canada—lived similar organizational experiences, in that they strove to transplant a similar model of research organization. Their histories have much in common for the three had abandoned a colonial state and pursued their march towards nationhood within a larger whole, the British Commonwealth.[111] That Canada broke her political, economic, and cultural ties relatively early and quickly in comparison with the slow and gradual detachment in Australia and New Zealand gives her a different colour. Yet the three societies, albeit menaced by the Americanization in native culture through broadcasting, maintained an unambiguous engagement in the Empire, and not simply as the result of self-conscious calculations of national interest. Evidently, significant variables such as size and geography, demography, industrial structure, and scientific tradition, impinged upon the mode in which radio evolved. But in general these variables could not hide what in the interwar period came to be a common and self-evident perception: while Dominions' dependence upon Britain for science, leadership, and ideas diminished irreversibly, their dependence upon the US for technology, capital, and commercial capacity augmented equally irreversibly.[112]

If the present discussion were straight forward, then one would only have to weigh the effect of such variables on the respective countries and to try to linearly extrapolate patterns and ideas. If, in an exercise of simplification, one extracted the principal constants which emerge from the framework of

[111]See the introduction in R. McLeod & R. Jarrell (1994), *Dominions Apart: Reflections on the Culture of Science and Technology in Canada and Australia, 1850–1945*. Special Issue of *Scientia Canadensis*, 17, 53–70, pp. 1–6.

[112]See the introduction by Smythe (1981).

variables defined, then one might see remarkable parallels between Australia and Canada. Both had modest populations and vast, practically empty territories (thwarting broadcasting), both concentrated most inhabitants on a few urban centres (centralizing stations), both sustained university research (albeit variously and often intermittently), both were beneficiaries of the plans of geomagnetic studies from the Carnegie Institution (feeding observational practice), both propitiated the proliferation of radio amateurism (often dispensing with government science), both wished to strengthen long-distance radio communications with the Empire (favouring short wave to the detriment of long wave), both housed a rich manufacturing radio fabric (while being contingent upon foreign capital), both held offspring of multinational corporations (vying with domestic firms in advantageous conditions), both stoically contributed to wartime radar development (as a prelude to radio astronomy). And yet, their histories differed enormously.

In reality, Canada's history is closer to New Zealand's than to any other country. In both countries, the reconnaissance of ionospheric physics and radio electronics was more tardy and less vigorous than in leading countries such as Britain, the US, and Germany, but also Australia (that is not the case of South Africa and India). The disparate nature of reconnaissance might be explained by an alleged overlap between ionospheric physics in both countries and that of their neighbours, with the ambition of government organs and researchers from the scientifically more advanced society (that is, the US and Australia) to usurp the contiguous field (that is, Canada and New Zealand, respectively). Much of this occurred in New Zealand, as has been seen. However, the deliberate strategy of exportation and transplantation, on neighbouring soil, of both experiments and scientific programmes can be questioned in Canada.[113] If the premise of tardy cultivation is accepted, if the complex combination of commercial, military, and geopolitical reasons is correctly understood, then the interference of institutions such as the DTM has only a secondary effect to exert in its evolution. Rather, the dominance of a free market based on commercial hegemony, the superiority of broadcasting stations, the control of patents, technological supremacy, and the exaltation of American progress as the ideal model to emulate, all point to invasive dynamics in which economic

[113] J.A. Fleming, Director of the Department of Terrestrial Magnetism, Washington, to R.W. Boyle, 22 May 1939: 'We don't as yet have definite plans for conduct of necessary ionospheric experiments in Canada. We were therefore most interested to learn at first hand of the advance of your work there'. In NAC [RG 77, vol. 173/ File 45–2-21], 'Ionospheric Radio Research, 1939–40'.

and cultural, not scientific, autonomy was at stake.[114] Canadian Radio League's proclamation, 'the question is the State or the United States', eloquently incarnates such perils to national identity.[115]

A description of radio development in those Dominions cannot overlook the disparities in the double pull of Britain and the US. In Canada's case, in so far as her exposure to American irradiation was mediated not only by language but also by geographical proximity and by the mutual lack of national existence upon which to build their national identity, one might deem her situation as unique. This pull seldom induced the paralysis of Buridan's ass, as historian Donald Fleming sarcastically noted, but led, among many other aspects, to a continual leakage of young Canadian scholars to the best American universities (over 600 in 1927, twice as many as in 1907).[116] What from the US was the incarnation of 'intellectual internationalism' in science, from the next-door neighbour it denoted a transfer of allegiance and intelligence, expressed academically as a 'kind of subinfeudation'.[117] The inverse phenomenon—that is, the permanent flow of American radio investigators to Canada—remained extremely rare, not to say non-existent.

The explanation of the *retardation* of radio development, and therefore of ionospheric physics, in Canada and New Zealand must be in part sought in the substrata of a technologically derivative radio industry. Effectively, as has been seen, among investors and shareholders there was no necessity, nor incentive, to endorse the notion of independent innovation as a business priority. If a large industrial laboratory with communication interests is required for the encouragement of radio R&D, then one could maintain that a state of technological dependence, which is usually defined by the preponderance of subsidiaries of foreign firms and by the oligopolistic hegemony on patents, would, in an unduly limited market, favour a small or at best medium-sized radio entrepreneurship that could only afford to manufacture and assemble. This would appear ineluctable unless an inheritor of any of the communications giants, with privileged—not to say unique—access to patents, was able to accumulate, through a strategy based upon diversified investment in

[114]See Mahant & Mount (1984, pp. 136–9); Hume Wrong, 1975. 'The Canada-United States Relationship, 1927–1951'. *International Journal*, 31, 529–45.

[115]Crean (1976, pp. 33–34).

[116]Donald Fleming, 1962. 'Science in Australia, Canada, and the United States: Some Comparative Remarks'. In *Proceedings of the Tenth International Congress of the History of Science, Ithaca, 1962*. Paris: Hermann, 179–196, p. 185.

[117]Ibid., p. 185. Metaphors such as 'a one-way street' reflect Canadians' view. See Berger (1976, pp. 142–43).

manufacture, broadcasting, and point-to-point communication (the triad for expansion), together with a vigorous nationalistic dose, substantial enough capital and resources to promote the conception of laboratory as an integral part of the company. This was especially so in the case of the Australian AWA, which ironically achieved self-reliance at the expense of accentuating reliance in other Southern Hemisphere countries. However, it is too easy to blame the parent companies and tactics of the more powerful neighbouring countries for this situation, notably for Canada's Americanization and New Zealand's Australianization, which were influential dynamic currents in the 1930s. Government bodies such as the Canadian ACRR and New Zealand RRC, overcome by Depression-induced shortages and excessively devoted to centralism (e.g. Ottawa Lab) and to short-term undertaking (e.g. the 1932 eclipse), contributed little to the sustained endeavour upon which the success of leading research depended. In this respect, the fecund relationships distinguishable between government and university in Australia (e.g. RRB and Madsen's Sydney group) starkly contrast with the unfruitful, isolated cooperation efforts between NRC's radio section and Canadian engineering schools and faculties. It is true that organizations involved in the management of public funds felt scant pressure from the radio industry to coordinate initiatives with the object of innovation. But, when that pressure was significantly exerted, notably from the armed forces, the balance of finances shifted unequivocally to defence issues, with a great marginalization of fundamental physical research. In Canada and New Zealand, radio R&D became identified with military-applied radar (not so defined in Britain and Australia). Yet, in neither case was it explicitly opposed to ionospheric physics (or later to radio astronomy). Rather, it was conceived as a practical response to an emergency situation in a very exceptional wartime context. Hence the responsibility for this phenomenon of retardation does not lie solely with entrepreneurs. Better still, one should turn the reasoning around: first-rate, sophisticated industrial research—and this must be italicized—*could only come successfully to fruition where there already existed first-rate university research*, *videlicet*, a factory of postgraduate, young investigators capable of nurturing the intellectual machinery of innovative venture. Research would not probably have been fruitful in an imaginary Canada or New Zealand with industrial laboratories but devoid of university research.

POSTCRIPT*

Overstating reality

> All great ***discoveries*** in experimental physics have been due to the intuition of men who made free use of ***models***, which were for them not products of the imagination, but representation of ***real*** things (Max Born)[1]

In one of the most imaginative and thought-provoking articles of his intellectual oeuvre on the history of ionospheric research historian C. Stewart Gillmor asserted that 'data produced by [engineers and physicists'] radio tools as well as their earlier *cultural* ideas reinforced a particular idea of the upper atmosphere as being vertically distributed in more or less discrete layers'. To explain how they 'overstated the reality of the ionospheric layers,' he contended: 'A concept can owe its existence in large part to the fact that people say it is real [...] I believe this was true in part because of the analogy of optics, in the minds of many, in which the radio waves 'bounced' or reflected off the ionosphere as if from a mirror...The idea of an ionospheric layer or layers was a useful idea and one subject to elaboration. The physical existence of the layers, however, came to be taken too literally by some. Once the phenomenon has been cast in terms of a metaphor the comparison and resemblance can widen.'[2]

Gillmor's contentions are highly suggestive. Coming from one of the most authoritative voices (first as an ionospheric physicist and then as a historian) in this discipline, they are absolutely worthy of consideration. But it must be added that Gillmor focused on the effects of radio technique upon physical concepts, and barely explained how this was *cultural.* Nor did he adduce almost any evidence to interconnect a disregarded community (such as geomagneticians), a technological innovation, and determinant socio-economic and geopolitical

*A shortened version of the postscript and the epilogue was published in *Studies in History and Philosophy of Science Part B*, 39, Aitor Anduaga, 'The Realist Interpretation of the Atmosphere', 465–510, Copyright Elsevier (2008). The author would like to thank Elsevier for permission to reproduce this work.

[1]Born (1953, p. 140), italics added.

[2]C. Stewart Gillmor, 1981. 'Threshold to Space: Early Studies of the Ionosphere'. In Paul Hanle & Von del Chamberlain eds. *Space Science Comes of Age: Perspectives in the History of the Space Sciences.* Washington, D.C.: Smithsonian Institution Press, 101–14, pp. 102–5.

circumstances, which concurred in the 1920s in the British Empire, the US, Japan, and part of Europe.[3]

We do not intend here to discredit Gillmor's propositions but rather to demonstrate that a crucial part of this story remains to be examined. In this postscript, we want to demonstrate *one* thesis: in the interwar years, most radiophysicists and some atomic physicists, for reasons principally related to extrinsic influences and to a lesser extent to internal developments of their own science, enthusiastically pursued, and fervidly embraced, a *realist interpretation* of the ionosphere. To this end, we shall treat the historical circumstances in which a specific social and commercial environment came to exert a strong influence upon upper atmospheric physicists, and in which realism as a product validating the 'truth' of certain practices and beliefs arose. To start, therefore, we shall confront the great paradigm of atmospheric sciences in the twentieth century: the discovery of ionospheric layers and the pursuit of atmospheric structure.

The antecedents of our story go back to the nineteenth century, when the use of sounding balloons allowed scientists to ascertain the lower regions of the atmosphere: the troposphere and the stratosphere. The 'discoveries' of the layers of the upper atmosphere took place in Britain and the US in the interwar years. The bare facts of the story are purportedly well known. In 1902, O. Heaviside and A.E. Kennelly suggest the existence of a conducting layer in the upper atmosphere (explaining long-distance propagation of radio waves), on purely theoretical grounds. In 1924, E.V. Appleton (Nobel Prizewinner in physics) and M.A.F. Barnett provide experimental evidence on its existence. A few months later in the US M. Tuve and G. Breit independently calculate by pulse-echo sounding the height of the reflecting stratum. In 1928, Appleton announces the finding of the fine structure of the ionosphere (the *E* and *F layers* and the substrata F_1 and F_2). Furthermore, French and British physicists determine the extent of a thin layer of ozone (*ozonosphere*) between the stratosphere and the ionosphere.

These crucial experiments, and their ensuing discoveries, have acquired a canonical status in physics textbooks and science encyclopaedias, and still remain embedded in the discipline of the history of science.[4] That the

[3]It must be said that Gillmor dealt with ionospheric physics in most of those countries in other papers—as will be seen subsequently. For prewar ionospheric research in Germany and Japan, see Maeda (1986) and Dieminger (1974, 1975). See also A.P. Mitra, 1984. *Fifty Years of Radio Science in India.* New Delhi: Indian National Science Academy; A.C. Brown, 1977. *A History of Scientific Endeavour in South Africa.* Cape Town: Royal Soc. South Africa, 362–5.

[4]See, for instance, Affronti (1977, pp. 293–5) and Schmidt (1988).

atmosphere and its structure were shaped by a continually advancing swell of 'discovery'—on the crest of the wave experiments antedated theory—was a common idea among postwar historians. Experiments accompanied transatlantic short-wave communications, and through their association with discoveries, became the end products of science, to the detriment of the social and commercial circumstances of their realization. Perhaps the most notable—without doubt, the most quotable—manifestation of formalization of this view came with the award of the Nobel Prize for physics to Appleton in 1947, for his physical investigations, 'especially for the discovery of the so-called Appleton Layer'.[5] The paradigmatic model, and the favourable response it evoked, were quickly and solidly established in the literature. In this respect, there is no question that an influential book by I. Bernard Cohen in 1949 and another by Donald H. Menzel from the celebrated series of *Harvard Astronomical Books* served as agents of popularization.[6] These authorized voices admirably demarcate our frame of preoccupations (and discordances): they show that the stratified structure of the atmosphere was unmistakably disclosed as a succession of experimental findings, with almost no connection with geomagneticians' perceptions on auroral regions.[7]

To many writers, the standard version reflected common sense. Its historical narrative resembled—albeit too relatively—the contemporary experience of atomic structure, which many upper atmospheric physicists took into account in construing the phenomena that were layering; indeed they meant to compare themselves to them and, if possible, to emulate them. The fact that one of the few treatises on the subject prior to 1939, *Exploring the Upper Atmosphere* by Dorothy Fisk, devoted the entire introduction to emphasizing the importance of the structure of the atom to atmospheric investigation is significant in the extreme.[8] Their perspectives were clearly idealist, at the time when many physicists adopted simplified mathematical models as essential operational tools in ionospheric physics. Hence it is counterproductive (and it would be

[5]E. Hulthén, 1964. Physics 1947. In *Nobel Lectures Physics: 1942–62*. London: Elsevier Publishing. Co., 75–8.

[6]I.B. Cohen, 1949. *Science, Servant of Man. A Layman's Primer for the Age of Science*. London: Sigma Books Ltd., 257–79; D.H. Menzel, 1953. *Our Sun*. Massachussets: Harvard University Press. Two well-documented and highly regarded sources of information about experimental practices (written by physicists) were A.L. Green, 1946. 'Early History of the Ionosphere'. *A.W.A. Technical Review*, 7(2), 177–228, and Harry R. Mimno, 1937. 'The Physics of the Ionosphere'. *Reviews of Modern Physics*, 1937, 9(1), 1–44.

[7]See also Dellinger (1947) and Peck (1946).

[8]Dorothy Fisk, 1935. *Exploring the Upper Atmosphere*. London: Faber & Faber Ltd., with an introduction by H.L. Brose, pp. 9–20.

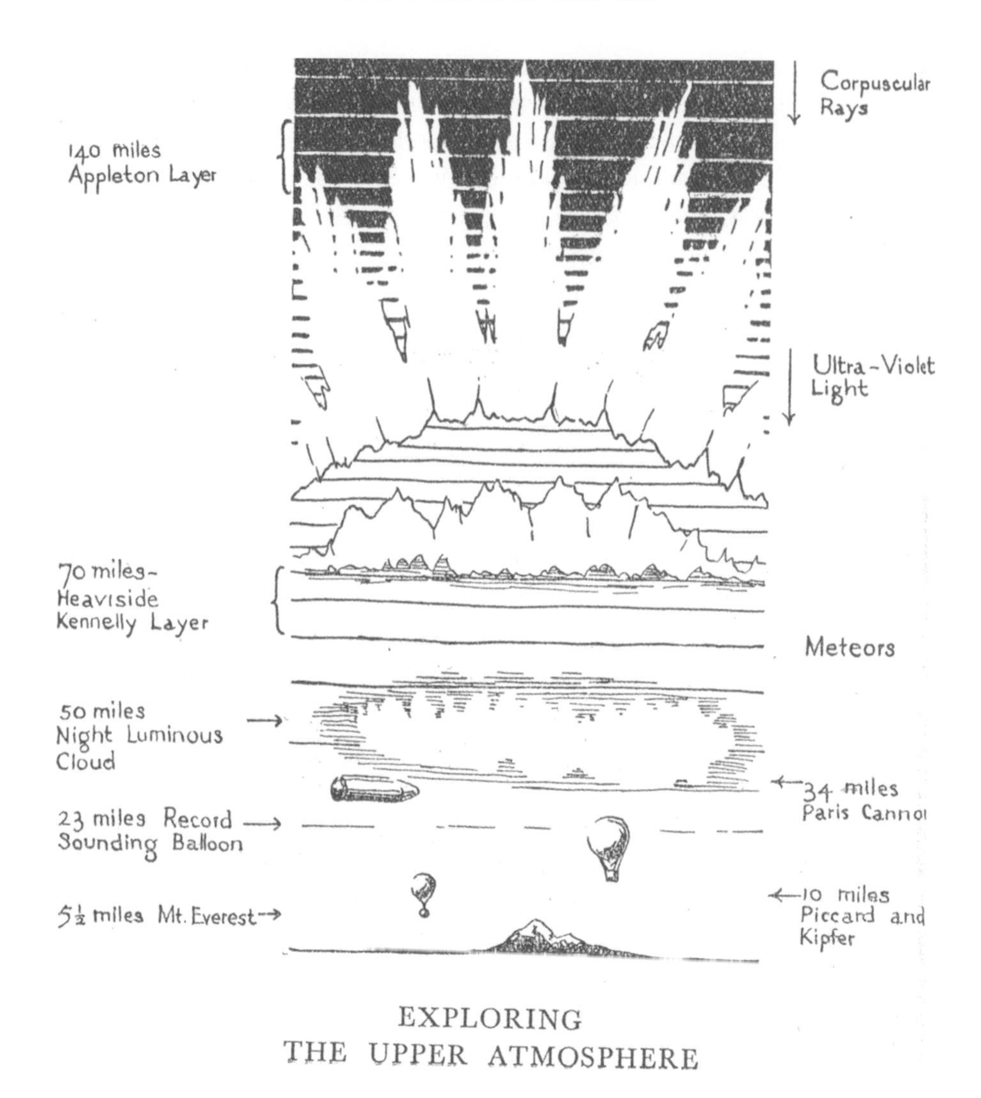

Fig. 24: The paths of radio and sound waves. From D. Fisk (1934), p. 94.*

* There are instances where we have been unable to trace or contact the copyright holder. If notified the publisher will be pleased to rectify any errors or omissions at the earliest opportunity.

misleading) to divorce the story of discoveries from that of the models of the atmosphere: the bonds of relationship are simply too close.

From a retrospective vantage point we can easily see why the lacunae of the standard version have passed practically invisibly with respect to historiography, especially in its geomagnetic neglect and in its asymmetry of 'physics–enginery' relations. One reason is the extent to which the figure of Appleton as a *discoverer hero* has propagated in the literature.[9] There is no lack of evidence of the high repute in which his contemporaries held him, especially, but not exclusively, those who regarded him to be the founder of ionospheric physics. Yet, by the 1970s, (coinciding with the fiftieth anniversary of the crucial experiments) Appleton's shadow was to become too long. Ronald Clark in his introspective biography on *Sir Edward Appleton* glorifies the event of *experimentum crucis* and scarcely mentions the confluence of scientific traditions in interwar Britain.[10] Perhaps more remarkable (certainly more influential) are the special monographs published in the *Journal of Atmospheric and Terrestrial Physics* and the *Philosophical Transactions of the Royal Society of London* in 1974 and 1975, respectively.[11] Here, in contemporary actors' eyes, the exceptional events of 50 years ago provoked a 'major breakthrough', and 'opened up a new and rich field of research.' Much the same can be said of the homage the *Journal of Atmospheric and Terrestrial Physics* paid to its founder, Appleton himself, in commemoration of the centenary of his birth.[12] Thus emerges the Turneresque figure of the conqueror of Nature, and of dominions beyond the stratosphere; and his discovery, 'the basis for a strategy for radio transmissions on a reasonably predictable basis'.[13]

It is possible that part of the reason for the invisibility of the current episode is in its general neglect in the historiography of science. As far as we have been able to determine from a survey of the papers published in the 20 leading journals in the last 40 years, *no* historian has focused attention on the exploration of the ionosphere in the interwar period. The most 'contiguous'

[9]On the functions of heroic stories of discoveries, see M. Terrall, 1998. 'Heroic Narratives of Quest and Discovery'. *Configurations: A Journal of Literature, Science, and Technology*, 6, 223–42.

[10]R. Clark (1971).

[11]The compilation of the *Jour. Atm. Terr. Phys.*, December 1974, 36(12), 2,069–320, includes, among others: M.A.F. Barnett, 'The Early Days of Ionosphere Research'; M.A. Tuve, 'Early Days of Pulse Radio at the Carnegie Institution'; J.A. Ratcliffe, 'Experimental Methods of Ionospheric Investigation, 1925–1955'; A.H. Waynick, 'Fifty Years of the Ionosphere: The Early Years—Experimental'; H.G. Booker, 'Fifty Years of the Ionosphere: The Early Years—Electromagnetic Theory'; E.O. Hulburt, 'Early Theory of the Ionosphere'; H.S.W. Massey, 'Theories of the Ionosphere—1930–1955'; A.L. Green, 'Early History of the Ionosphere'; J.A. Ratcliffe, 'The Formation of the Ionosphere. Ideas of the Early Years (1925–1955)'. The issue of the *Phil. Trans. Roy. Soc. Lond.,* 1975, 280, 1–130, includes, for example, J.A. Ratcliffe, 'The Early Days of Ionospheric Research'; and A.H. Waynick, 'The Early History of Ionospheric Investigations in the United States'.

[12]See *Jour. Atm. Terr. Phys.*, 1994, 56(6), 693–731.

[13]Excell (1994, p. 697). Even Charles Süsskind's article on Appleton in the *Dictionary of Scientific Biography*, vol. 1, ed. by C.C. Gillispie (New York: Schribner, 1981), 195–6, is biased considerably towards his role of patriarch of the discipline and of explorer of a New Continent.

seem to have been absorbed by the apparently more attractive whirlpool of atomic physics and quantum mechanics (over a hundred). For the peripheral dominions of physics, this suction has constituted a permanent brain drain, which has certainly left its mark in the still (rather shaky) foundations of the tiny historiographic edifice of ionospheric physics.

Fortunately, within the last ten years, there have been several attempts to revise these—in retrospect, rather vulnerable—constructions. Our indebtedness to recent studies on the ionosphere and radio propagation by historians such as Dominique Pestre, E.C. Jones-Imhotep, Olav Wicken, and C.P. Yeang is evident; thanks to them, ionospheric physics has acquired a three-dimensional character and presence, in which the ultimate reference is cultural, geopolitical, and epistemic.[14] Nevertheless, they have focused attention on postwar experience and we are still very far from appreciating the place of atmospheric geophysics in interwar science and society. In particular, Jones-Imhotep has incisively examined knowledge production in cold-war sounding techniques and its relationship to the graphic record known as the ionogram.[15] Although the details of the process cannot be explained here, he concludes that the visual instantiation of the propagation of radio waves yields reliable knowledge: the ionogram's virtue resided in rendering static the dynamic behaviour of the ionosphere. The parallel between the reliability of the 'eye' as the basis for warranting knowledge and the direct visual apprehension of reality is thus established. For Jones-Imhotep, this latter effect is the result not only of convention and of construction but also of craft: through discipline and practice, the 'layered' structure of the ionosphere was captured by 'the trained eye of the radio scientist'. With the standardization of images and the devices producing them (ionosondes), with the careful execution of the craft of 'reading' (that is, of testing physical models of layer profiles against the contours of the graph), ionograms acquired the status of mirrors of reality. This analysis is, we believe, correct for postwar instrumental practice. Yet, it does not seem the explanation of the realist impulses we are seeking for the interwar period. The reasons are rather obvious: the ionosonde was developed in 1933, this and its resulting ionograms were systematized in the late 1930s, and our 'crucial experiments' date from the mid-1920s.[16]

[14]Pestre (1997); Wicken (1997). On the other hand, Yeang (2003) has examined the prewar period.

[15]Jones-Imhotep (2000, 2001, esp. pp. 87–107).

[16]The deep-rooted predisposition towards the visual apprehension of reality through instruments has been noted by several writers; see, for example, Don Ihde, 1991. *Instrumental*

Pervasively, historians of science have drawn upon the visual representations as the basis of their causal accounts of the apprehension of reality.[17] Whether describing Feynmann diagrams of physical processes of particles, the Dutch descriptive paintings in seventeenth-century art, or stratigraphical columns in eighteenth-century geology, representations that had been often introduced as mnemonic devices then came to be seen by practitioners and observers as pictures of the real world.[18] Images (photographic, pictographic, diagrammatic, or inscriptional) emerged thus as a condition for realism. In the events with which we are concerned, however, the dotted lines on the oscilloscopic screen of radiation frequencies vs. virtual height (whose gaps and crinkles were construed as discrete layers) were observed in the late 1920s, several years after the definitive papers of Appleton, Breit, and Tuve.[19] This presupposition of the negligible effect of both instrumental and representational realism is essential here; they form the basis of our attempt to provide an answer to the *crucial* question—in its broadest sense crucial to all history of science—: how then did realism arise?

In sum, we can now better appreciate the symptoms of a historiographic situation that has at the very least nourished the ostracism of the episode we are to recount. We have an almost pathological neglect (of which Gillmor is an illustrious exception), the rudiments of a discovery-based explanation (which implicitly hide the neglect), and an unreflective (not to say non-existent) treatment of realism and of models. Firstly, retrospective narratives—in particular, the standard version of the 1940s—not only eclipse the complexity of the actual historical process, they also metamorphose it: 'a complex enterprise, accessible to historical and sociological understanding, generates objects which are then labelled as discoveries. Subsequently, the story of that process is rewritten. The lengthy enterprise is telescoped into an individual moment with

Realism: The Interface between Philosophy of Science and Philosophy of Technology. Indianapolis: Indiana University Press.

[17]See, for example, George Levine ed., 1993. *Realism and Representation: Essays on the Problem of Realism in Relation to Science, Literature, and Culture.* Madison: University of Wisconsin Press.

[18]On images and visual languages in science, see Rudwick (1976) and Lynch (1985). For an example of realist ascription in representational diagrams in twentieth-century physics, see David Kaiser, 2000. 'Stick-Figure Realism: Conventions, Reification, and the Persistence of Feynman Diagrams, 1948-1964'. *Representations*, 70, 49–86. For a broader outlook, see Levine (1993). For a valuable study about art history for historians of science, see Svetlana Alpers, 1983. *The Art of Describing: Dutch Art in the Seventeenth Century*. Chicago: Chicago University Press.

[19]It was not until 1930 that the pulse-echo technique (manual and photographically equipped) enabling the measurement of the height of ionospheric echoes continuously through time and over a band of variable frequencies was in widespread use.

an individual author.'[20] Secondly, the historian has often taken the *corpus* of the discovery story and has used its causal explanation as a product of his own analysis (Schedvin's account in *Shaping Science*...is a good example).[21] Thirdly, it is established that the current events are the commencement of a new enterprise, not the confluence of traditions and of practices. And, finally, while many authors have treated *vertically* the structure of the atmosphere and internal developments of the discipline, we shall strive for broadening its confines through a panoramic study, undertaken *horizontally*, across multicontextual frontiers.

Since here we shall dispense with the category of 'discovery' and the historical narrative associated with it, the approach that will be adopted next in the epilogue is indicated. Firstly, our tone will not be descriptive; it will be demonstrative. We intend neither to evaluate nor prescribe: we intend to offer a causal analysis that explains the circumstances under which, and the reasons by which, physicists embraced the conception of stratified atmosphere. Now, there is a relationship between the espousal of this conception and the overstatement of the 'reality' of layers. Discoveries are associated with realism, and a stratified atmosphere *must* have a real reference. We have explained why we shall dispense with the visual power of inscriptions and of pictures as the *force majeure* in the apprehension of natural reality. Neither shall we situate the origin of these realist impulses at the level of individual decisions; a 'psychological' analysis in which conditioning experiences and previous scientific practices appear as determinant of present attitudes, we believe, is not the most appropriate. Instead, we shall adopt a 'sociological' approach: we shall treat physicists' mental posture as a reaction socially conditioned, and strongly influenced, by the climate of excitement that followed short-wave transatlantic communications *circa* 1924. However, our purpose is not to go deeply into the sociology of scientific knowledge (in fact, such a possibility might be debated at length). Our purpose is to ascribe a strong influence of the social and commercial environment upon actors' dispositions and their scientific discourse (here we agree with Forman's view).[22] We shall be interested in perceptions, sensitivities, sentiments of resentment towards 'radio invasion', rather than in the conditions of evaluation of scientific knowledge. This said, we shall be adopting a

[20]Simon Schaffer, 1986. 'Scientific Discoveries and the End of Natural Philosophy'. *Soc. Stud. Sc.*, 16, 387–420, p. 397. See also Pickering (1984) and Brannigan (1981, esp. 120–62).

[21]Schedvin (1987).

[22]See Paul Forman, 1971. 'Weimar Culture, Causality, and Quantum Theory, 1918–1927: Adaptation by German Physicists and Mathematicians to a Hostile Intellectual Environment'. *HSPS*, 3, 1–115.

stance in which the objections to the canonical authority of 'discovery' seem rational, reasonable, and credible. In this respect, we shall propound a reinterpretation of the pursuit of atmospheric structure from new historiography perspectives. Our intention is not to effect a 'true judgement' of history: most early concepts such as conducting surfaces, mirrors, and reflectors were subsequently rejected by the scientific community itself.[23] Our objective is to analyse the aura of realism surrounding the exploration of the upper atmosphere by means of radio techniques, and which eclipsed previous physical conceptions of the geomagneticians. To this end, we shall suggest that the vertical structure of the atmosphere and its partition in layers was a *conceptual representation*, rather than the result of a cascade of discoveries. We shall try to prove that there was nothing uncontestable or unproblematic in the series of historical declarations on those 'discoveries', which produced a scientific consensus in favour of a sharply stratified atmosphere. Even better, what we shall find is a *reification* of metaphors (such as the reflecting layer) and a strong tendency to ascribe material existence to theoretical entities. We shall seek the explanation of these realist impulses in the very excitement of the most immediate social environment. Implicitly or explicitly, the radiophysicist was the target of the continuous exhortations to the elucidation of revolutionary communications. We shall construe the eagerness, the anxiousness of those men to erect the foundations of their science as a reaction to their environment. In *the realist interpretation*, realism arises neither from the habit nor from pictorial representations, nor is it a product of craft or of convention; it arises in a subtle way: 'short wave' incarnated all that was amazing in long-distance communications—astonishing audibility, miraculous instantaneousness, extraordinarily economical, cogent refutation of diffraction. To explain such attributes, there *must* be overhead a physical reality independent of perception and substantiation. Reflecting layer acquired thus *ab incunabulis* the status of reality.

The current account delivers us a proposition of realist interpretation by upper atmospheric physicists and radio engineers. But it also provides a model with historiographic value, we think, for the conditions under which such an interpretation is likely to crystallize. One may think that when an immediate commercial and military environment badly requires a necessity (e.g. long-distance communication) and a scientific community is incapable of providing convincing solutions, then their members are unduly inconvenienced

[23]Even the term *layer* proved unsuitable and ambiguous with the use of sounding rockets in the 1940s, being substituted by the words 'region' and 'zone'. See C. Stewart Gillmor, 1976. 'The History of the Term *Ionosphere*'. *Nature*, 262, 347–8.

by this situation. Devoid of any consistent explanation, they feel captive to social pressure, to their relative ignorance—which usually means reluctance to embrace doctrines and proliferation of conjectures. When, almost suddenly, there appears a technological intruder (*short wave*) which marvellously meets that need and which 'renders' self-evident in the eyes of the society the existence of the metaphor assumed (a reflecting layer, whatelse?), then scientists feel obligated to supply evidence. The same scientists who previously perceived discomfiture are now experiencing a boost from a highly excited milieu around them. Their responses will generally be attempts to accommodate the public image of science (the positive image of radio as a revolution) so as to gain prestige (a prestige still to consolidate). In a climate of excitement, their endeavours will imply an overemphasis on the reality of the metaphor socially accepted by virtue of the authority of our commercial and social achievement (transatlantic communications). Such circumstances may even impinge upon the concepts and doctrinal contents of the discipline itself—as we shall try to demonstrate.

It is a commonplace to say that reducing this history to a simple discussion about discovery is to misrepresent historical complexity in conceptual terms. The paradigm of 'layered' atmosphere was never a linear succession of unproblematic findings, nor was it necessarily heroic. On the contrary, it was driven by the mechanical action of conceptual linkages, in which the categories of 'realism', 'model', and 'interpretation' were the three 'sails of a windmill'. We shall deal with these thorny questions centrally, but in a manner somewhat different from that which characterizes much philosophy and sociology of science. To start with, 'realism' will be treat as a historical product, as process achieved, as category that requires a complex reading—not merely a theory about the nature of science but rather a space, a complex of impulses, a combination of motives, and a central role in our conceptual representation of the atmosphere. We shall be interested in explicit statements about how theoretical constructs corresponded to objective reality, but such realism statements will be analysed in connection with scientific and the engineering contemporary practices.[24]

Now, if the curiosity of the reader is sustained essentially by the philosophical doctrinal fundamentals, he or she will be surprised by a notable paradox:

[24]Important philosophical surveys of realism that offer helpful resources to the historian of science include Craig Dilworth, 1990. 'Empiricism Vs. Realism: High Points in the Debate During the Past 150 Years'. *SHPS*, 21(3), 431–62; Valeria Mosini, 1996. 'Realism Vs. Instrumentalism in Chemistry: The Case of the Resonance Theory'. *Rivista della Storia della Scienza*, 4(2), 145–68. See also Ian Hacking, 1983. *Representing and Intervening*. Cambridge: CUP.

this historical episode of intense realist impulses was not characterized by the steadfast espousal by these men of an ideological-philosophical doctrine such as realism—at least not explicitly. Before this apparent paradox, one might be tempted to attribute such impulses to a naïve realism, to regard them as a projection into the scientific domain of a common-sense attitude, to resort to the frequently mentioned (albeit controversial) practising scientist's philosophy of science. But such inferences would be rather impetuous. Here the essential question is the nature of the reaction of the physicist to the excitement of his most immediate environment. And by adducing 'environment' we seek the relations between the responses of the upper atmospheric physics community and the conduct of private firms' radio engineers and of amateurs in general.

By alluding to 'model' it is commonly understood that one is referring to a copy of an original or to an idealized or abstract version of an object or system. In both cases, the concept is ambivalent in the extreme, even polyvalent in meaning in science, logic, and philosophy ('impossible' to be properly defined, according to L. Apostel).[25] Here, we shall not be concerned with the mechanical models that were built up for experimental research (such as Birkeland's *terrella*),[26] or with the representative models having a close analogy with the original system (e.g. the billiard-ball model for the kinetic theory of gases).[27] Rather, we shall deal with theoretical physical models, which were later elevated to the status of a theory (a familiar case is Niels Bohr's atomic model).[28] However, our purpose is not to evaluate their heuristic function but to explain and describe the loose way in which physicists used them. We shall be striving to broaden our understanding of what models meant for upper atmospheric scientists, and to show how their use in the interwar years related to practical interests in fields of enginery and in the wider society. The way we shall do this by situating models, and interpretations about them, in their social context.

To this end, we shall borrow from J. Audretsch and S. Hartmann the notions of 'preliminary physics' and of 'model' as the typification of this.[29]

[25]Apostel (1961).

[26]For mechanical models, see, for example, Thomas P. Hughes, 1988. 'Model Builders and Instrument Makers'. *Science in Context*, 2, 59–75.

[27]For a classification of types of models in physics, see Groenewold (1961, pp. 98–103). Cushing's classification has numerous similarities with Groenewold's. See Cushing (1982, p. 8). See also Michael Redhead, 1980. 'Models in Physics'. *British Journal for the Philosophy of Sciences*, 31, 145–63.

[28]For Bohr's semiclassical model of the hydrogen atom as the cornerstone of the old quantum theory, see Heilbron (1985, pp. 33–49).

[29]Audretsch (1989, pp. 373–92); and Hartmann (1995).

We aim to approach the use of models as the result of *pragmatic attitudes*. Just as for Audretsch and Hartmann physicists' models are frequently not justified and inconsistent approximation schemes which are explicable as part of a human activity, so we shall treat the use of models as patterns of activity which are exposed to external influences and guided by practical ends. We shall suggest that the 'models of electron density profiles' for the ionospheric layers in the 1930s were examples of preliminary (ionospheric) physics reflecting various views of scientific method. In this respect, we shall try to prove that there existed a hierarchy of attitudes based on ontological commitments about models and that physicists' stances on the function of models were embedded within practical needs and within aspects of social order. In this hierarchy of stances—from purely instrumental to strongly ontological—at one extreme there lay those scientists for whom models simply provided possible mechanisms of how natural systems could be operating, while at the other end there lay those for whom the 'real atmosphere' was practically like the entities in the model. That is one of the questions that *The Realist Interpretation of the Atmosphere* will try to demonstrate.

Interestingly, the notion of 'interpretation' now acquires the status of link between the different uses of models.[30] For if we accept that model relates to the idea of approximation, if we allow its latent metaphorical content, then the degree to which models were taken as exactly representing reality (or as merely calculating devices) has an essential role to play in our enquiry. In broaching issues of interpretation we shall take seriously the historical dimension of theoretical constructs. For us, models and metaphors associated with them will not be treated as a set of formal and unproblematic entities capable of variously representing reality, and not at all from a preceptive view of a realist ontology of science.[31] We shall be concerned, however, with explicit ontological assertions about how contemporary physicists suggested interpreting them, but such realist sentiments will always be analysed in relation to the precise environment in which they were incubated, with a view to enquiring into the reasons by which they were produced, and in accordance with the real nature of scientific practice. More important still, our purpose will be an examination of interpretation regarded as practical and human action, and subject therefore to influences from the immediate environment. This approach

[30]For a valuable collection of essays addressing the interpretation of models in science, see Churchland & Hooker (1985).

[31]See Michael Bradie, 1980. 'Models, Metaphors, and Scientific Realism'. *Nature and System*, 2, 3–20.

raises several further questions: how did the pursuit of the knowledge of the ionosphere become a part not only of government-funded research policy but also of industrial development of private companies? To this we might add: how did the introduction of models of layer profiles prove for the prediction of most optimum frequencies so decisive, and at once scientifically and metaphysically so divisive? And finally, what happens when scientific, military, and commercial relations in radio communications become so interlaced that theoretical models prove vital for the reliability and audibility of transmissions? Can environment then affect ontological interpretations?

EPILOGUE

THE REALIST INTERPRETATION OF THE ATMOSPHERE

Pre-1924 conceptualizations of the upper atmosphere

As formulated by geomagneticians

In the early 1920s, geomagneticians regarded with pride and satisfaction the massive accumulation of observational data they had collected in magnetic observatories all over the world.[1] They, certainly more than meteorologists and probably more than any other sector of the physical scientific community, also felt confidence in contributing to the expansion of the knowledge of the atmosphere, hitherto confined to the troposphere and stratosphere. Looking back over their discipline, they steadfastly and self-confidently claimed that Balfour Stewart was the first to infer, in 1882, that 'the upper atmosphere must include a region with significant conductivity' (Chapman, 1956, p. 1385). Stewart, in effect, had suggested that the solar daily geomagnetic variations (known as 'S') arose from electrical currents induced in a conducting layer in the upper atmosphere (Stewart, 1882).

In 1908, the professor of physics at the University of Manchester, Arthur Schuster, gave mathematical form to Stewart's dynamo theory of diurnal variations. He pointed to the sun as the main cause: 'We may reasonably retain the view that the powerful ionization of the air...is a direct effect of solar radiation' (Schuster, 1908). The dynamo theory that Schuster formulated was extended and perfected in 1919 'with extraordinary rigour' by his pupil Sydney Chapman, who introduced a lunar component into the daily variations of the earth's magnetic field (Chapman, 1919). Drawing upon spherical harmonics he was able to explain the daily solar (S) and lunar (L) variations at any time of the year and to predict the total conductivity required in the upper

[1]A comprehensive history of geomagnetism in the twentieth century is yet to be written. Apart from the literature discussed below in connection with the early dynamo theory, accounts are almost exclusively in the form of brief surveys and memoirs. Probably the most thoroughgoing study is by Barraclough (1989). Valuable collections are also Chapman & Bartels (1940, pp. 898–936); Chapman (1967a; 1967b, pp. 3–28); Good (1998, pp. 350–7); and Dudley Parkinson (1998, pp. 357–65).

atmosphere.[2] Moreover, the then recent observations of Norwegian geophysicists Carl Störmer and Kristian Birkeland on the height of aurorae led Chapman to think that magnetic storms and aurorae might originate 'in similar regions of the atmosphere'.[3]

The argument for the existence of a conducting layer on the side of terrestrial magnetism was not wholly theoretical. The strongest reason, in the opinion of the director of Kew Magnetic Observatory, Charles Chree, was straightforward observation.[4] The association of geomagnetic variations with aurora prompted many geomagneticians, if not all, to accept the upper atmosphere as 'the seat of electrical currents'. However, magnetic observations may have indicated the pattern and intensity of electric current flow, but they did not fix the height *h* at which it flowed. And according to geomagneticians, this failing undermined the credibility of the atmosphere–geophysical picture.

Geomagnetic theorists agreed that conductivity and ionization were essential characteristics of the high atmosphere. In a discussion on atmospheric ionization held at the Physical Society of London in 1924, Chapman summed up the picture of the upper atmosphere from the geophysical viewpoint:[5]

> There are two independent regions of high conductivity, ionized by independent solar agencies. One of these regions is a layer extending nearly or quite over the whole earth...[whose] ionizing agent is ultra-violet radiation; it seems likely that the ionization is associated with the production of the layer of ozone...in the upper atmosphere [at a height of about 40 or 50 km. for the ozone layer].
>
> The other ionized regions are the auroral zones round each pole.... Measurements of the height of aurorae indicate that [charged particles coming from the sun] penetrate the atmosphere down to about 90 km. above the ground; the aurorae are observed up to a height of several hundred kilometres.

It is thus not surprising that the representatives of this, to a certain extent, 'British geophysical school'[6] thoroughly agreed with its founder, Sir George

[2]Unlike wave reflection theories, as we shall see later, Chapman's theory contained a systematic means for providing predictions: the total conductivity integrated throughout the thickness of the layer was 2.5×10^4 siemens. For further information, see Vestine (1967, pp. 19–23).

[3]Chapman (1919, p. 47). On Birkeland's and Störmer's auroral research, see Friedman (1995).

[4]Chree (1927, pp. 82–3): 'The observed large universal increase of [S and L] as we pass from sunspot minimum to sunspot maximum is, through their association with aurora, a direct consequence of the electrical currents overhead.'

[5]Chapman (1924–25, pp. 44D–45D). The meeting was held on 28 Nov 1924.

[6]To a certain extent and with reservations; for, even if Chree was Darwin's student, he criticized his method. Kushner (1993, p. 217).

Darwin, that the earth should be viewed and treated as a planet in space. Following the nineteenth-century Cambridge tradition of applied and mixed mathematics, Chapman developed mathematical-minded models which, unlike those of radio physicists, had no ontological pretensions: firstly, approximating the complex atmosphere with the best possible mathematical model; and then, reducing the number of parameters to facilitate analysis. And, even better, where, as with the radio reflecting layer, there was great scientific interest in specific, mainly commercial demands, that interest—in contrast with their non-commercial, therefore 'worthier' one—was often, to my knowledge, construed by the geomagneticians as an affront, showing disrespect for their discipline.

When at the end of 1924 radiophysicists suddenly 'discovered' the hypothetical Kennelly-Heaviside layer, most geomagneticians felt that their picture of the atmosphere had been drastically eclipsed, a fact implying a negative valuation of their contribution. That, without doubt, was their impression before the new reality.[7] One may perceive it, or at least sense it, in the disaffected (and often circumspect) tone of the opinions and replies of geomagneticians in different forums of discussion. Whereas during this time of euphoria the expressions of radio engineers and experimental physicists transmit excitement, those of geomagneticians exude caution, prudence in the interpretations and claims for their merits. And, although it is not easy to reflect these sensations on paper, the following fragments may serve to give some idea of the credits to which the author clearly believes to be entitled in all fairness. Thus, in January 1927, in view of the constant omission of geomagnetic precedents in *Nature*, Chree felt impelled to remind readers that 'estimates of the altitude of a stratum of high electrical conductivity were made [by Störmer] long before the times of wireless communication'.[8]

[7]An important aspect underlying the relatively high degree of reticence towards radio observations among some geomagneticians—definitively not in Chapman's case—was, of course, reputation, especially the idea of radio-based evidence as a substitute for magnetism-based evidence. In 1932, in the prime of radio effervescence and in spite of having been identified with several different layers, the German geomagnetic theorist Julius Bartels (1932, p. 616) still concluded that the most valuable information came from terrestrial magnetism. He deprecated the lack of coordination, and stated: 'The hypothesis of the high conducting layer, deduced...by Balfour Stewart in 1878... had to be re-deduced, from data of radio propagation, more than 20 years later!' According to Bartels (1939, p. 387), 'the pioneer experiments in 1925' verified, not discovered, 'in the most direct manner the existence of a conducting layer'.

[8]And Chree (1927, p. 82) added: 'We may hope to learn much from wireless which it might be difficult or impossible to derive from auroral observations'. Earlier, Schuster (1922, p. 325) had observed in *The Electrician* that testimony from magnetic studies had 'only been lightly touched upon'.

Only two unambiguous tokens of recognition of the preceding achievements of geomagneticians before audiences of radio engineers have come to light.[9] One can feel sympathy for geomagneticians when reading Victor Hess' influential *The Conductivity of the Atmosphere* (1928) which makes no mention of the evidence provided by terrestrial magnetism regarding the ionization of the upper atmosphere.[10] Or how, to mention another example, a manual on short-wave wireless communication by two engineers from the Marconi Company labelled the information elicited from meteors, aurora, and magnetism with the opprobrious appellations 'meagre' and 'inaccurate'.[11] This tendency to disparage was so pervasive among radio experimentalists that Appleton, always considerate and respectful, usually made reference to geomagnetism, as if to counter the indecorous attitude and the hubris of his colleagues.[12]

As formulated by radio engineers and physicists

Guglielmo Marconi epitomizes an empirical observational attitude that was widespread among early wireless telegraphers and experimentalists.[13] In December 1901, Marconi achieved transatlantic communication between Poldhu in England and Newfoundland in Canada, a distance of 1,800 miles, the equivalent of a wall of sea water 100 miles high interposed between the

[9] Sir Frank Smith, 'How Radio Research Has Enlarged our Knowledge of the Upper Atmosphere', delivered at the Institution of Electrical Engineers, London, on April 1933 (*The Electrician*, 5 May 1933, 581–2); and 'Note by Dr. Chree on the Present Information as to the Properties of the Upper Atmosphere', 5 Dec 1925, DPA.

[10] Hess (1928). And even where there was a certain interest from radio engineers and experimentalists in specific studies of terrestrial magnetism, that interest was seldom interpreted by them as evincing acclamation and appreciation of their endeavour. Rather, it appeared as a means to an end: the validation and confirmation of their ideas.

[11] Ladner & Stomer (1932, p. 41). Compare this case with the favourable treatment given by another Marconi's engineer in 1915, Dowsett (1915, p. 282).

[12] See, for example, 'Geomagnetism and the Ionosphere', EUA [D40], inaugural meeting of the Physical Society, London, 1949, in which Appleton describes how geomagnetic and radio-physics were converging. See also chapter one of Appleton's projected book (1941) on the ionosphere, EUA [D17].

[13] The early history of the ionosphere and radio wave propagation is reflected in a vast secondary literature. The most comprehensive contemporary summations are by Green (1946), Mimno (1937), Kenrick & Pickard (1930), Tuska (1944), and Mesny (1926). For an invaluable and unknown bibliographical repertoire (with 474 references), including many published in the less easily accessible journals, see Sacklowski (1927). For a general history of geophysics in the twentieth century, see Bush & Gillmor (1995). Among more recent studies, see Oreskes & Doel (2002) and Yeang (2003). For bibliography, Manning (1962) and Bureau of Standards (1931), which contains 620 ref. and short abstracts from 1900 to 1930.

stations (Beynon, 1975; Hong, 1996). The achievement symbolized his particular battle against the hegemony of cable companies. Wireless was a victory of technology over logic: 'at distances of over 700 miles the signals transmitted during the day failed entirely, while those sent at night remained quite strong up to 1551 miles' (Marconi, 1902).

Among the first to respond to the theoretical challenge was a brilliant and introspective theorist, Oliver Heaviside, an expert on cables and wireless, who in an introductory article on telegraphy in the *Encyclopaedia Britannica* propounded what he called the 'guidance hypothesis'. He had in mind a mechanism comparable to that of wave transmission along a conducting telegraph wire. 'There may possibly be a sufficiently conducting layer in the upper air. If so, the waves will, so to speak, catch on to it more or less. Then the guidance will be by the sea on one side and the upper layer on the other' (Heaviside, 1902, p. 215). The speculation that waves could be confined in their propagation as if along a two-wire transmission line was in fact unprecedented.[14] But Heaviside offered no justification for this 'guidance' nor did he provide a source for the upper layer or an explanation of the behavior of the waves.

A few months earlier Arthur E. Kennelly, a professor of electrical engineering at Harvard University, had nonetheless proposed a reflection model on an equally flimsy basis. Drawing upon J.J. Thompson's ideas on the electrical conductivity of air at low pressure, he inferred that 'At an elevation of about 80 km, or 50 miles, a rarefaction' existed. He went on: 'There is well-known evidence that the waves of wireless telegraphy, propagated through the ether and atmosphere over the surface of the ocean, are reflected by that electrically conducting surface'.[15]

Far from being unconnected conceptions, the contributions of this group of, mainly British, Maxwellians (Heaviside, Fitzgerald, Lodge), signalled the existence of a by no means marginal current of scientific thought, one with strong links to scholarship on telegraphy and late Victorian physics. These

[14]Heaviside was very likely influenced by Irish physicist G.F. Fitzgerald (1893, p. 526), who wrote in 1893: 'The hypothesis that the Earth is a conducting body surrounded by a non-conductor is not in accordance with the fact. Probably the upper regions of our atmosphere are fairly good conductors.... If the Earth is surrounded by a conducting shell its capacity may be regarded as that of two concentric spheres'. And in a footnote Fitzgerald mentions the dielectric *layers* of spherical condensers!

[15]Kennelly (1902, p. 473). Anyone acquainted with contemporary studies on gases would be surprised that Kennelly (like Henry Poincaré) believed that air was rendered conducting by 'its extreme rarification', that is, simply by being at low pressure. For at that time it was thought that air conducted if, and only if, it were ionized by some external agent. See Ratcliffe (1974, pp. 1,034–5).

conceptualizations, whose sources lay in the British electrical theory (which was strongly influenced by telegraph technology), but which welled up following Marconi's feat, continued to dominate the academic intellectual milieu until the early 1920s.[16]

After reflection came diffraction. During the same 20 years, a group of European mathematical physicists and mathematicians strove to explain Marconi's experiment in terms of wave diffraction. Hector M. MacDonald (1903), Henri Poincaré (1904), John W. Nicholson (1910), Jonathan A.W. Zenneck (1907), Arnold Sommerfield (1909), and George N. Watson (1919) developed rigorous mathematical theories that represented 'substantial contributions to physical optics' and to mathematical physics (Kenrick & Pickard, 1930, p. 650). Their ethos, methods, and aims differed from those of experimental physicists and electrical engineers. The former opted for what the latter had renounced: wave transmission according to Maxwell's equations, mathematical tractability, quantification instead of physical intuition, and theory as the instrument and justification for prediction. Yet they never evolved their own physical picture of the atmosphere.[17]

It was not until 1912, when William Henry Eccles—another self-confessed Maxwellian—investigated the causes of wireless disturbances ('atmospherics', 'strays' or 'statics'), that he made known his physical theory for long-distance wave propagation.[18] Eccles' *ionic* theory contained a qualitative, metaphorical, vivid model: the earth is surrounded by 'a permanently conducting upper layer which is somewhat sharply defined...we may call it Heaviside's reflecting layer' and by another located between it and the earth in the middle atmosphere (Eccles, 1912, 1913). The then professor at the University of London distanced himself both from Kennelly's mechanism of reflective propagation and from Heaviside's enigmatic guidance channel. In his model, the bending of radio wave results from an ionic refraction in the middle layer and from pure reflection in the Heaviside layer (Ratcliffe, 1971, pp. 200–1). But in his efforts to shed light on the refractive character of wave propagation,

[16]Bruce J. Hunt (1991a, 1991b) has argued that British pre-eminence in submarine cable technology promoted the field theory of electromagnetism by focusing attention on the propagation of signals and the behaviour of dielectrics.

[17]Although the diffraction accounts are much more complex and richer, here we only want to highlight their scant contribution to the structure of the atmosphere. For further details, see Green (1946, pp. 183–4); and Yeang (2003).

[18]Eccles is a curious acolyte of the intellectual tradition of late nineteenth-century microphysics, a Maxwellian yet clearly also an experimentalist in his vision of physical constructs. Buchwald (1985).

dispensing with 'wave surface' and with 'ground conditions', he fails to provide predictions consistent with the empirical law of long-wave propagation, the so-called Austin-Cohen formula.[19] This stated that wave intensity decayed exponentially with the inverse square root of wavelength and with distance. Since 1910, this formula had become the empirical basis for the engineering design of radio stations.[20]

During the 1910s, the reflection theory was synonymous with confusion and controversy.[21] In seeking the reasons for this situation, the Austin–Cohen empirical formula looms large. Its acceptance compelled them to use extremely long wavelengths and powerful stations for trans-oceanic communications. High-power stations and long waves responded to the Navy's imperatives, but their ineffectiveness for long-distance transmissions cast doubts on the credibility of the reflecting model inferred from Eccles' theory.

Yet, however much that formula erred for short waves (as proved in the early 1920s in wavelengths shorter than 500 m), the important fact is that, after the war, engineering practice and fundamental investigation coalesced. This proved decisive. Now, in 1920, with a larger availability of vacuum tubes, we find the most conclusive proof of the existence of a stratum in the upper atmosphere. The man responsible for the proof was a lucid mathematical-physicist from the Marconi Company, T.L. Eckersley, who worked on the bearing errors in direction-finding apparatus. Eckersley attributed such errors to 'irregularities of the Heaviside layer' There was no question: 'The existence of a ray reflected at night time from some upper conducting layer... may be

[19]The importance of the Austin–Cohen formula as a technical specification for transmission stations has been stressed by Yeang (2004).

[20]Apart from Eccles, one of the first to observe an intimate connection between ionization in the upper atmosphere and variability in wireless signals was John Ambrose Fleming (1912, 1914, 1915), a scientific adviser to Marconi, to whom the sun was the 'likely source' of the ionization required for communications. W.F.G. Swann (1916, pp. 1–8), a professor at the University of Minnesota, was perhaps, the first to consider both lines of approximation when he precluded the possibility of ultraviolet light as the principal ionizing agent on the basis of *all* existing evidence, both geomagnetic and radiotelegraphic.

[21] See, for example, the controversy unleashed in the pages of *The Electrician*, then the *agora* of wireless par excellence. Here, in a crossfire of correspondence, Eccles, Lee de Forest, E.W. Marchant, J.E. Taylor, and others confronted their stances fluctuating from the stalwart sentence 'clearly demonstrative proofs of the existence of the Heaviside layer' to the tempting invitation 'to get rid of [such] superfluous assumptions'. See 'Correspondence: The Heaviside Layer', *The Electrician*, 7 May 1915, 169; 14 May, 209; 21 May, 251; 'The Heaviside Layer: Some Further Correspondence', *The Wireless World*, 3 (1915), 7–8; and Merchant (1916). Another discussion, less strident but almost certainly important, took place at the Institution of Electrical Engineers in London. See Merchant (1915, pp. 329–44).

considered to be beyond doubt' (Eckersley, 1921, p. 248). What was undeniable was the existence of a descending and abnormally polarized ray, a *sine qua non* of any successful demonstration, the response that is expected from a reflecting stratum. But Eckersley gave no height. And in this, he definitely lost cogency (Green, 1946, pp. 187–91).

Although Eckersley's experiments did not obtain unanimous approbation, the tide in favour of the reflecting-refracting layer augmented.[22] Between 1920 and 1922, the issue still remained unsettled.[23] But perhaps, most important of all, a substantial shift in the metaphorical language regarding the conceptual representation of the atmosphere took place for reasons of simplicity and engineering practicality. Where in 1915 Eccles et al. surmized electromagnetic constructs such as *conducting surfaces* and *reflecting strata*, seven years later, a Marconi radio engineer, E. Bellini, drew upon a typically optical analogy, 'a horizontal *reflecting mirror*', to explain the errors of direction-finders (see Figure 25).[24] And such optical modelling remained subliminally but firmly attached to the substrata of experiments and theories on wave propagation after 1922.

The revolution of short wave: from excitement to discovery

In the next sections we shall examine from diverse angles the commercial and military conditions pervading the milieu of radio engineers. To ascribe extrinsic influences upon the dispositions and scientific discourse one should recognize not only the perceptions and conceptions, the reactions and the social approbation, but also the accepted vision of the socio-economic and cultural reality surrounding the scientific community. In this section, however, we shall

[22]H.J. Round, G.W.O. Howe, and A. Hoyt Taylor had also attributed the night errors to a reflecting effect. See, for example, Round (1920, pp. 224–57).

[23]Once again, *The Electrician* reflected the situation. Professor Elihu Thomson (1922). 'A Short History in Wireless'. *The Electrician*, 11 Aug 1922, [and replies, *89* (1922), 326, and 415], deemed the 'so-called Heaviside layer' as an unnecessary invention and a superstition. Oliver Lodge, writing about 'On Earth Transmission in Wireless', *The Electrician, 89* (1922), 206–7, retorted: 'It may be argued that the conducting layer...is too gradual to give sharp reflections...too irregular, too corrugated and uneven, to act as anything like a reasonably good mirror. [But] a conducting layer in the upper air is inevitable' (p. 207). Howe (1922, pp. 260–61) quite agreed: 'Everyone realises that it must be a very imperfect reflector, subject to rapid and enormous changes of position or contour or character, but causing a certain amount of reflection and refraction'. Eckersley's replied, ibid., 242–3.

[24]Bellini (1921, pp. 220–2). Again, there was still a persistent tendency, especially among radio engineers, to confound rarefaction with ionization as the cause of conductivity. See Bellini (1921, p. 220).

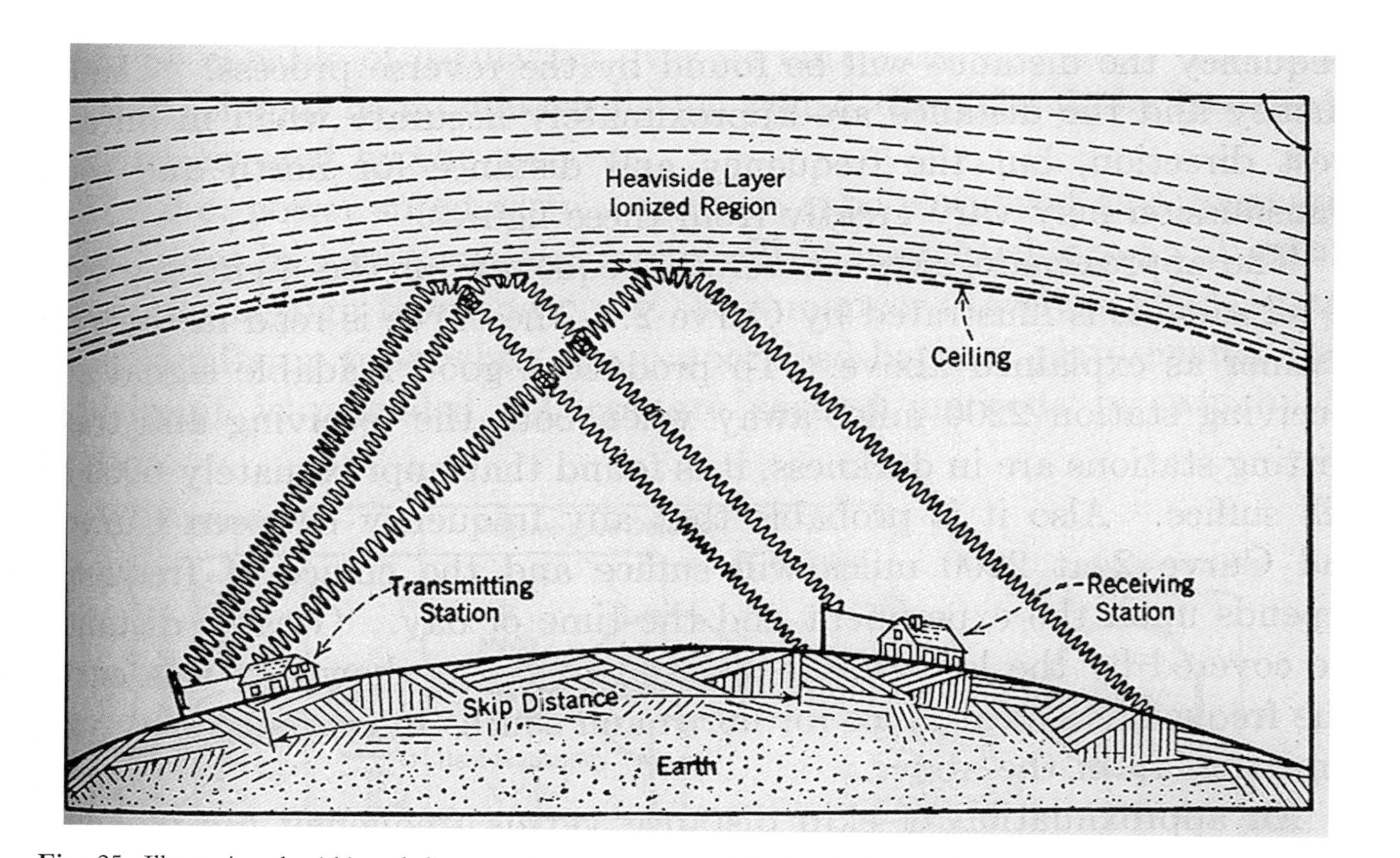

Fig. 25: Illustrating the 'skipped distance' features accounted for by the Heaviside-Kennelly layer theory of the ionized condition of the upper atmosphere. From Duncan & Drew (1929), p. 757.

identify merely the collective spirit, the mood surrounding radio technology, in order to understand the conceptual development of the atmosphere.

The advent of short wave, the exploration of a virgin band of the spectrum of wave-lengths, claimed by both radio amateurs and Marconi's engineers, began a period of innovation with profound repercussions. Physics, the military, industry, society, the British Empire, and the US were the principal beneficiaries. Short-wave transmission, previously restricted on military grounds to lengths under 200 meters, was 'startling', 'wondrous', 'unexpected'.[25] Besides its inherent advantages over long wave (more economical in power, more immune to disturbances), short wave bore the stamp of the allied victory in the war.[26] The revolutionary and quasi-visionary optimism it inspired set the tone for postwar euphoria.[27]

Circa 1924, in the prolegomena of the classical experiments, one perceives great unanimity regarding the following fundamental feature of the socio-cultural environment: in the commercial and military spheres most closely related to the radio field, but especially among radio investigators, a generalized state of excitement prevailed. And inherent in this excitement was the expansion of the radio communications industry. Far from being a hermetic realm, this phenomenon was felt to be a discovery of a new world, a unique opportunity not only for communications and broadcasting but also for science. In March 1924, Walter S. Roger, American adviser to the Peace Conference in Paris, wrote about 'Air as Raw Material':

> In a certain sense the development of radio has opened up a new domain comparable to the discovery of a hitherto unknown continent. No one

[25]To take but one of the countless testimonies: Hiram Percy Maxim, the president of the International Amateur Radio Union, in an address on 'The Radio Amateur', in Codel (1930, pp. 141–57), depicted 'the passionate intensity of purpose that had become stirred' by the idea of communicating without wires. 'There is no way to hold down such passionate interest' (p. 141). 'It was positively thrilling to pick this out of the air and to realize what it all meant.... Two-way communication of quite good reliability was effected over distances that were almost unbelievable. The greatest enthusiasm for long-distance communication was aroused.... Astounding results were achieved.... The accomplishments of the amateurs on the short wave awakened the commercials to the possibilities'. (p. 148).

[26]Kintner (1925, pp. 423–4). For a discussion on the participation of the amateurs and Marconi's, see Armstrong (1951, pp. 21–8)

[27]Professor of physics Low (1924) allegorically characterized this magical episode: 'One can imagine broadcasting of the future linking up every city from China to London; one can see special wave lengths for men, and equally special wave lengths for women' (p. 38). 'Undoubtedly, we shall see wireless controlled tanks, submarines, and torpedoes.... Power will one day be transmitted by wireless' (pp. 71–2).

> can foresee with certitude the possible development of the transmission of energy through space. Really great stakes are being gambled for. And private interests are trying to obtain control of wave lengths and establish private property claims to them precisely as though a new continent were opened up to them and they were securing great tracts of land in outright ownership. (Rogers, 1924, p. 254—quoted in Childs, 1924, p. 520)

Although the excitement associated with the labels 'discovery' and 'revolution' had been palpable on previous occasions in wireless, it appeared as an intense widespread sentiment only after the advent of short wave.[28] 'The dawn of a new era' prophesied amateurs in February 1923 at a meeting of the Radio Society of Great Britain (quoted in Eccles, 1930, p. 16). So it was according to Appleton: 'Much of the fascination of wireless may be attributed to the fact that the use of different wavelengths leads to very widely divergent results.... That this is so is a modern discovery—and one that has surprised even the experts' (Appleton, 1930, p. 710). And, in the early 1930s, Eccles still exalted the rapture of 'a bloodless revolution' that had led to 'an improved knowledge of the atmosphere' (Eccles, 1930, p. 17).

The quasi-religious belief in the revolution, in the 'conquest of space', became a common phenomenon in the American radio amateur community between 1922 and 1924.[29] As if beguiled by a new continent to explore, amateurs hastened in their transatlantic communications to demonstrate the inoperability of the Austin–Cohen formula for high frequencies and underlined problems that physicists were obligated to elucidate.[30] By emulating Marconi's engineer C.S. Franklin's experiments of 1919 in England, radio amateurs such as J.L. Reinartz, C.D. Tuska, K.B. Warner, D.C. Wallace, and J.J. Lamb made short-wave radio a passion, but also a scientific research tool.[31]

Also of major importance for our present enquiry is the entry of American companies into the short-wave broadcasting industry. Between July of 1923, when the first short-wave programmes were broadcast, and the end of 1924, this waveband became a prime objective of domestic commercial interest—especially

[28]The ascription of the epithet 'discovery' was common in the contemporary press. See the editorial 'Who Discovered the Long Range of Short Waves?', *EW & WE, 3* (1926), 715; and the German radio engineer of the Telefunken Co., H. Rukop (1926, pp. 606–12), rhapsodizing about 'the discovery of the extraordinary effectiveness' of short wave for trans-oceanic traffic.

[29]The story has been thoroughly described by De Soto (1936, pp. 88–105), Douglas (1987), and Aitken (1976).

[30]See the transcription of the first transatlantic communication on 28 Nov 1923 by Deloy (1924, p. 40).

[31]McNicol (1946, pp. 200–2).

to Frank Conrad of Westinghouse, to David Sarnoff of the Radio Corporation of America (RCA), and to the engineers at General Electric (GE) (Phipps, 1991, pp. 215–27). A *New York Times* editorial of July 1924 compared the elimination of the high-powered, long-wave 'super station' by short wave to the action of David's 'well-directed little pebbles in his contest with Goliath' (*New York Times*, 20 Jul 1924, p. 4). And the fundamental characteristic of the mood that was generated was 'excitement', 'amazement', 'ecstasy'.[32] Over and over again, the general press as well as the specialist journals identified short-wave radio with revolution, discovery, and progress, and glorified it as 'the wave of the future'.[33] And thus we see how the rhetoric surrounding short wave was inflated to gigantic proportions.

In Britain, it was especially the Marconi Company's engineers who promulgated this enthusiastic version of the revolution of short wave. Its purest commercial expression was the so-called 'beam system', in which waves focused by a parabolic reflector were beamed towards the distant target. It was to place radio 'on a par with cables for the first time'.[34] It obliterated the past and the obsolescent long wave by a 'new technique' that promoted audibility, economy, simultaneousness, propinquity, confidentiality, and, above all, the *unity* of the Empire. Marconi himself trumpeted 'the demise of long wave', of which the revolution convulsed the very theoretical foundations: 'The whole theory and practice of long-distance wireless communication is just now undergoing a most important and radical change' (Marconi, 1924, p. 62—quoted in Beynon, 1975, p. 663).

However, the change in wavelength did not affect the physical model of the atmosphere: implicit in this notion of revolution was the unanimous acceptance of a reflecting ceiling (for those feats would be 'practically impossible' without the assistance of 'a conducting layer').[35] And, furthermore, implicit in it was a positive valuation of the modus operandi and practices of contemporary radio amateurs and professionals. If the lay listener realized that the waves, almost inaudible at short distances, emerged 'at great strength in another continent thousands of miles away!', (Appleton, 1930, p. 710), then it is reasonable

[32]Illustrative of the strength of excitement generated in the US is Archer (1938, p. 369).

[33]See, for example, 'Marconi Foretells a Radio Revolution', *Current Opinion*, Sep 1924, 352–3; *World Wide Wireless*, Apr 1924, 15–16; *New York Times*, 20 Jul 1924, 15; 20 Nov 1924, 15; Morecroft (1924, p. 296); Phipps (1991, pp. 217–9); Archer (1938, pp. 329–30).

[34]British Information Services (1963); Barty-King (1979, p. 193). The revolutionary nature of the beam system is vigorously trumpeted by Marconi and his engineers, for example, in Morse (1925, pp. 88–91).

[35]Marconi's Co. (n.d., p. 15).

to assume that the prestige of those who had investigated this discipline for both academic and commercial purposes would be significantly augmented. For the same reason, in the layman's eyes, it is also reasonable to assume that the endeavours of those who had explored the atmosphere by other methods, such as geomagneticians, without offering any solution to wave propagation, were seen as merely academic.

Reactions of radio scientists to short wave, circa 1924

By the autumn of 1924, the revolution of short wave generated in physicists' minds certain cogitations and preoccupations regarding the treatment of ionic refraction. While forcing the revision of Eccles' theory, short wave reinforced the physical model the theory incorporated.

The movement in favour of the reflecting layer, in effect, reaches its highest level of popularity between 1923 and 1924. There are numerous testimonies from this period which indicate that the certainty of its existence is reinforced. Such is the case with M.P. Lardry, a member of the Office for the Coordination of Amateur Scientific Observations of the URSI, who maintains that only *one* hypothesis is capable of explaining the strong diurnal variations of short-wave strength: 'that of the Heaviside layer'.[36] Similarly, the contemporary physicist G.W.O. Howe confessed in September 1924 that the idea of a conducting layer in the upper atmosphere 'is now generally accepted'.[37] Whereas in 1922, he had recognized the experimental obstacles in the path of explaining 'the character of the layer', now he thought it possible to ascertain its 'exact height and characteristics' from the attenuation of short waves.[38]

Once again, in seeking the reasons for this gradual fermentation of a widespread conviction of the existence of a reflecting layer *circa* 1924, one cannot help but think of the recent achievements in long-distance communications. It is only by reference to a generalized sentiment of excitement and bewilderment that one can justify the generalized assent to the reflecting hypothesis among

[36]Landry (1924, pp. 449–510): 'Au milieu de ce chaos, une seule hypothèse paraissait sérieuse : celle de la couche d'Heaviside'. A. Hoyt Taylor (1924, p. 13), from the US Navy Department: 'To me this [the high degree of intensity of the signals received] would indicate that there is so complete a reflection of these waves at some upper and probably ionized layer of atmosphere'.

[37]Howe (1924, p. 548): 'The wonderful results obtained by amateurs…have completely taken a spill our preconceived ideas' on waves propagation.

[38]In September 1922, Howe (1922, pp. 260–1) emphasized in his diatribe against Elihu Thomson that 'one is sometimes asking too much of the Heaviside layer', and that 'there is still much to be explained'. In September 1924, Howe (1924, pp. 282–3) placed emphasis on short-wave transmission as a means of the exploration of the upper atmosphere.

radio scientists and amateurs. For it was not until the autumn of 1924 that Eccles' ion-refraction theory was extended and perfected, and it continued to be as debated then as it had been before the irruption of short wave. In reality, it was not so much a mathematical treatment as a sophisticated, idealized physical model: a vertically ionized atmosphere with gradually varying refractive indices but no magnetic field. It was a precursor of the atomic and molecular physics of the late 1920s and 1930s, a vague indication of how, embracing a reflecting-refracting mechanism, one could formally explain the interaction between electrons and radio waves, between ionization and propagation. Ironically, the successful short-wave transmissions (and their peculiar behaviour) persuaded physicists both of the insufficiency of Eccles' ionic theory and the 'reality' of the visualizable atmospheric model the theory incorporated.[39]

In fact, in the midst of all the flurry of transatlantic communications, we find only two serious attempts to offer an alternative to the Heaviside-Kennelly layer. The most solid we owe to a radio engineer of the National Telegraph Engineering Bureau of Germany, M. Baeumler, who was interested in the diurnal and annual variations of wave intensity. He supposed that 'the waves are refracted, absorbed or reflected at the boundary surfaces of air masses of different densities', and that 'the electric turbidity of the atmosphere' caused 'the diminution of field intensity'.[40] The most scathing, however, comes from a German electrical engineer, A. Meissner, who discredited radio engineers' attempts 'to explain everything by the assumption of a condition of the upper atmosphere which may not exist at all' (Meissner, 1924). Both views passed practically unnoticed.[41]

Although the irruption of short wave provoked in physicists certain doubts about the mathematical treatment of refraction, the 'objectivity' of the concept of a reflecting layer remained intact, as Appleton described on 21 November, one month before his crucial experiments:[42]

[39]Short wave forced Eccles' theory, to be revised but reinforced the reflecting model that it incorporated. This is the case with Howe (1924). Similarly, for O.F. Brown (1924, pp. 595–7), the technical secretary of the RRB, the effects of fading of short-wave signals supported the existence of the Heaviside layer, though little was known of the cause. He proposed 'certain astrophysical hypotheses' to account for 'the production of such layers'.

[40]Baeumler (1925, p. 26). While published in 1925, the paper was received in July 1924.

[41]Reviewing the theories on the propagation of waves, Mesny (1926, p. 456) stated that 'les arguments favorables [à l'hypothèse d'une couche conductrice] se soutiennent d'eux-mêmes.... [Baeumler and Meissner's] argumentation se borne à quelques affirmations non étayées. Si difficile que l'on puisse être sur les justifications des théories, la vraisemblance d'une haute atmosphère conductrice s'impose'.

[42]'Beam Wireless', manuscript, 21 Nov 1924, *Papers*, Appleton Room, EUA [D5].

> In 1924 the last of a series of wireless links between Great Britain and the Dominions has been completed.... Some years ago, results...were obtained by wireless amateurs and by the engineers of the Marconi Company.... By that time it was becoming clear that...wireless waves can reach any particular spot not only along the ground, but also by a kind of overhead path via the upper atmosphere. There is, about *fifty miles* above the ground, a layer free of electricity which sends back or reflects the wireless waves. When such waves come back to the ground they can produce a signal in just the same way as can the waves which travel along the ground all the way. It was therefore not a difficult matter to explain the results of the short wave experiments in terms of these sky-waves and ground waves.
>
> More recent experiments carried out with all the resources of the commercial companies have shown that these short waves seem to travel enormous distances with very little loss of strength.... Four years ago an account of [shortwave experiments] using reflectors to produce a wireless beam was published by Mr. Franklin.
>
> I have mentioned that we think that the signals received at great distances are due to the sky or overhead waves and this may have made you wonder whether better results might not be got by deliberately projecting the waves upwards at the sending station. To test this experiments have been made using a horizontal aerial with point.... As was expected the signals were much stronger when the beam was tilted upwards so that the waves were projected into the upper atmosphere. It is quite likely that many more of the points which puzzle us in connection with short wave wireless may be solved using similar methods.[43]

We have quoted Appleton so generously for several reasons. Firstly, no assertion regarding the existence of the reflecting layer in such a convinced, steadfast, not to say 'ontological', tone would be likely to be found before 1925. Secondly, the commercial milieu and its experiments are not a mere backdrop. Appleton is fully aware of both the pressure—n.b., not leverage—that private firms are exerting in their quest, and the urgent need for solutions to long-distance communication.[44] He has resolved to rehearse for short-distance

[43]Here, an observation is necessary: the frequency at which contemporary physicists and radio engineers regarded the wavelength to be short—i.e. the dividing line between medium and short waves—was continuously shifting. In his historic papers of 1925, Appleton considers the wavelengths used (within the BBC's broadcasting range, i.e. about 350 m.) as short waves. And frequencies of 1,500 kc, corresponding to 200 m., were labelled as very short waves.

[44]It is necessary to add that at this time, in addition to the research engagements at Cambridge and at the British RRB, Appleton acted as an industrial consultant on wireless subjects to the Pye group, Trippe and Philips, and the Post Office, and had close contacts in the radio

experiments, basically because he has not the resources of a large company, but also because he believes that the effects between sky and ground waves can be much more easily interpreted in this way.[45] And, last but not least, he has adopted the interference (strangely enough, as Eckersley did it in 1921) and the vertical projection of waves as the ideal method of elucidating the 'puzzle of shortwave'.[46]

In this case, the commercial milieu begins to provoke reactions among radiophysicists. And when we see how in October 1924 the aged and venerable professor of physics at Cambridge University, Sir Joseph Larmor, extended Eccles' ionic theory to explain the 'mystery' of short-wave transmissions, admitting their influence upon academia, then it is reasonable to conclude that the explanation of the reactions of the time must be sought in the excitement of the commercial and social environment.[47]

It seems pretty clear that the British physical community involved in radio research was infected by this heightened mood, that their perception of the insufficiency of the theory of wave propagation was affected by the echoes of revolution. But here we must allow that this infection was deeper than mere rhetoric. For, as we shall see, radio experimentalists, more than mathematical-physicists, tended to make use of the very rhetoric surrounding short wave when publicly announcing their results. Just as the mood of excitement was a focus of contagion, so it was also to become a culture medium, a seedbed propitious to the 'reification' of layers, to the substantiation of accepted metaphors. By conferring the title of 'discovery' on their experiments, Appleton

industry—such as the brothers Eckersley, who were working for the BBC and for Marconi's. See Clark (1971, p. 31, pp. 68–9).

[45]'Some Wireless Methods of Investigating the Upper Atmosphere', *Papers*, Appleton Room, EUA [D1].

[46]On 28 Nov 1924, Appleton (1924–25, pp. 38D–45D) adumbrated his classic experiment, and mentioned Eckerley (p. 18D): 'The variations of bearing errors in direction finding were first described by Hoyt-Taylor and Eckersley. If we attribute these errors to the interference of the two rays and assume that the reflected one is the variable one, we should expect the errors to be noticeable at shorter distances over land, where the direct ray is strongly attenuated, than over sea, where little absorption takes place'.

[47]Larmor (1924, pp. 1,025–36), abstracted in *Nature, 114* (1924), 650–1. Larmor's class lectures in the Mathematical Tripos in February 1924 in which he explained his theory were not well attended. In this respect, he said: 'The attention now excited by long-range free electric transmission, the most wonderful sudden practical evolution since the telephone, may attract the interest of a wider audience' (p. 1,027). Again, Green (1946, p. 199): 'The value of the Eccles-Larmor theory of ionic refraction is best seen in combination with the success of short-wave communication over very long distances'. See also Fleming (1925, pp. 123–4).

and his colleagues were not only to share in the spirit of their most immediate environment but were also to establish a communion with radio amateurs and engineering audience. Over the course of the next years, their endeavours were to shed light upon all those questions which their *confrères* found most imperative—height of layers, most optimum frequencies.

In sum, it is correct to infer that physicists, being immersed in a highly exhilarating climate, saw the 'revolution of short wave' as an intellectual challenge, something which inspired them to comprehend the 'new-discovered continent' But by then (early 1925) the reflecting layer had already been invented, solidly manufactured, metaphorically moulded..., and socially accepted.

The realist interpretation as confirmed by ionospheric physicists

Historians, like E.H. Carr, remind us how useful it may be to learn from the testimony of contemporary observers when one needs absolute confirmation that one is not going astray. Carr urged us to view the voice of secondary actors as a window on the past. With a certain apprehension and inability to undertake an exhaustive reconstruction of the intellectual universe of radiophysicists at that time, we have opened that window—first through contemporary ionospheric physicists, and later through modern textbooks—in order to seek support in their opinions and reflections. In doing so, we shall see how they themselves debated—in retrospect—about the reification of the layers.

For the purposes of this work, the retrospective analysis by D.R. Bates, H.S.W. Massey, and W. Dieminger—published in commemoration of the fiftieth anniversary of the finding of the Kennelly-Heaviside layer—[48] are of special interest. Although these ionospheric physicists are more interested in the truthfulness (or falseness) of early theories, their appraisals of the then physical picture—and sometimes their characterizations of the commercial and technological milieu—in fact throw much light on our question. In particular, the proliferation of studies as a result of the *in situ* exploration of the upper atmosphere with rockets discloses many of the inconsistencies and cultural beliefs which pervaded the radiophysicist community during the interwar years. Notwithstanding the variety of their nationalities, theoretical or experimental credentials, and technological preferences, they agreed on the same diagnosis of the conceptual representation of the atmosphere: hypostatization of the term layer; excessive predisposition to regarding an abstraction as if it had concrete or material existence; overemphasis on the physical existence of discrete

[48]See the special issue of the *Jour. Atm. Terr. Phys., 36* (1974), 2069–2320. Some of the next quotations are extracted from Gillmor (1981).

layers; literal, unjustifiably realist interpretation of a conceptual entity, with a longing for the immediate apprehension of reality rather than for the functional adscription of concepts, according to which theoretical constructions are epistemological instruments for predicting observations.[49] This attribution of existential nature, from which a clearly stratified atmosphere emerged, was but the solution, in Bates' view from the University of Belfast, 'to a problem which can scarcely be said to have existed. We undoubtedly attached too literal an interpretation to the word *layer*. Perhaps some of us were even misled by the parabolic model used by radio scientists and failed to appreciate fully that they are without real physical significance' (Bates, 1973, p. 1,942).

H.S.W. Massey of University College, London, pinpointed instrumentation as the main cause, and inspiration, of this realist point of view: 'The techniques of ionospheric sounding used to study the properties of the reflecting layers naturally tended to convey the impression that the maxima were unique and sharply defined. If we look at electron or ion concentration altitude profiles obtained from rocket flights this impression is very much less clear'.[50] Likewise, Dieminger noted the often inconspicuous parallelism between the breakthrough of instrumentation and the change in the way the physicist conceived the phenomenon, or in other words, the scientist's interaction with nature through technology: "It sounds strange that in those days the F-layer was termed a *regular* layer in contrast to the *irregular* E-layer. This was a result of the technique of observation, since on fixed frequencies the reflections from the F-layer showed a more or less regular variation of height with the time of day, whereas reflections from the E-layer appeared and disappeared irregularly" (Dieminger, 1974, p. 2,086).

Although these retrospective perceptions by contemporary ionospheric physicists provide us with evidence of a realist attitude towards science, they

[49]One aeronomist went beyond those considerations: 'The term *ionospheric physics*', he wrote in 1972, 'is rather meaningless. The charged particles of the upper atmosphere form a very small percentage of the total and are largely at the mercy of the neutral atmospheric components which form the vast majority. It is true there is a cosy, well-knit community of *ionospheric physicists*...but their very narrowness explains their lack of progress in the past 40 years'. Quoted in Gillmor (1981, p. 111).

[50]And he continues: 'Because of this the interpretation factors distinguishing the ionospheric layers sometimes presented a major problem which would perhaps have not appeared so prominent if the basic ionospheric data had come from space flights. Sometimes in the initial development of a theory it is a disadvantage to know too much and at least the early ionospheric theorists did not suffer in this regard'. Massey (1974, p. 2,141). Or further: 'One of the few advantages which the early theorists enjoyed in discussing the ionosphere was the belief that the ionospheric layers...were more clearly defined that in fact they are'.

fail to shed too much light on the motivations and inspirations which may have provoked such impulses. The following disclosures, nonetheless, not only seem to suggest a valuation of the external environment but also invite us to recognize the decisive role played by the expansion of commercial and military radio communication.

As a result of the success of short wave in long-distance transmissions, the use of graphs representing the virtual height of layers versus the frequency of radio waves became pervasive. In analysing the validity of these curves in 1937, F.H. Murray and J. Barton Hoag of the University of Chicago clearly detected the roots of the reification of the layered model: 'Curves such as [those], while *possessing great value for communication purposes* [italics added], in that they express the delay time of the sky wave pulse over that of the ground wave pulse, give a greatly distorted picture of the true altitudes to which the waves rise and the locations of the various ionized strata in the upper atmosphere' (Murray & Barton Hoag, 1937, p. 334). A more penetrating and insightful argument of this tendency towards compliance within a highly demanding commercial and military context is that formulated by S.A. Bowhill and E.R. Schmerling in 1961. Here, the phenomenological approach appears as a generalized, if inappropriate method symbolizing the somewhat submissive attitude of radio-physicists to the pressure and exigencies of radio communication. 'Although the ionosphere is a very active research field, its study has been to some extent hampered by the fact that the practical applications of radio propagation have led to a phenomenological type of approach to ionospheric behavior. The early availability of large amounts of numerical data on critical frequencies, obtained at the world-wide sounding stations set up for the purposes of long-distance radio propagation prediction, contributed to this approach'.[51]

According to the textbooks, the bias introduced by instrumentation (whose most expressive icon was the ionogram) limited the spectrum of possible agents responsible for the realist tendency. In 1960, Y.L. Al'pert, examining the structure of the ionosphere and its influence upon radio wave propagation, attributed to technology the powerful (and perilous) faculty of shaping the physical picture of the atmosphere. This was not an error, nor was it an illusion or a figment of imagination, but simply a distortion that was only disclosed when it was supplanted by another technology.

> Radio pulse methods make it possible to observe directly the ionospheric electron concentration maxima N_M. As a result of the fast increase in the group delay time as the maximum N_M is approached, and the frequently

[51] Bowhill & Schmerling (1961, pp. 265–326),—quoted in Gillmor (1981, p. 111).

> observed sudden changes in the ionospheric altitude-frequency curves, one gains the impression that there exist ionospheric layers with sharply defined electron concentration maxima. This circumstance has led to the notion of a well defined, stratified structure of the ionosphere.... However, with the development of methods for the analysis of the altitude-frequency characteristics of the ionosphere, which yield the true distribution of the electron concentration with altitude, both above and below the electron concentration maxima, it became clear that if not always, then at least frequently, the electron concentration maxima are apparently not sharply defined. The layered model of the ionosphere has gradually changed, as new though indirect data on the altitude variation N (z) became available. (Al'pert, 1960, p. 21)

Eloquent in this sense is the introduction to the excellent treatise, *The High-Latitude Ionosphere and its Effects on Radio Propagation*, by J.K. Hargreaves and R.D. Hunsucker. For, besides the feasibility, suitability, and potentiality of propagation prediction, the authors underscore the distorting effect of instrumentation in visualizing data. They hold that an understanding of ionospheric mechanisms is vital to efficient radio communication, but also that 'the identification of the regions was much influenced by their signatures on ionograms, which tend to emphasize inflections in the profile, and it is not necessarily the case that various layers are separated by distinct minima' (Hargreaves & Hunsucker, 2000, p. 13). In a treatise on the *Fundamentals of Aeronomy* (1971), which was highly regarded in the 1970s, R.C. Whitten and I.G. Popoff accept the reification of the layers with the ambiguous epithet 'inherent': 'The notion that layers with unique properties exist in the upper atmosphere was inherent in the development of ionospheric studies' (Whitten & Popoff, 1971, p. 15). Observational technique also deserves mention: 'in early days... layer formation theories appeared to be very explicit, relatively simple, and very relevant because radio sounding detected what appeared to be specific layers'.[52]

In light of some of the aforementioned assertions, one might be tempted to think that radiophysicists and engineers were misled by a mere optical effect, and that the substance of debate is reduced to a misinterpretation, a

[52]Ibid., 17. See also Whitten & Popoff (1965, p. 7) characterizing the outlook in the space age: 'It has become more commonplace to refer to *regions* of the ionosphere rather than *layers*. This practice has arisen since the advent of rocket sounding experiments which do not show the well-defined *layers* that seemed to exist as a result of earlier interpretations of radio sounding experiments. Both models and experiments indicate that layers are simply large gradients of electron concentrations.... The boundary altitudes are certainly not sharply defined'. Similar viewpoints were upheld as early as 1942 by Barton Hoag (1942, p. 54): 'It is not clear just why there should be layers instead of a continuous distribution of ions and electrons'.

misapprehension, or even to a misguided practice. *De facto*, one might object, most ionospheric physicists did not go much further in their reflections, and imputed it to instrumentation and to data presentation on the basis of nebulous arguments. But all these issues will be elucidated later. Here the question that concerns us is what picture of the atmosphere and its vertical structure the radio scientists devised. And this was without a doubt the picture of a layered, sharply stratified atmosphere, the result of a realist interpretation.

Reification of invention

Contributing his chapter 'Radio and the Ionosphere' to the posthumous tribute to Clark Maxwell in 1963, Appleton began with a memorable pronouncement with profound epistemological and philosophical implications. 'The ionosphere is a realm of the universe', he asserted, 'which was both *invented* and *discovered*' (Appleton, 1963, p. 70). In essence, the notion of a natural and objective structure to which he alluded is neither outlandish nor unorthodox. Most radiophysicists in all probability subscribed to it *circa* 1925.[53] It was *vox populi* and *credo communis* in academia. If one peruses the definition, however, one encounters two utterances which seem either antithetical (is not *discovery* the finding of a new 'natural phenomenon' previously hidden—by no means invented?) or at least contentious (is it possible to *invent* an entity inherent to nature itself?). It is precisely these apparently irreconcilable and controversial expressions which point to matters Appleton implicitly deemed as unproblematic and indisputable with regard to the interaction between experiment and experience. Due to its complexity, we shall not analyse here the first of these—the replication and authorship of discovery as a complex issue of negotiation inside the scientific community.[54] It is the second issue, the reification or hypostatization of

[53]The dichotomy between discovery and invention was explicitly evinced by the German theoretical physicist Felix Auerbach in 1923 when he asserted that X-rays were not discovered by Röntgen but invented by him. 'Es ist ein entscheindender Charakterzug der Physik, dass in ihr das Experiment die Beobachtung fast völlig verdrängt hat' [Auerbach (1925, p. 3)—quoted in Otto Sibum (2004, pp. 60–1)]. Sibbum maintains that Auerbach's reflections on the experiential basis of physics are an integral element of a long historical debate on the epistemological status of experiment and experience.

[54]There are two noteworthy aspects in this process: firstly, a consensual attribution of primal authorship by reason of an overemphasis on the 'first direct proof' (to Eckersley's detriment), which is reflected in Appleton's reiterated citation on the lack of 'adequate experimental evidence on the existence of the Heaviside layer'—delivered by R.L. Smith-Rose and R.H. Barfield [see *EW & WE, 2* (1925), 373] in September 1925; and, secondly, an immediate and generalized replication, also vigorously sustained in realist terms, of Appleton and Barnett's experiments and of Tuve and Breit's with such disparate results that made it necessary to negotiate the identity

an invention—that is, to construe a conceptual entity such as 'layer' as if it had material existence—which we shall discuss and examine in detail.[55]

But what was in essence Appleton discovery? To what extent did he extricate himself from his predecessors? And, perhaps more importantly, upon what evidence did it rest? That 'finding' was, in a word, the *reflecting-refracting layer* (previously manufactured by Eccles and Eckersley).[56] And the evidence offered, the equivalent, rather than actual, height; that is, the height 'that a wave would reach if it travelled in a straight line through the ionosphere and was then reflected as though from a mirrorlike surface' (cf. Bellini's 'optical analogy').[57] In effect, in December 1924 Appleton and M.A.F. Barnett, a research student from New Zealand, devised an ingenious experiment for determining this height. In Figure 26, radio waves travel via two paths—one direct (ground ray) and one indirect (sky ray). By varying the wave frequency from the sender, they artificially produced an interference phenomenon in the intensity at the receiver. Since the sky ray path is an isosceles triangle of altitude h and since the path difference between the rays is easily measurable, they geometrically inferred an equivalent height of 100 km, which they construed as an actual layer height of 80–90 km. Here, their readiness to regard this datum as an inherent feature of nature did not originate from reading the empirical graphs but from acceptance of metaphors already approved by the relevant community.

and reality of such events, and which we shall only touch upon in passing. Both aspects figured prominently in the emergence and consolidation of the scientific discipline in the late 1920s. We do not, however, attempt here to treat this intricate and inextricable question. On discovery and the fixing of scientific practices, see Woolgar (1976), Pickering (1984), Collin (1985, pp. 38–46), and Schaffer (1986).

[55]As often happens in modern historiography, and even more in the literature of scientific divulgation, when paradigmatic discovery stories are considered, the complexity of the real historical process—that is, the mutual articulation of experiment and theory—is diluted in chronological narratives. In this respect, Schaffer (1986, p. 397) points out that 'a complex enterprise, accessible to historical and sociological understanding, generates objects which are then labelled as discoveries. Subsequently, the story of that process is rewritten. The lengthy enterprise is telescoped into an individual moment with an individual author'. This analysis is germane to ionospheric layers: the story of their finding has been related over and over again as a sequence of crucial discoveries. An excellent example is *Mirror in the Sky: The Story of Appleton and the Ionosphere*, a 20-minute film produced by Mullard Ltd. in conjunction with the Educational Foundation for Visual Aids, London; and Harrison (1958).

[56]Appleton & Barnett (1925, pp. 621–41). Although Appleton & Barnett (1925, pp. 333–4) start from the postulate of a reflecting layer, they add that 'the term *reflection* used for convenience must be taken as meaning *ionic deflection*', especially in the case of short waves.

[57]Terman (1943, p. 718).

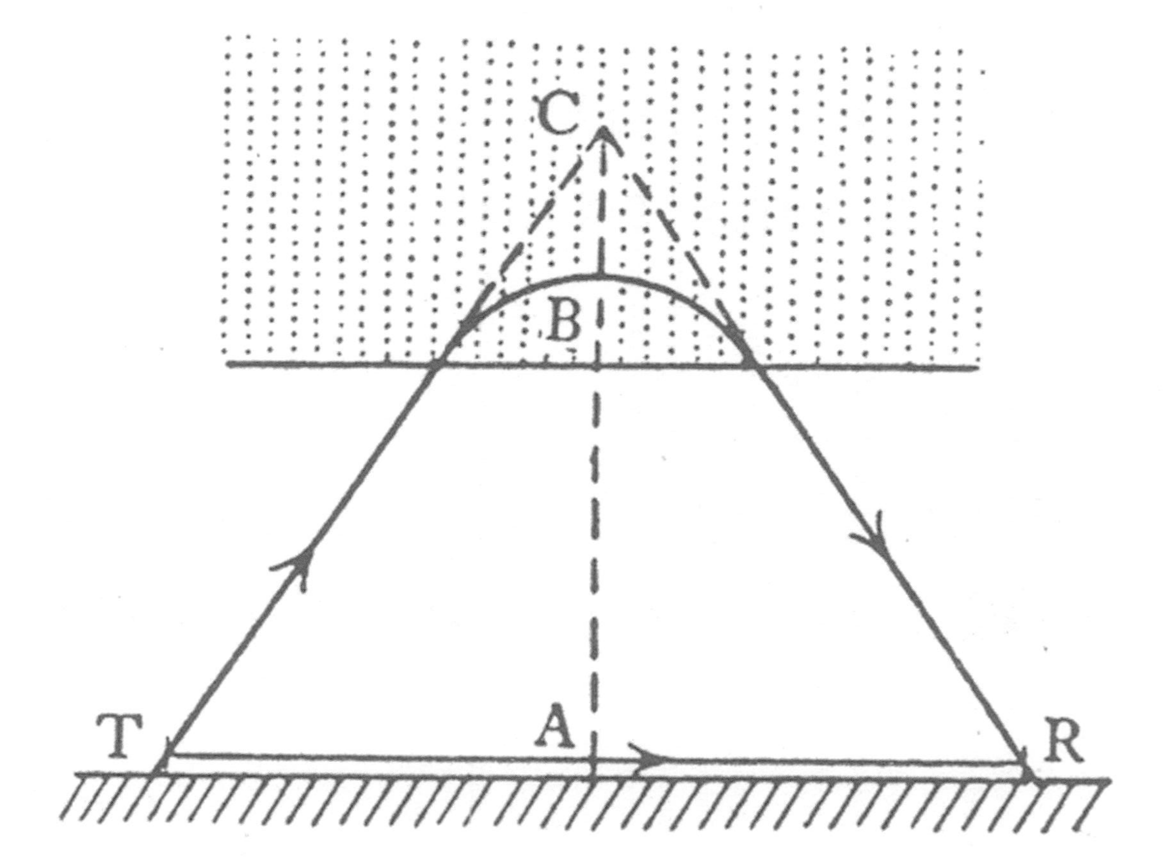

Fig. 26: Radio waves travelling through two paths—one direct (ground ray) and one indirect (sky ray).

This very notion of an essentially reflecting layer in accordance with optical analogy was that adopted by Gregory Breit and Merle Tuve at the Department of Terrestrial Magnetism (DTM) of the Carnegie Institution of Washington in mid-1925. Now, instead of a gradual frequency change, Breit and Tuve emitted pulses at a given frequency—a method known as 'pulse-echo sounding', the precursor of the ionosonde. They recorded the amplitude and time delay of the echoes photographically at a receiver several miles distant. The time delay was proportional to the height of the pulse echoes and thus, to the height of the reflecting layer. The time delays measured corresponded with equivalent heights between 88 and 225 km.[58] But, as Tuve acknowledged, 'the true heights to which the waves travel must be obtained indirectly from the equivalent heights which are actually measured', which, clearly, led 'to an important difficulty' (Tuve, 1932, p. 161).

Resorting to a fictitious reflector in order to disclose the vertical structure of the atmosphere, radiophysicists faced the problem of ascertaining the relation between the 'equivalent' and 'actual' height, that is, of filling the gap between artifice and nature. However, this *gap* (prelude to reification) took on different aspects according to whether the reflection occurred from a discrete layer or from a diffuse region. If from the former, the propagation mechanism could be likened to a total internal reflection, regulated by a relatively sharply defined low boundary, whose equivalent height exceeded the actual one by only a few per cent. If from a diffuse region, however, the appropriate analogy was a refracting medium governed by the magneto-ionic theory, stretching out indefinitely, whose equivalent height had no precise meaning (indeed, it often exceeded the actual height by at least 25%!), but which seemed necessary to approximate to 'reality'.[59]

To elucidate these issues, radiophysicists resorted to vertical incidence ionospheric observations, which were substantially improved by introducing oscilloscopes. Screens enabled visualization of the height-versus-time graphs (replaced in 1930 by the crucial height-versus-frequency curves) of radio echoes. Between 1927 and 1930, Appleton and then many others construed the

[58]Breit & Tuve (1926); Breit & Tuve (1925, p. 357). In 1925, at least four other methods were deployed by G. Munro (in New Zealand), J. Hollingworth (at the British RRB), R. Bown, D.K. Martin, and R.K. Potter (at the American Telephone & Telegraph Co.), and R.A. Heising (at the Bell Telephone Laboratories) to determine the layer height. Green (1946, pp. 219–25), Tuska (1944), Kenrick and Pickard (1930).

[59]Turner (1931, p. 56): 'Since the return of the ray…is effected rather in a refracting region than at a reflecting surface, the meaning of the effective height of the Heaviside layer has not always been clear'.

discontinuities of these graphs as the morphological traces of the fine structure of the ionosphere.[60] The typology comprised the *D* layer (associated with the ozonosphere), the *E* layer (discrete), and the diffuse *F* region (composed of two substrata, F_1 and F_2) (Figures 27 & 28).[61]

However, if such was the radiophysicist and engineer's notion of layer, with its thorny postscript of equivalent height, how indeed could they hypostatize so emphatically, almost dogmatically, this conceptual artefact? The question was partially elucidated in 1933 by Frederick A. Lindemann, the future Viscount Cherwell and a meteor connoisseur (among many other virtues), at the conclusion of a discussion on the ionosphere held at the Royal Society of London. 'It might well be, for instance, that the two main layers instead of being separated by a hundred kilometres or so, as is generally assumed *sub silentio*, are really quite close together and merely represent more or less typical changes in the ionic density gradient. Questions such as these must be carefully considered and determined before we expend too much time or effort in endeavouring to relate the existence of these layers with other physical phenomena.... In discussions such as these..., one tends subconsciously to identify equivalent heights with real heights'.[62] Lindemann's last consideration is a thought-provoking one, and his reservations ought to have, without doubt, stimulated ruminations from the audience. For if one finds radiophysicists

[60]Builder (1932, pp. 667–72). See also 'How many ionized layers?' *EW & WE, 8* (1931), 463–4.

[61]For an exhaustive review on the indications of substratification in this period, see Mimno (1937, pp. 27–30). The early radio evidence almost immediately provoked reactions in favour of the use of the term 'region' instead of 'layer', especially by the leading radiophysicists. But not cogently. If in general the radio engineers and physicists did not go much further in their insistence upon this issue, that was in great measure because their stances in this respect were often ambiguous. Appleton himself urged his colleagues to promote such 'conversion' [see Appleton to Ratcliffe, 24 Sep 1932—quoted in Gillmor (1981, p. 108); and *Proceedings of the URSI, London, Sep 1934, Vol. 4* (pp. 46–50). Brussels] but at the same time he seemed to feel comfortable with a narrative language tending to accentuate the stratified nature of the ionosphere—note that the F region was baptized as the *Appleton layer* [see, for example, Appleton (1963, 1964)]. Even in 1938 the delegates at the URSI Meeting in Venice, *Proceedings of the URSI* (Brussels, 1938), p. 28, had to resolve that 'les termes *région* (region) ou *couche* (layer) soient considérés comme équivalents lorsqu'ils se rapportent à des parties individuelles de l'ionosphère'.

[62]Lindemann had previously asserted: 'In order to correlate our information about the ionosphere with our geophysical knowledge, it is essential to know the real heights at which reflection of the wireless waves take place, for all our observations and calculations of temperature, density...refer to real heights. It is perhaps worthwhile emphasizing that there does not seem to be any conclusive evidence that the equivalent heights are closely related with the actual heights'. See 'Meeting for Discussion on the Ionosphere'. *Proc. Roy. Soc. Lond., 141A* (1933), 697–722, pp. 720 and 722.

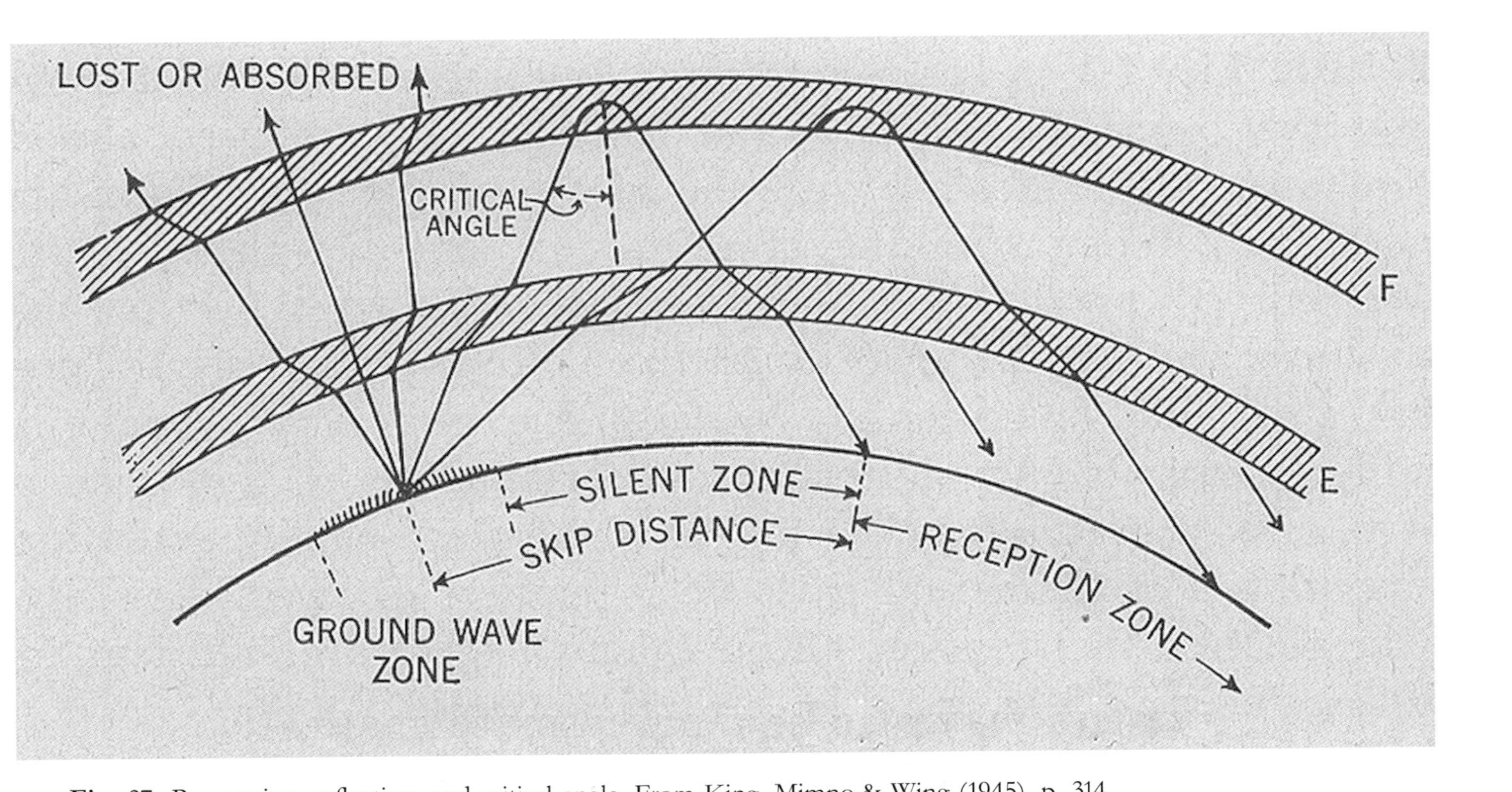

Fig. 27: Penetration, reflection, and critical angle. From King, Mimno,& Wing (1945), p. 314.

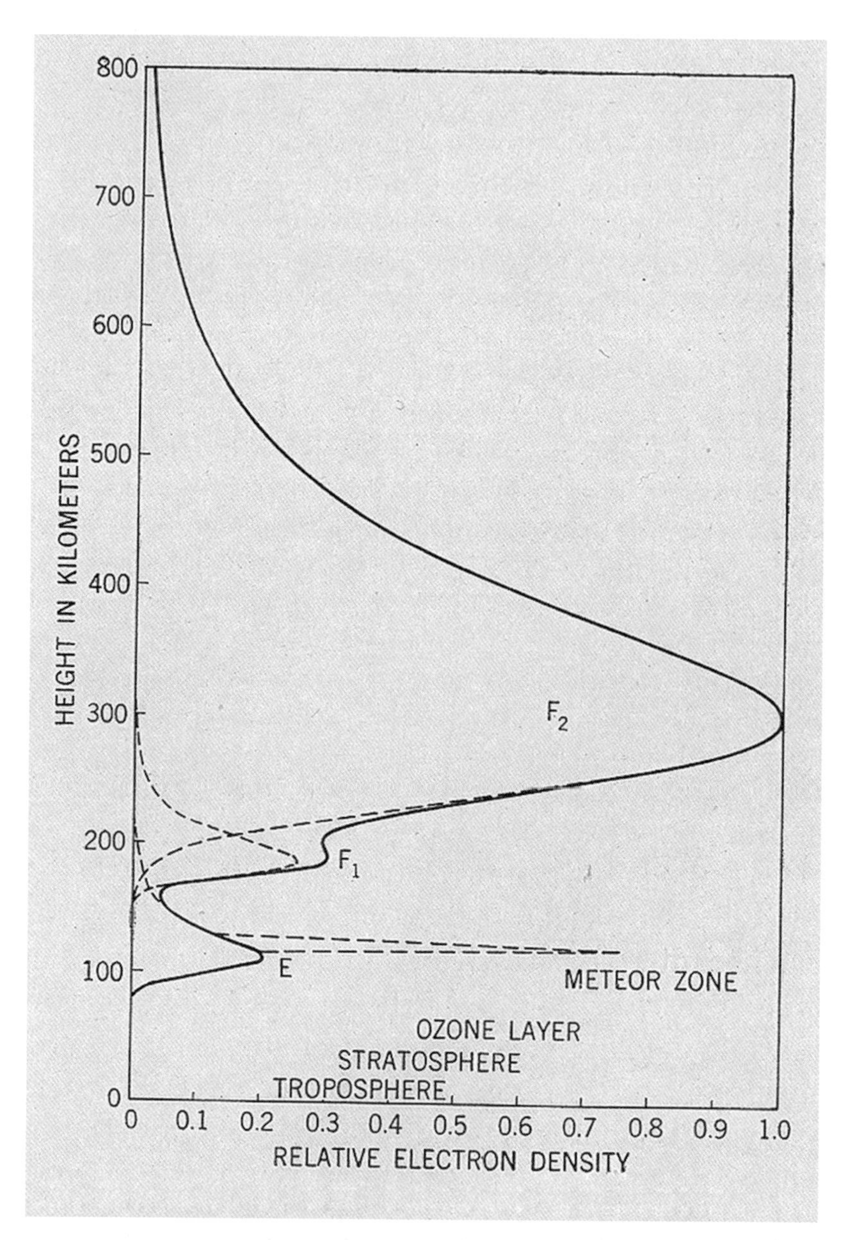

Fig. 28: Layers of ionization in the atmosphere. From King, Mimno, & Wing (1945), p. 311.

being apt to ascribe real existence to operational concepts—as a result of an unconscious exercise—without realizing the varnish of realism with which they coloured their notions and beliefs, then it is reasonable to assume that we can construe such impulses as a sort of reflex action or stereotyped response that was intuitively and automatically generated in light of a specific stimulus and a science deeply imbued with forms of radio engineering. And yet, Lindemann's appreciation is not cogent enough for our purposes.

Although sporadic, reservations such as Lindemann's regarding the ontological nature of layers did exist. Such reactions generally reflected the geophysicist's stance, and bore the unmistakable, if more temperate hallmark of his predecessors at the beginning of the century. Whereas Chree felt bitterness (and a fair degree of indignation) at the disparagement of the historical geomagnetic memory by radiophysicists, the moderate fringe around the theoretical-mathematic geophysics advanced in the 1930s towards mutual consensus. For example, Chapman urged recognition of the two realms 'not as overlapping and contradictory, but as complementary, giving information as to different aspects of the problem', as he asserted in his memorable Bakerian Lecture in 1931 (Chapman, 1931, p. 368). He promulgated the communion of forces: there was no mutually exclusive and competitive evidence but rather different pieces of knowledge coalescing into *one single* substance. Taking the variance in Chapman's sense, it is not surprising that his caveat upon 'the difficulty in interpreting the equivalent heights' strongly attracted attention from radiophysicists.[63] In the same reasoned vein, L.V. Berkner and his colleagues of the DTM at the Carnegie Institution of Washington considered the several layers not as physically separated 'with non-ionized regions between them', but rather as 'one ionized region whose ionization varies with the height in such a manner that the retardations of radio waves fall into fairly definite groups'. Here, clearly, the authors dissent from Appleton's postulate 'equivalent height-leaps = sharply discrete layers'.[64]

Meanwhile radio engineers worried increasingly about 'that necessary evil', the unpredictable dependence of radio propagation upon the upper atmosphere.

[63]Chapman (1934, p. 908): '[The powerful and valuable radio methods] afford clear evidence, which probably few workers on the earth's magnetism expected ever to gain, of the decrease of the [magnetic] field with height.' Berkner (1941) wrote in a similar conciliatory tone.

[64]Yet the dissidence is not radical: 'These retardations which determine the virtual height are due, first, to the actual height of the reflecting layer and, second, to the reduction in group velocity of the pulse caused by passing through lower ionized regions. The resultant virtual heights fall into fairly definite groups, but the real heights of the layers and the ionization between them is not known'. Kirby, Berkner & Stuart (1934, p. 18).

Between 1925 and 1930, the period that saw the crucial experiments and the multiplication of layers, there are a good number of examples of attempts to subdue the variable and capricious state of the ionosphere. To refer only to those initiatives in which the ionosphere was an object of search and research for commercial and military purposes, we can mention T.L. Eckersley and K. Tremellen's 'ionic density charts' at Marconi's in 1929, A. Hoyt Taylor's 'graphic representations of skip distances' at the US Naval Research Laboratory (NRL) in 1926, and R.A. Heising's 'transmission curves' at Bell Telephone Co., New York, in 1928.[65] Broadly speaking, each tries to resolve the same enigma: the prediction of wave behaviour. But, manifestly, in their attempts, each applies a different modus operandi. Yet they had one characteristic in common: all are 'based upon an *equivalent reflection* that introduces no inaccuracy' for skip-distance calculations.[66] And this fact powerfully indicates that the assumption of a sharply discrete stratification, far from being extraneous and inconsequential in the commercial and military milieu, was not only welcomed but also nurtured by them.[67]

Can one observe this chain of incursions and not surmise that there are close parallels between the reification of layers by radiophysicists and the commercial success in long-distance communication, that Eckersley, Taylor, and Heising's emphasis on the realism of idealized strata was accentuated by their perceptions of predictive achievements, that possibly the rhetoric of discovery (and therefore the omission of invention) was not only nourished by, but contingent for its existence upon, a highly exciting atmosphere on the threshold of a burgeoning radio industry? Can one ignore engineer L.B. Turner's words in 1926, when he alludes to the very fact that in short wave 'the Heaviside layer really does behave as a *good reflector*...confers a *good measure of reality*'? (Turner, 1926, p. 43). Here perhaps it is worthwhile remembering once again what Murray and Barton Hoag asseverated in 1937 in this respect: 'Curves such as

[65]Eckersley & Tremellen (1929), summarized in *The Marconi Review, 17* (1930), 1–17; Hoyt Taylor (1926); Heising (1928).

[66]Taylor (1926, p. 521). This conception is essentially that depicted most vehemently by Terman (1938, p. 339): 'The skip distance for a particular ionosphere layer can be calculated with fairly good accuracy by assuming that the wave undergoes a mirror-like reflection in the ionosphere at a height corresponding to the point of maximum electron density'. For Taylor's practical communication charts, see Taylor (1948, pp. 111–12).

[67]Calculations of skip-distances and silent zones were performed by Captain S.C. Hooper of the US Navy to ascertain the viability of high-frequency communication. For the use of 'working schedules' purveying the most effective frequency for stated hours by American engineers and military, see Duncan & Drew (1929, pp. 756–9). For the skip-distance project at the US Navy, see Hevly (1987, pp. 23–39).

[those of "equivalent height vs. frequency"], while possessing great value for communication purposes,...give a greatly distorted picture' of the substantive physical nature of the ionosphere (Murray and Hoag, 1937, p. 334).

It must be acknowledged that in the years which followed the crucial experiments, Eckersley, Heising, and other radio engineers were pushing for formulations more consistent with their utilitarian viewpoints. This fact was most eloquently reflected at the meeting of the URSI held in London in 1934, whose commission of wave propagation discussed several questions of nomenclature.[68] At issue here was not the extirpation of ontological overtones from layers but the acquiescence, approbation, and normalization of a definition of 'ionosphere' as an international reference. Appleton introduced the matter.[69] 'The ionosphere', he propounded, 'is the sphere of air whose predominating physical characteristics is ionization'; on the other hand, Bell Lab's engineer Heising urged for 'that part of the upper atmosphere which is ionized *sufficiently to affect* [italics added] the propagation of wireless waves'.[70] The question was neither an inanity nor a mere caprice on categorization or terminology; it was a physical conception versus an engineering one. 'After much discussion', Heising's version carried (Beynon, 1975, p. 51). And thus, by the late 1930s, this definition had became a quasi-religious epigraph in the reviews and textbooks of radio engineers—indeed, it was ratified by the IRE in 1950 and widely subscribed to even in the 1960s—so that the 'ionosphere', stripped of almost all physical attributes, was deemed equivalent to an operational construct.[71] Strangely enough, one might be forgiven for thinking a physical fact

[68] *Proceedings of the URSI, London, Sep 1934, Vol. 4* (pp. 46–50). Brussels. The commission comprised several engineers and radiophysicists including Eckersley and Heising.

[69] The existence of a subterranean current tending to accentuate engineering practicality might be sensed by some public reservations to espouse physicists' notions. Thus, addressing the commission, Appleton acknowledged that some agreement was 'desirable as to the acceptability of the term *Ionosphere*', which is, in Beynon's words (see note below), 'a little surprising [for] it had been fairly widely used from about 1930'. Here it is important to note that Appleton not only admitted an engineering component in ionospheric physics but seems to have perceived an interconnection between both in a spirit of a mixture of approbation and resignation: 'The physicist sought to elucidate the phenomena of nature, while the engineer tried to utilize them in the most efficient way for our practical benefit. [However], this distinction...is not of universal application..., and we see [in radio communication] how closely allied and interdependent is the work of [both]'. (See 'Electrical Communication and its Indebtedness to Physics', manuscript, 1931, EUA [D13].

[70] *Proceedings of the URSI, London* (1934, p. 46).

[71] See 'Standards on Wave Propagation: Definitions of Terms, 1950'. *Proc. IRE, 38* (1950), 1,264–8, on 1266. To mention but few examples: Rishbeth & Garriot (1969, p. 3) define the ionosphere as the part of the atmosphere 'where ions and electrons are present in quantities

is defined in terms of the way in which scientists observe and explore it; unusually, one might conclude, the very technology (in the broad sense of the term) is part of the essential content of the concept itself. Such a conception followed not evidently from the electromagnetic or magneto-ionic theories but rather from the leverage exerted by a commercial environment.

Electron density profiles: the parabolic model, 1935–1939

We have now come to the very epitome of the realist interpretation of the atmosphere, the parabolic model of ionization profile. Next we shall attempt to prove that there are reasons to believe that such an interpretation does not result ineluctably from an optical effect of instrumentation or derive inexorably from a theory of ionospheric propagation. With the introduction of Chapman's theory of layer formation in 1931 and of Appleton-Hartree's magneto-ionic theory in 1932, upper atmospheric physicists realized that radio data interpretation could no longer rest chiefly upon speculative inferences. They perceived the calculation of electron density profiles as the avenue to an exact knowledge of atmospheric structure. And with the development of T.R. Gilliland's automatic ionosonde in 1933 and its ensuing iconic pictogram, the compass and dimension of visualization were substantially altered (Gilliland, 1933, 1935). We shall not try here to examine exhaustively the momentous repercussions of all these changes, but only to underscore how disposed, not to say anxious, some of the leading radiophysicists were to embrace a stance congruent with their previous 'discoveries'.[72]

It is tempting to suggest that during the 1930s advances in radio technology appeared more likely to confirm radiophysicists' optimism regarding the interpretation of layers in realist terms than to vindicate the more moderate positions of geophysicists like Chapman. In several papers published after the invention of the ionosonde, Appleton predicted a brilliant future for the elucidation of layers (e.g. Appleton & Naismith, 1935, p. 688). But even if ionosondes provided valuable information on critical frequencies, it was not obvious to radiophysicists that their experimental ionograms conclusively

sufficient to affect the propagation of radio waves'. In the same vein, see Jouast (1936, p. 286): 'la partie de la haute atmosphère...qui intervient dans las propagation des ondes radioélectriques'; Darrow (1940, p. 455): 'a region from which radio signals are reflected', and, more pompously (p. 458), 'a canopy of ions overarching the earth'; and the influential Dellinger (1939, p. 803): 'the entire ionized region...which affects the transmission of radio waves'.

[72]The abovementioned changes are reflected in some noteworthy papers. On the magneto-ionic theory and the crucial role of an unknown figure, as was W. Altar, see Gillmor (1982). For a convincing analysis on the use of ionogram as a realistic representation of the ionosphere, based on Canada's postwar experience, see Jones-Imhotep (2000, pp. 87–107).

showed the fine structure of the ionosphere in the way they understood. In a comment that helps us understand the limitations and doubts of the realist interpretations, Ratcliffe stated that he and his colleagues paid more attention to the penetration frequency than to the shape of the ionogram. This seems surprising, for the shape must be related somehow to the height distribution of the electrons, and thereby to the morphology of the ionosphere. In retrospect, he said, 'it seems a little strange that the relationship was not explored more energetically by the early workers: indeed it was not until about 1950 that it was fully investigated' (Ratcliffe, 1970, p. 83).

Or, in other words, the function of the experimental ionogram was not demonstrative but corroborative. In the 1930s, physicists and radio engineers did not attempt to deduce the electron distribution from the ionograms; quite the contrary, assuming a distribution from theoretical models and data, they deduced the ionogram that fitted best.[73] Let us see why. Here an explanation might help. An ionosonde emits a series of pulses of gradually increasing frequency. The returned echo provides information regarding the electron density over the path in the form of height versus frequency graphs. However, to extract the information, one must solve complicated integral equations practically unsolvable before the advent of digital computers.[74]

There is another difficulty. The observer receives *only* echoes from layers up to points of maximum electron density. If the electron concentration decreases above a maximum, no echo will return from those regions (for the pulse has been previously reflected); and therefore there is no way to know its density. The problem is aggravated when two layers (such as the *E* and *F*) lie superimposed in such a way that the electron density falls to a minimum between them—the so-called *valley ambiguity*.[75] The problem can be appreciated

[73]Ibid., 84–5. In contrast to this reading 'backwards' (through theory to the physical preconditions responsible for the appearance of profiles), postwar communications engineers instead would seek to project the image 'forwards' through a series of propagation theorems. As Jones-Imhotep (2000, pp. 64–5) has made clear, they read propagation conditions—and saw the ionospheric structure!—directly in the shadow of the graph.

[74]In the simplest form of the pulse-sounding method, the measured time delay can be converted into an equivalent height of reflection $h'(f)$ given by $h'(f) = \int \mu'[N, f]\, dh$. Here $\mu'[N, f]$ is the group refractive index that can be derived from the magneto-ionic theory as a function of electron density N and of a wave frequency f. In 1930, De Groot demonstrated that, if the magnetic field effect is ignored, this integral equation assumes the form of Abel's equation and thereby admits an analytic solution. But, in a general situation, one must resort to some form of approximate solution.

[75]The mathematical difficulties of N (h) analysis are emphasized by Wright & Smith (1967, p. 1,120): 'A region in which the density is less than that at a lower height cannot produce virtual-height data; such a region or *valley* is often present between the E and F layers.... Moreover,

from Figure 29. Any of the densities shown in (b) might correspond to the ionogram shown in (a). Radiophysicists had tried to attack the ontological connotations of this problem by asserting that a strongly ionized intermediate region could exist between *E* and *F*. In fact, in 1933 and 1934, Appleton and Ratcliffe discussed such a possibility. Even in 1934, Appleton thought he had measured a transition in the reflection coefficient, which would have substantiated the foregoing thesis.[76] But however much they endeavoured to introduce intermediate ionized spaces, the essential point in this discussion is that the *valley ambiguity* hit the core of any reading in realist terms, for the limitations inherent to sounding technology produced regions inaccessible to observation. Might the various lower layers E and F_1 be only inflections, or ledges, in a continuous gradient of electron concentration? And, consequently, might the well-defined 'layers' that seemed to exist as a result of earlier interpretations be simply realist excesses?

Limited, on the one hand, by some seemingly insuperable computational obstacles, but obliged, on the other, to construe the experimental ionograms, radiophysicists resorted to mathematical models of electron density profile. Earlier in 1932, the Chapman function and the normalized curve derived from it had showed the form of the ideal distribution. But, since 1933, with the new automatic ionosonde operating continuously over a wide frequency range and with an ideal profile that did not fit well with the data, radiophysicists' response was the pursuit of approximations to the Chapman layer (Kaur, Srivastava, Nath & Setty, 1973, p. 1746). Now, in a model devised by Appleton, D.R. Hartree and Henry Booker (and quickly championed by Appleton itself), any uncertainty that might have existed due to the tardiness in the processes of numerical computation, was remedied with the introduction of analytical expressions being amenable to integration and differentiation.[77] The model was predicated upon an electron distribution described by two superimposed

the lowest part of the ionosphere is likely to remain unexplored by pulse-reflection techniques because of overpowering absorption, equipment limitations, and failure of ray theory.'

[76]Between 1933 and 1935, Appleton and Ratcliffe exchange over 50 letters, many of which treated, or touched upon, the controversial issue of the ionization between the *E* and *F* regions. Cf. Appleton to Ratcliffe, 23 Feb 1935, EUA [1985/218a].

[77]On this point, Appleton did not act alone. As Gillmor notes, his parabolic layer model was shared by the young Henry Booker, who treated linear and parabolic models in a course at Cambridge with the title 'Ionosphere', sending his work to Appleton. Likewise, D.R. Hartree exchanged his ideas on ionization density profiles with Appleton in several letters and in the manuscript 'Notes on the Propagation of Electromagnetic Waves in a Stratified Medium'. See Hartree to Appleton, 22 Feb and 6 Jun 1936 and Booker to Appleton, 4 Jun 1936, EUA—quoted in Gillmor (1981, p. 109).

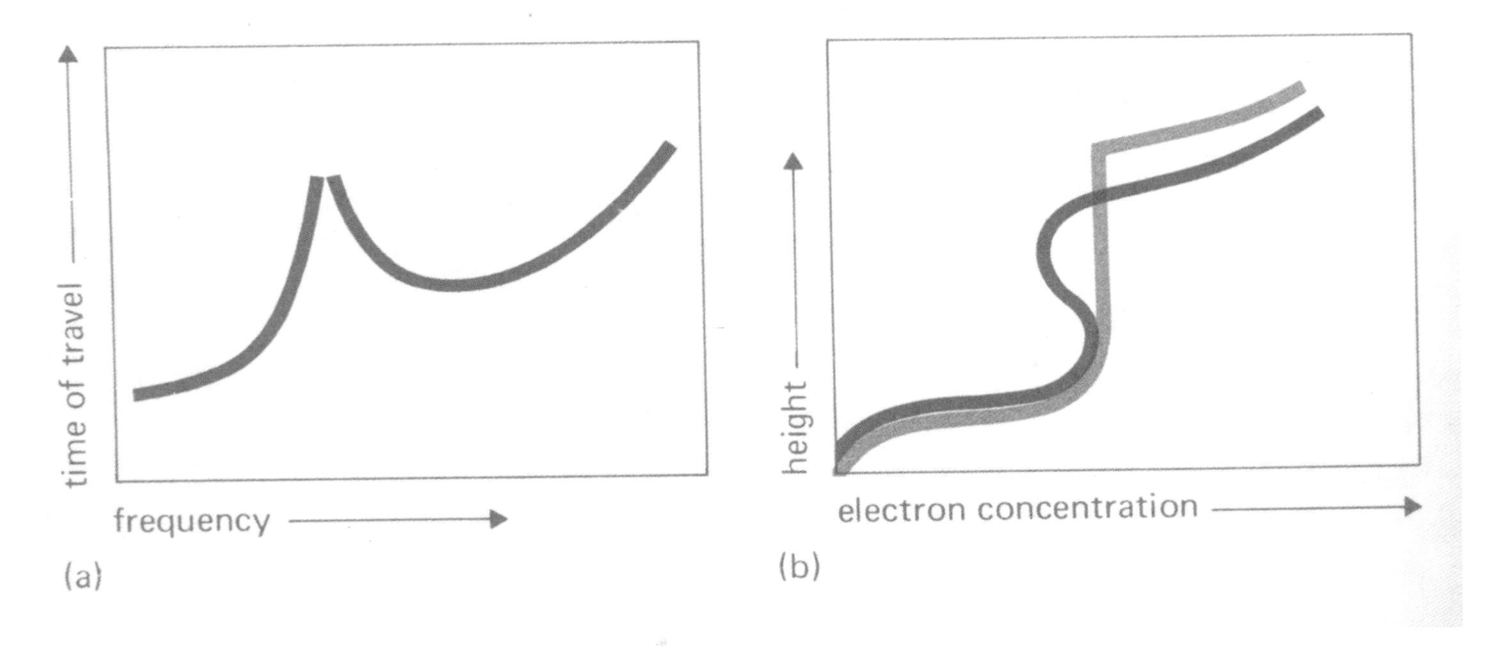

Fig. 29: (a) An ionogram representation showing the time of travel or delay time as function of wave frequency. (b) Theoretical curves of height as function of the electron concentration. The two assumed electron distributions might correspond to the ionogram shown in (a). From Ratcliffe (1970), p. 82.*

parabolas (E and F), in which both the data and the ideal Chapman layer fit satisfactorily.[78]

Although Appleton always enjoyed his assistants' collaboration in seeking ammunition for his attack on the intermediate region, one of his former pupils, J. Hollingworth, seems to have been the first to publish a research article against the physical separability of layers and in favour of a continuous ionization distribution (Hollingworth, 1933, pp. 229–51). In this work we discover that the different orders of retardation of echoes do not necessarily indicate a stratification of the ionosphere. For, argues the author, if the ionization in the purportedly intermediate region is comparable with that of the top of the *E* layer, as the evidence strongly seems to suggest, then it raises the question 'as to whether the "shelves" have a *real existence* or merely arise as points of inflexion on a group-velocity curve'[79]—a point which escaped Appleton and Ratcliffe at that time. Hollingworth's corollary is the audacious assertion that the grouping of echoes into certain heights is the result of the mode of exploration, rather than an ontological attribute of nature.[80]

In this respect, the objections of F.H. Murray and J. Barton Hoag of the University of Chicago are also especially interesting, for they not only disclose the interpretive difficulties which theorists encountered in ionospheric physics but also include a *prima facie* demonstration of a hitherto unidentified relation between the hypostatization of layers and the discontinuities observed in experimental curves. In their article published in *Physical Review* in March 1937 (but received in September 1936), the plurality of 'regions of *virtual* heights' is emphatically described as hypothetical and controversial. 'A sharp discontinuity [in the virtual height vs. frequency curve], observed as the frequency was increased, was interpreted by Appleton as due to the existence of two *physically separate* regions' (Murray and Hoag, 1937, p. 333). And the essential

[78]Mathematically, the parabolic profile (as the linear, exponential, quadratic, devised later) was easily derivable from the Chapman model—indeed, they were approximations. See Davies (1965, pp. 134–9). For the radiophysicsts, however, it functioned as an approximation to the experimental ionogram, legitimizing their realist interpretations.

[79]Hollingworth (1934, p. 462). The existence of an inter-layer ionization is most fully and forcefully held by Hollingworth (1935, p. 844): 'It must therefore be assumed that the ionization at the top of the E layer persists with only a slight diminution in value until the F layer is reached.'

[80]The hypothesis of a continuous ionization is also suggested by other (if few) research works in the intervening period. In December 1934 Ionescu & Mihul (1934, p. 1,303) explained the jumping of echo on the assumption of electron collisions, and regarded all variation in electronic density as continuous: 'Les discontinuités que l'on observe expérimentalement ne sont qu'apparentes et les niveaux de réflexion réels varient d'une façon continue.'

inconsistencies of this interpretation lead them to develop a method of calculation for the true heights of reflection. After analysing the only three possible situations, namely that the electron concentration increases, decreases, or remains essentially constant with altitude, they enquire into the connection between the observable and the observed. And the conclusion acquires the character of *sine qua non*: 'It appears that mathematically a discontinuity... is a necessary but not a sufficient condition for the existence of two distinct layers' (Ibid., 333).

Not everyone, however, agreed. The overlap between ideal and real became obvious in Appleton's Bakerian Lecture to the Royal Society in June 1937 (Appleton, 1937). In this lecture, he praised the Chapman model before introducing a simple model of parabolic distribution for the regions E and F.[81] Neglecting the effect of the magnetic field, he developed a mathematical expression for height as a function of frequency (clearly ideal), which was then 'scaled' by adjusting certain parameters (height, thickness, peak concentration, etc.) to 'fit the experimental ionogram as closely as possible'[82] (Figure 30). It is hard to explain why Appleton showed plots of two clearly separate parabolic layers, with an ionization *tending to zero* on each side of the curves. To what extent his representation aimed to persuade the audience of the existence of two well-defined layers (instead of two diffuse regions being a function of latitude, season, etc.), or to what extent it was an attempt to evade the problem of 'valley ambiguity' (where, paradoxically, his own conclusions seemed to lead him to an ionization in the intermediate region) remains a mystery. The effect, however, was unmistakable: the figure supported the idea that the maxima were unique and sharply defined.[83]

Adaptation of knowledge to the necessities for ionospheric prediction

But can this account of the parabolic theoretical model represent the whole truth? The examination of the reasons for its adoption by most radiophysicists cannot stop at that elusive and porous borderline where science and the quest for knowledge end and engineering motivation and external pressure begin.

[81] Appleton expressly underlined that his parabolic distribution followed the formula given by Lenard in 1911 (relating ion production to the angle of incidence); thereby, he seemed to relegate Chapman's theory of layer formation to the background. Ibid., 452.

[82] On the F layer Appleton categorically stated 'that the ionization is fairly well represented by a region with a lower boundary at 270 km and a half thickness... of 100 km'. Ibid., 472.

[83] Gillmor (1981, p. 109) asserts that this figure 'had considerable influence on ionospheric theorists'.

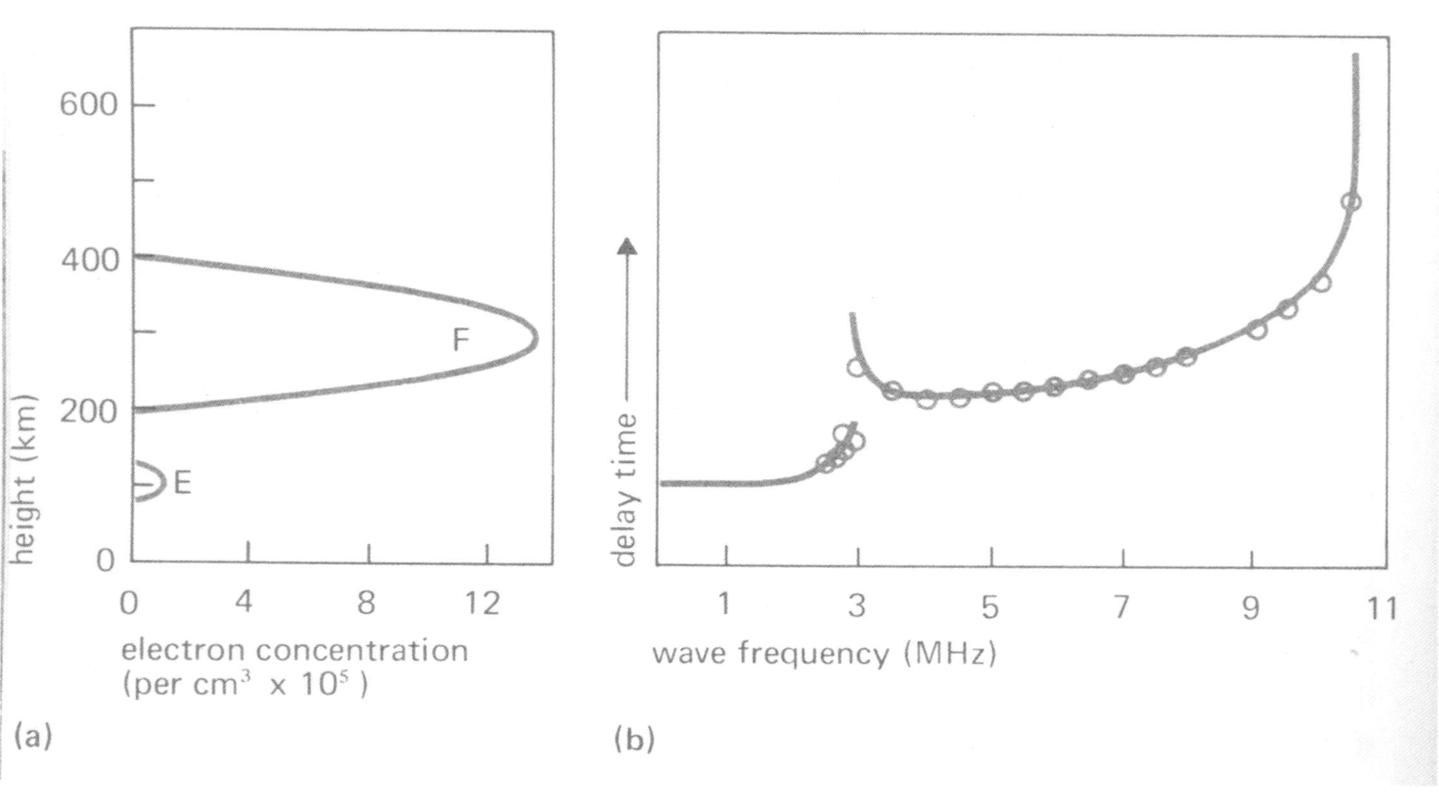

Fig. 30: Theoretical relation between equivalent height and frequency for two 'parabolic' layers. The assumed ionization distribution is shown in (a). From Ratcliffe (1970), p. 84.*

* There are instances where we have been unable to trace or contact the copyright holder. If notified the publisher will be pleased to rectify any errors or omissions at the earliest opportunity

In analysing the 'realist interpretation', should not ionospheric prediction and the practicability of radio communications, both commercial and military, be regarded as its typical expressions? And might not these factors have been inherently congenital to theoretical discourse and the achievements of ionospheric physics? Thus, then, we return to the substrata of radio industry and Empire, to the existence of a highly competitive and exigent environment, which was an essential constituent of the enterprise of radiophysicists before World War II.

Certainly, the large radio companies, with their objectives of imperial communication networks, had a receptive attitude towards investigation. In fact, the policy of commercial expansion was predicated upon the amalgamation of communication and broadcasting services with manufacture. And ionospheric research was a component in this amalgam. But how important was the knowledge of the upper atmosphere for them? The response lay in waves. By the mid-1930s, radio engineers had realized that the successful and profitable frequencies for long-distance transmission were determined, on the one hand, by the F_2 critical penetration frequency (the outermost layer), and, on the other, by absorption in the lower layers. The lower the frequency the greater the absorption, they believed. Hence practical radio communication required a *compromise*. Practically all radio experts subscribed to two maxims. Firstly, propagation prediction was an *art* (the art of determining the optimum frequency for given ionospheric conditions); and, secondly, an understanding of ionospheric mechanisms was *basic* to efficient communication.[84]

One need only glance at the minutes and memoranda of the Radio Research Board (RRB), the civil-military body responsible for coordinating radio research in Britain, to realize that ionospheric prediction was increasingly necessary.[85] The RRB's *Memorandum* in respect of its new committee on '*Propagation of Waves through Ionosphere*', dated 1937, reads: 'Its objective [is]

[84]One can find prewar testimonies of this connection between ionosphere and radio industry in many papers and business addresses. So, for example, Millington (1938, p. 801): 'The problem of determining the characteristics of long-distance transmission through the ionosphere, as regards the maximum usable frequency and the absorption en route, is a very important one from the point of view of the engineer who has to allocate wave-lengths for any projected service.'

[85]In 'Minutes of the Fifth Meeting of the Committee *Long Distance Propagation*, Held on 8 May 1936', DPA, under T.L. Eckersley's chairmanship, it is recommended to explore 'the possibility of predicting the future expectations of transatlantic propagation', and 'the desirability of carrying out measurements on the properties of the ionosphere for oblique angles'. Furthermore, the validity of the extrapolation from vertical incidence data is regarded as 'doubtful in the presence of a magnetic field'.

the application of the results to the improvement of radio communication'.[86] This committee arose from the merger of a committee on 'Propagation of waves', chaired by Appleton and concerned with vertical exploration, with one on 'Long-distance propagation', headed by Eckersley and related to oblique incidence studies for long-distance propagation. The memorandum betrays the stagnation of the work at vertical incidence led by Appleton, and the commencement of 'a gradual extension...to oblique incidence in comparatively small stages, over moderate distances'.[87] Addressing this Committee in October 1937, Colonel J.P. Worlledge, President of the Royal Engineer Board of the British War Department, urged its members to obtain information 'on the angles of incidence in Britain'.[88] But this missive was but the expression of a long-running demand repeatedly solicited and exhorted by the fighting services: the quest for optimum frequencies. A letter from the Colonel himself to the secretary of the Australian RRB in October 1935 is an example:[89]

> We are anxious to obtain all the information we can upon the optimum frequencies for use when communicating over ranges of from 50 to 1,000 miles; and any direct information in this respect that you could let us have would be very much welcomed. Furthermore we imagine that, apart from such direct information, you must be in possession of much valuable data as to the variations in the ionosphere...; and from such data we could probably deduce much of the information we require. The paper by T.R. Gilliland [Sept. 1935 *Proc. IRE*] gives a quantity of data of this kind for the conditions experienced in the U.S.A., and we should very much like to be put in possession, if possible, of similar data.

Far from being a marginal practice, the oblique incidence approach and its apologists (in Britain, mainly Marconi's engineers) dominated the commercial realm between 1936 and 1937. They placed special emphasis on the predictability of long-distance communication rather than on the objectivity of physical knowledge. This was the case, for example, with G. Millington, who, as early as 1932, drawing upon Chapman's layer formation theory, had constructed ionization charts 'to predict the behavior of short waves' (Millington, 1932, p. 580). And, now, in 1937, by adopting the technique of Newbern Smith of

[86]'Memorandum on the Constitution of the Committee *Propagation of Waves through Ionosphere* and the Program of Work', by R.L. Smith-Rose of the National Physical Laboratory, 27 Sep 1937, DPA.

[87]Ibid., 5.

[88]'Minutes of the First Meeting of the Committee *Propagation of Waves through the Ionosphere*, Held on 14 Oct 1937' (chaired by Appleton), 4, DPA.

[89]Colonel J.P. Worlledge to G.A. Cook, Oct 1935, in Evans (1973, p. 268).

the American National Bureau of Standards (NBS), he derived transmission curves from which the optimum frequency could be determined (Millington, 1938). The fact that in 1937 the British Broadcasting Company (BBC) regularly used these curves prepared by Marconi's for its imperial transmissions led Naismith to propound similar procedures within the RRB to meet the Company's needs.[90] *Even better*, in November 1937, the BBC itself appealed to the RRB for 'the regular publication of available ionospheric data' similar to those the NBS was publishing weekly in Washington.[91] And so we see how the pressure upon the RRB increased.

The American NBS's case is somewhat different, for, unlike Marconi's, now the pressure upon the RRB came from a foreign rival. Thus, in this case, one must ask, rather, if the new overseas methods and their policy of disseminating ionospheric data constituted a 'serious menace' to the scientific status, international prestige, and practices of the RRB. The competition between British and American radiophysicists had been increasingly overt since the NBS' initiation of radio prediction forecasts worldwide in 1935 compelled the RRB to redefine its strategies.[92] But it was Smith's prediction system that epitomized the rivalry.[93] When one examines the influence exerted by the NBS (and particularly by this system) in Britain, one cannot help but surrender to its magnitude. It owes much to the fact that the essential notion of Smith's method cogently impinged upon the foundations of radio communication. With the empirical transmission curves, data elicited from vertically incident waves were *transformed* into information about their propagation over oblique paths

[90]'Memorandum on Information Available from the Present Program of Measurements on the Ionosphere and its Possible Use for Communication Purposes', RRB, Committee P2, 22 Nov 1937, DPA.

[91]'Minutes of the Meeting with a Representative of the Radio Research Board to Discuss Information Required from the P2 Committee, Held in H.L. Kirke's Office at the Research Station of the British Broadcasting Corporation', on 16 Nov 1937, 1, DPA.

[92]The circular letters were prepared by the US Dept. of Commerce under the heading 'The weekly radio broadcasts of the National Bureau of Standards on the ionosphere and radio transmission conditions'. Evans (1973, Append. 16). Four years later a monthly bulletin of ionospheric data was issued by the RRB. 'Minutes of the Fourth Meeting of the Committee P2, Held on 16 Mar 1939', DPA.

[93]Some antagonism, not to say animosity, towards American radio attitudes is explicit in the letter that Ratcliffe wrote to Appleton on 4 Apr 1935, EUA [E101]: 'The Americans work at problems without ever reading what is being done elsewhere, and their papers never refer to *foreign* work. It is the characteristic of the English work that we read widely, and try to connect what we are doing to what is being done in the rest of the world. I think in English papers (largely as a result of your own excellent example) we rather take a pride in mentioning anyone who has worked on the same subject.'

(Gilliland, Kirby, Smith & Reymer, 1937, 1938). While the former was relatively simple, the latter paved the way to prediction. Hence, when Smith, following Gilliland and Kirby's research line, introduced his curves to determine maximum usable frequencies (or MUF) in 1937, their dissemination overseas was almost instantaneous.[94] Millington immediately adopted them to elaborate the *P′f* curves which the BBC then utilized (Millington, 1938, p. 809). Even the Royal Aircraft Establishment in Farnborough had to use the American scheme (at least, since 1936) owing to the lack of a national system.

Having assayed the pressure, particularly strong within the RRB, to adopt predictive tactics and formulae, we can better appreciate the great steadfastness with which Appleton and the radiophysicists in general held to specific models of electron density profiles *circa* 1937. In this connection, the lecture that Appleton delivered at the Institute of Electrical Engineers of London in October 1939 is worth examining, for it illustrates the resonance that the parabolic model had in engineering academia, and points to the motives which may have induced the speaker to make that 'unexpected' espousal. The subject was 'Wireless'. Appleton emphasized the value of *equivalent height* and *critical frequency*, the two keystones of ionospheric physics.[95] 'From the highest penetration frequency', he stated, 'one can derive the maximum electron density'; and from this, 'one infers the structure of the ionosphere'.[96] The outcome, he categorically asserted by showing graphs, were two parabolas with the limits tending to zero; 'the representation of electron density vs height curves for the E and F layers' (see Figure 30). More importantly, the bulk of the talk discussed maximum usable frequencies (MUF), which Appleton treated with a considerable realist emphasis involving a thin layer for trajectories of obliquely incident waves.[97]

[94]Smith's method, published in July 1937, was presented in part at the joint meeting in May 1936 of the IRE and URSI at Washington. See Smith (1937).

[95]IEEA 'Personal notes by T.D. Meyler, a student at the City and Guilds College, in London, on the lectures given by E.V. Appleton on 30 Oct 1939'. Appleton was—alongside R.L. Smith-Rose—the most prominent lecturer in radiophysics in Britain.

[96]Ibid. According to the magneto-ionic theory, the refractive index of ionized medium (μ) is a function of the electron density (N) and the radio waves frequency (f_0): $\mu^2 = 1 - Ne^2/\pi m f_0^2$. The reflection of radio waves at vertical incidence takes place when $\mu = 0$, that is, when $N = \pi e^2 f_0^2/m$, or in other words, when $N = 1.24 \cdot 10^{-8} f_0^2$ (for ordinary wave). Now, if the values of f_0 refer to critical penetration, and therefore maximum frequency, the values of N refer to electron density peaks. By finding $f_{0\,max}$ it is possible to ascertain the electron concentration N_{max} of layers, and thereby the structure.

[97]For the MUF curves and the choice of suitable frequencies for communication services, see Smith, Kirby, & Gilliland (1938, pp. 127–33).

Compare now with the parabolic model as it appears a few months later, in May 1940, in an article published by Appleton and W.J.G. Beynon in which the authors announce a method of calculating the MUF for long-distance communication (Appleton & Beynon, 1940). After repeating the foregoing analysis of the reflection of obliquely incident waves by a thin layer (suitable for the abnormal E-layer), they take on thick layers (*de facto* the usual practical case), and then deduce the MUF. Thus, they now consummate what had been adumbrated a year earlier: the parabolic-layer treatment is the most appropriate method for the precise determination of MUF for practical communication. Appleton thereby establishes implicitly the association 'parabolic model ↔ ionospheric prediction'.[98] In this essay, as in the aforesaid course, there is, of course, an overemphasis on critical frequencies and no mention of true reflection heights at all. Unlike the frequencies, however, the heights were not indispensable for prediction.[99]

Clearly, the defence of a specific layer model was of more than mere symbolic or academic importance. During World War II, analysts of the ionosphere determined the most efficient frequencies from models based on layer profiles.[100] Understandably, if Appleton and his colleagues at Slough wanted to rise to the occasion they had first and foremost to undertake the task of predicting MUFs—that universally coveted feature in the art of communication. They developed their own predictive method based on the parabolic model. They feared the competition of a predictive technique partly supported by their Australian colleagues, but which had not aroused the same sympathies in the

[98]After the war, Appleton adumbrated the fundamental data (among them the MUF, critical frequencies, and reflection constants) and characteristics required for an effectual radio prediction service. See 'Memorandum on Ionospheric Research and Prediction of Radio Propagation,' by E.V. Appleton, typewritten document., appendix 1, DPA [D23].

[99]Beynon (1967, p. 1,118) analyses in retrospect the neglect of true heights: 'The readily available experimental data on critical frequencies provided an accurate and unambiguous measure of peak electron densities, and these data enabled a great deal of ionospheric research to be carried out without requiring accurate knowledge of true reflection heights. [Moreover], it was clear that a really reliable determination of true height might involve elaborate and lengthy calculation, and it was appreciated that even when this was done the final result would not be completely free from any uncertainty.'

[100]Several organizations with their own predictive methods proliferated in parallel with, and largely as a result of, the fortunes of war. Thus, the American NBS drew upon Smith's approach at the Interservice Radio Propagation Radio Laboratory; in Germany, another method was developed at the *Zentralstelle für Funkberantung*—improved after the war by the French *Service de Prévision Ionosphérique Militaire*. For a comparison between Appleton-Beyton's parabolic model and Smith's transmission curve method, see Rawer (1958, pp. 152–60) and Evans (1973, pp. 300–9, 339–58).

US, where the NBS was exporting Smith's transmission curves to Canada and New Zealand. In this respect, the fact that the British military themselves constituted in 1942 an Inter-Service Ionosphere Bureau at Great Baddow headed by Eckersley and Millington, which had the power to enact and implement Marconi's predictive schemes, served to accentuate the rivalry (Millington, 1948, in foreword). It widened the competition in ionospheric prediction and persuaded Appleton even more that a reading as close and faithful as possible to reality was one on which there could be no discussion.[101]

The explanation of the adoption of the parabolic model, with its addendum of realism, must therefore be sought, at least in part, in the pressure and competitiveness of the commercial and military environment in the years prior to the war. There are firm indications that Appleton altered, in all probability deliberately, his own postulates on the intermediate region to bring it into closer conformity with the needs and priorities of the time. His readiness to introduce a model of parabolic layers upon which his prediction method then rested must be construed as a response to an aggressively exigent environment. This reaction, however, is not by any means irrational; when radiophysicists (and scientists in general) are subject to external competitive pressure, they feel impelled to adopt measures to counter that effect and thereby assuage the tension. Perhaps the most unusual component here is that the countermeasures adopted by Appleton and his colleagues undercut the substantive doctrinal corpus of ionospheric physics itself, and accentuated the already marked propensity towards a reading in realist terms.

The legitimization of the realist conception

If one peruses the manuals and textbooks on wireless telegraphy and radio engineering in the interwar years, one finds an overwhelming percentage of treatises which include a section on the upper atmosphere in their syllabus. More remarkable, however, is the almost total omission of evidence derived from terrestrial magnetism.[102] Normally, these are practical compendia and *vade mecums* on wave propagation, the summations of radio engineers rather than

[101]The ever-increasing rivalry was evidenced at the International Radio Propagation Conference of Washington, in April 1944, and at a meeting held in London in March 1944, in which Appleton firmly endorsed the parabolic method, for it 'always gave answers within 3% of the American transmission curve method'. Gillmor (1981, p. 110); 'Minutes of Discussion on Ionospheric Problems', 25 Mar 1944, 1–2, NAC [RG 24, vol. 4058, File NS-1078–13-8]. For a detailed review of the conference, see Evans (1973, pp. 358–62).

[102]Of over 60 textbooks examined, only *one* unambivalent, explicit acknowledgment of such evidence from geomagnetism has been found: Brown (1927) does not mention any 'Heaviside

the cogitations of academics. Yet, although the intellectual compass of most books is restricted to practical engineering enquiries (e.g. design specifications for antennae, stations, etc.), this restriction did not preclude their contents from embracing a profoundly realist picture of the atmosphere.[103]

During the interwar period, the City and Guilds of the London Institute in Britain and the Institute of Radio Engineers (IRE) in the US introduced examinations in radio communications. This practice gradually spread to other entities. In 1929, the British IRE initiated its Graduateship Examinations, which helped stimulate more comprehensive and specialized curricula. The rise in demand for qualified staff in the radio industry during the late 1920s prompted an increase in the number of textbooks endeavouring to meet the requirements of examinations and national certificates.

It was also during this same period that the first definitions of radio terms officially established by the International Electro-Technical Commission and by national commissions, such as the British Standards Institution, appear in dictionaries and glossaries.[104] In 1926, while compiling a glossary of technical expressions for electrical engineers, S.O. Pearson depicted the reflection of waves as follows: 'Ether waves striking a plane conducting surface induce eddy currents therein, and these in turn send out ether waves, that are partially reflected in much the same manner as light is reflected from a mirror'.[105] Here the author is obscurely construing the reflection via an arcane mechanism of electrical induction, the Heaviside layer regarded as a conducting surface comparable with a mirror (cf. optical analogy).

But even more interesting than these utterances themselves is the association of categories which the definitions of layers disclose: concepts are defined in terms of the instrument of observation. This occurs toward the mid-1930s, immediately following URSI's precepts of 1934 in the matter of nomenclature and terminology. Thus, a handbook based upon the British Standard Glossary

layer' and asserts that 'evidence of the presence of such a conductivity shell can be found in the variation of terrestrial magnetism and in auroras' (p.198).

[103]A list of early English and American textbooks dealing with wireless and radiocommunication can be found in 'Catalogue of Books on Wireless Telegraphy'. *The Wireless World*, *1* (1913), 591, *1* (1914), 655; 'Publication of the Wireless Press, Ltd'. *The Wireless World*, *10* (1922), suppl. 27 May.

[104]Roget (1924, 1931, 1938) and Stranger (1933) are good examples of this new pattern.

[105]*The Wireless World* (1926, p. 180). At that time reflection and refraction were interchangeable: 'The waves [are] reflected and refracted back to the surface..., just as a beam of light is reflected from a mirror or refracted through a prism', 'Dictionary of Technical Terms', *The Wireless World*, *17* (1925), 716.

of 1935, which was in high demand among students and radio engineers, defined the Heaviside layer as 'the ionized layer about 100 km. above the Earth's surface *which reflects* [italics added] long waves', and the Appleton layer as the one '*that reflects* [idem] short waves'.[106] Here the boundary between conceptualization and means of exploration dissolves.[107]

Such association between physical concept and experimental procedure is also explicit in many manuals and textbooks for radio and electrical engineers used in universities and technical colleges. However, where in 1924 the authors of these regarded *mirror-like reflection* as the cornerstone of the lawfulness of radio propagation, a decade later the same metaphor clearly had operational connotations. One might object that these two conceptions are not incompatible, and that a radio engineer of 1924 could have also used the metaphorical language in a strategic sense. The espousal of a (weak) stance on operationalism would not be outlandish, a stance in which the notion of truth would ultimately pertain to actions. This possibility was not completely unknown to radio engineers in the years following the formulation of a magneto-ionic theory which paid no heed to this feature. Nevertheless, here the fundamental question is that every such intimation of a reading in operational or commercially strategic terms was seen as a reinforcement, and legitimization, of the realist interpretation. In fact, if one examines carefully the definitions of concepts and the explanations of physical processes, one will find as a rule the ionosphere and its properties being depicted through a multifarious, but always operational prism—as if they were tied to the functions of transmitter and receptor, subordinate to the behaviour of waves, wedded to fading and interference phenomena, subject to experimentalist's whims and bound to commercial imperatives.[108] And, in the majority of cases, these characterizations of the ionosphere were delivered in

[106]Starr (1935, pp. 3 and 9). Likewise, in a series of articles that appeared in *World Radio* under the title 'A Wireless Alphabet' reprinted as Decibel (1937, p. 6) defines the Appleton layer as 'a layer of ionized gases...which *acts* as a reflector to wireless waves'.

[107]Roget (1924, p. 192) defines the Kennelly-Heaviside layer as 'the lower section of the ionosphere...between which and the earth's surface the *waves used in wireless communication* [italics added] follow the curvature of the earth, owing to being reflected or refracted thereby'.

[108]An admirable and extraordinary pictorial instance of this remarkable operational style is given by King, Mimno, & Wing (1945, p. 314): 'Each ray acts as if it had intelligence and purpose. It bores into the layer, seeking an electron density sufficiently great to turn it back (by total internal reflection).... If such a density does not exist in the E layer at the time, the ray passes through the F layer, where it repeats its search. If again unsuccessful, the ray passes out into interstellar space'.

conjunction with an approbation of the validity of the realist interpretation of the atmosphere.[109]

Embracing an oversimplified version of the magneto-ionic theory of wave propagation, the authors of these texts reinforce the conception of a discrete-layered atmosphere held by the majority of radiophysicists.[110] Radio engineers demarcate between theory and practice. It is practicality that compels them to resort to overhead mirrors; it is neither an abandonment of current theory nor a diatribe against refraction. Furthermore, any operational utilization of the magneto-ionic theory, they argue, requires the interchangeable use of the terms 'reflection' and 'refraction' since the electron distribution cannot be known in detail. 'When speaking of refraction', contends a manual on radio communications, 'one has in mind the *actual* curved path ABCDE (Figure 31). When speaking of reflection one introduces the concept of an *equivalent reflection* which would produce a similar down-coming wave. The equivalent height is readily measurable by means of a determination of the angle of arrival or of the time lag of the down-coming radiation. [On the contrary,] computation of the actual path BCD requires detailed knowledge of the electron distribution'. The corollary to this reasoning is unequivocal: 'Equivalent heights are sufficient for most engineering purposes'.[111]

The importance that the textbooks and manuals attach to reflection, the conversion of layers into sharply defined mirrors, and radio engineers'

[109]E.g. Terman (1938, p. 358): 'The layer heights, obtained from radio pulse signals by making calculations on the basis of the velocity of light, are *virtual* or apparent heights. These are always greater than the actual height reached by the wave.... Because the layers of the ionosphere are fairly sharply defined, the difference between the virtual and actual height is small.' Similarly, Ladner & Stoner (1932, pp. 42–3): 'Experimental evidence suggests the presence of at least two layers... the ionic density and gradient... rising to a first peak with a sharp final gradient, then falling away to rise to a greater peak value at a greater height, forming the second layer.... From the wireless point of view, therefore, we may picture the upper atmosphere as two ceilings concentric with earth.'

[110]Apart from the above-mentioned literature, examples concerning the simplification of the theory for practical reasons occur predominantly in the form of calculations of the variables 'frequency', 'skip-distance', 'relative heights', and 'electron densities'. Such are, for instance, the treatises by Glasgow (1936, pp. 491–502), Terman (1943, pp. 709–58), and Turner (1931, pp. 56–61).

[111]King, Mimno, and Wing (1945, p. 313). Likewise, Glasgow (1936, p. 491) vigorously states the aptness of such approximation: 'The existence of not one, but of several ionized layers has been demonstrated by experiments. Instead of the wave being reflected from the conducting layer, as light from the surface of a mirror, it enters the medium and is bent back to earth again by refraction. However, it is convenient for purposes of calculation to regard the process as one of simple reflection.'

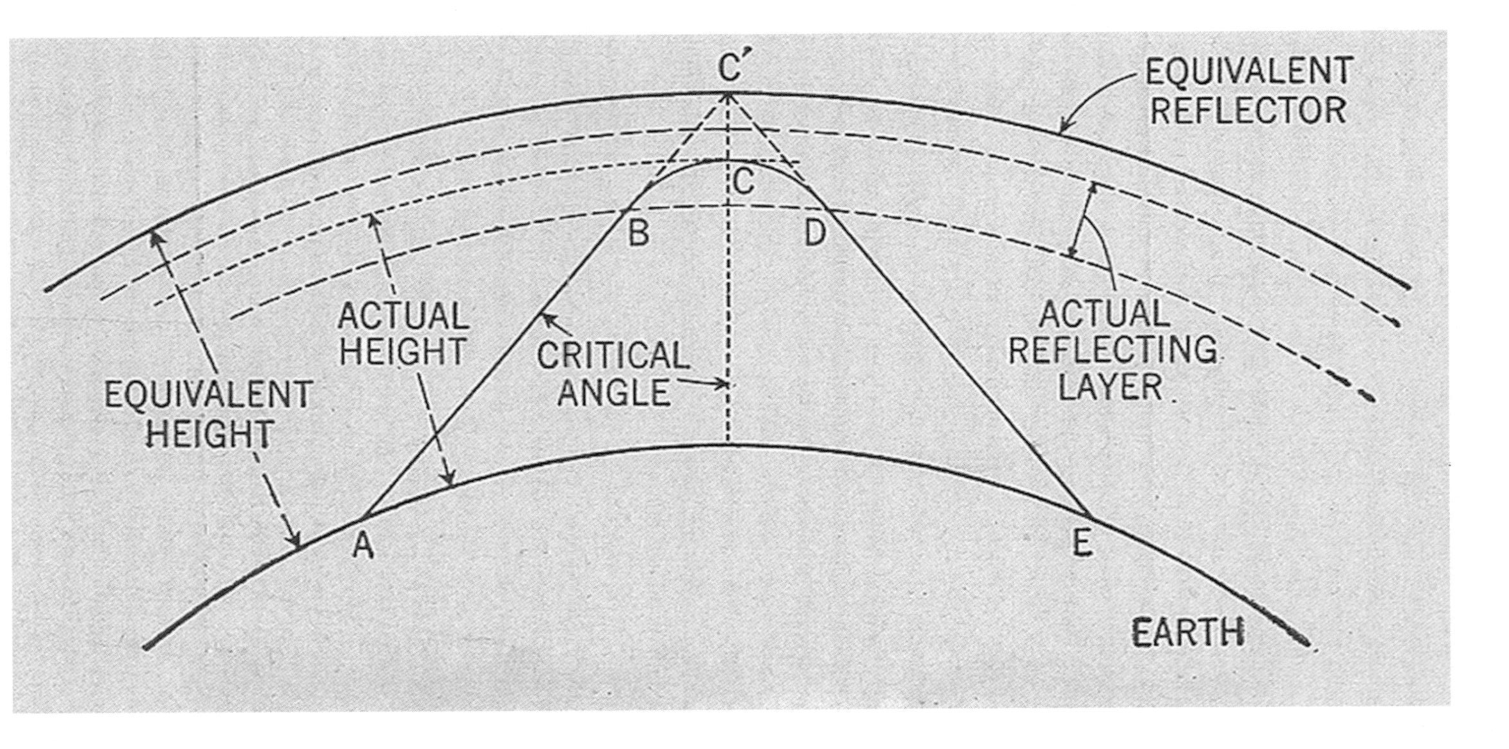

Fig. 31: Actual path and path of equivalent reflected ray. From King, Mimno,& Wing (1945), p. 312.

sustained endeavour to simplify the magneto-ionic theory—as a functional construct in which the *equivalent height* and the *critical frequency* are the defining characteristics—in the interests of optimization and effectiveness in radio communication, all indicate that the radio commercial environment not only facilitates but also nourishes the maintenance of a generalized conviction of a realist interpretation of the upper atmosphere.

Conclusion

We began this work with some of Gillmor's thought-inspiring assertions about the essential nature of layers in the ionosphere, and it is with some more of his thought-inspiring assertions that we shall finish. Gillmor was concerned with the influentiality of technology, which he—correctly, we believe—adduced to have inoculated with realist sentiments among upper atmospheric physicists since the advent of the ionosonde in the mid-1930s. In this respect, he said:

> Change in technology of instrumentation and data presentation can cause a change in the way the scientist conceives of the phenomena. Much of the jargon in any technological field is related to man's interaction with nature through his instrumentation. [And he concluded:] What I wish to stress [with these reflections] is that not only had the term *layer* been in use for decades, now the ionospheric sounder produced a height/frequency plot on which the ionosphere worker could *see* the layers. This has certainly been so in my own experiences in ionospheric physics, and I still have this impression as I examine ionograms [Figure 32]. (Gillmor, 1981, pp. 105–6)

Gillmor's contention regarding the inescapability of the influence of instrumentation is without a doubt worth considering. But, as we have tried to show here, much in the realm of the radio industry, geopolitics, and technical education escaped the net. While Gillmor's contribution (as well as Imhotep's rigorous study) in this matter is the specific identification of the apprehension of reality as the momentous effect of instrumentation, he fails to analyse how the commercial environment at the very least facilitated the precipitation of a generalized conviction of the reality of a sharply stratified atmosphere. Or, *scilicet*, how there was a strong predisposition among physicists towards the interpretation of nature in realist terms, arising as a form of interaction with their most immediate environment.

Now, as the foregoing evidence has shown, this predisposition is inherent in the advent of short wave—or its commercial congener, the *beam system*. Previously, electrical engineers and experimental physicists had fruitlessly endeavoured to explain long-wave transatlantic communications by means of

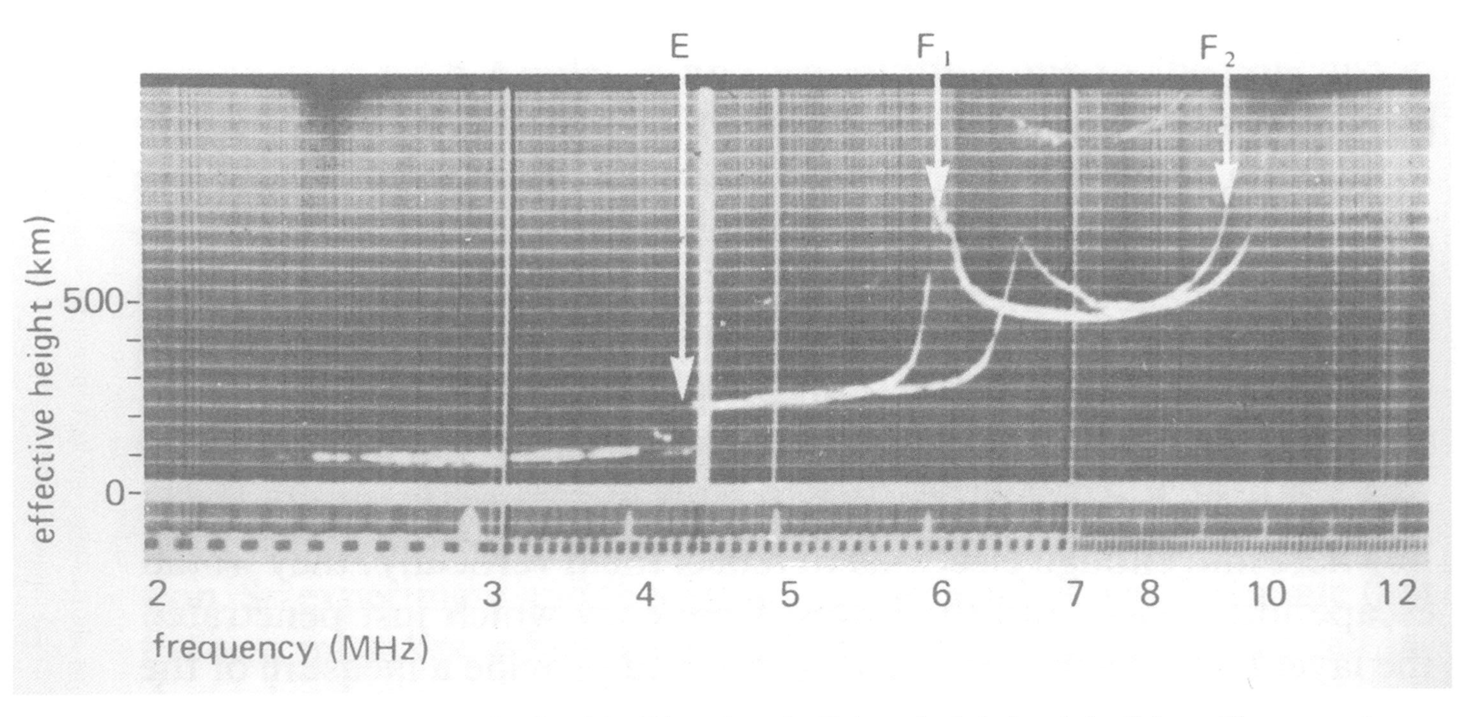

Fig. 32: An ionogram showing partial split of the F layer into the F_1 layer (or ledge) and the F_2 layer. The penetration frequencies are marked at E, F_1, and F_2. The trace is doubled by magneto-ionic splitting. From Ratcliffe (1970), p. 72.*

a reflecting layer. The fact that this had never been directly observed had not constituted an epistemologically significant reason for refusing to posit its existence, but had confined it to the sphere of conjecture, of *theoria*, of *postulatum*. Simultaneously, geomagneticians advocated conducting regions upon rather more solid evidence. But short wave changed everything. The compunction of some geomagneticians about the flagrant disrespectfulness towards their historical memory demonstrates an overshadowing on the part of the conceptions of the upper atmosphere originating from radio engineering. This is hardly surprising if one remembers that the rationale underlying these ideas with which radio communication was so intensely imbued after the war was predicated upon mirror-like reflection, upon commercial triumph, and upon the demise of long wave and diffraction. Indeed, the physical discovery—the epitome of a reality independent from the observer—was pertinaciously pursued as the most direct way of substantiating a generalized conviction and, thereby, of vanquishing the unfruitfulness shown hitherto. *A fortiori*, it was the idea of apprehending all that embodied the most exciting aspects of commercial dynamism—astonishing audibility over long distances, amazingly economical, entrepreneurial excitement, and the Dominions' exhilaration—which conferred upon the physicists' undertaking the substantive, magnified character of reification. It is precisely this evidence on the interrelationship between 'layer hypostatization' and 'excited environment' which indicates that the hypostatization was a reaction to the excitement, that the discovery was the *realization* of an invention. As Appleton stated in explanation and justification of his crucial experiments: 'If the ionosphere had been invented as an essential feature of theories of both geomagnetism and radio propagation, it was necessary to prove, by direct experiment, that it really existed—in other words to discover it' (Appleton, 1963, p. 72). And in all of this sensitiveness there is an amazing sincerity, an ingenuous espousal of the excogitation known as 'inference to the best explanation'.[112] It denotes a natural, unproblematic ebullition among the physicists themselves of fundamentally—and often subliminally—realist impulses.

In these circumstances, one can see then how realism arises neither from the habit nor from pictorial representations, nor is it a product of craft or of convention; it arises in a subtle way: 'short wave' incarnated all that was amazing in long-distance communications—astonishing audibility, miraculous

[112]The 'inference to the best explanation' is the style of reasoning utilized by the realist by which the Heaviside-Kennelly hypothesis was argued to be more likely to be true than any rival theory on the base is that it provided the best available explanation. See Newton-Smith (1981).

instantaneousness, extraordinarily economical, cogent refutation of diffraction. To explain such attributes, there *must* be overhead a physical reality independent of perception and substantiation. Reflecting layer acquired thus *ab incunabulis* the status of reality.

There are, moreover, indications that this realistic propensity flourished elsewhere in the atmospheric sciences. The concept of 'atmospheric wave' representing barometric fluctuations, which John Herschell, aided by American meteorologists, propounded in the early 1840s, came in fact to be constructed at first as a real entity. Herschell's proposal was suited to the commonsensical image of the atmosphere: the barometric curve was 'a simple result of direct observation [and therefore] *must have a meaning*'. 'By the sudden alteration in the Mercury it appears that the atmosphere...must have been greatly agitated and proceeded over the earth in vast waves' (Jankovic, 1998, p. 34). And I would emphasize here that much of the realist excesses announced with great fanfare by atmospheric scientists (more clamorous, certainly, in the 1920s than in the 1840s) owed much to specific problems of observation and inaccessibility in conjunction with, and aggravated by, the growing pressure to provide a useful image of the atmosphere.

In order to demonstrate our thesis, we have underscored three fundamental facts. Firstly, the metaphor of a stratified and reflecting atmosphere had been invented and manufactured credibly before 1925. Secondly, this was almost unanimously reified by radiophysicists and engineers before being 'vindicated' by the magneto-ionic theory of wave propagation in the late 1920s. And, thirdly, the propensity to hypostatize concepts and to read the height vs. frequency graphs in realist terms was patent and generalized before the invention of the ionosonde in 1934 and the production of the first ionograms.[113] It is, therefore, reasonable to conclude that substantive theoretical developments in upper atmospheric physics and the visual persuasiveness of imagery were minor factors in the nascence and incubation of this realist persuasion. Rather,

[113]Such a 'credential of realism' was still valid in the 1960s, although with a loss of credibility. See Lied (1962, p. 2): 'Observations of the virtual height's frequency dependence are then used to determine the arbitrary parameters.... Such considerations lead to a description of the ionosphere in terms of layers, and the parameters usually measure some property of a layer [such as critical frequency, height of the layer maximum, total electron content, etc.] Although such methods have been widely used, there has always been doubt as to their accuracy, particularly if some of the layers are not fully developed.... Comparisons of the profiles deduced by [models and numerical] methods frequently show marked differences in critical frequencies or the height of a layer maximum. In fact the very concept of a critical frequency for a layer is sometimes found to be of doubtful significance.'

the social influence exerted upon the upper atmospheric physicists from a given environment appears the most embryonic determinant.

And, on this point, it is perhaps worth reiterating that all indications suggest that it was not due to a steadfast intellectual espousal of any contemporary philosophical doctrine. However, this very circumstance does not imply the non-existence of latent stances of realism in different degrees of identification and sensitivity. Indeed, one finds here fundamental differences between those physicists (and engineers) who hastened to hypostatize the construct of layer and those who maintained more diffident and cautious views. For Appleton, Eccles, Ratcliffe, Tuve, and Breit, the conception of a reflecting-refracting stratum (or region) was essentially a feature of physical nature; for these authors, the supposition that concepts as useful as 'layer' and its acolytes 'equivalent height' and 'critical frequency' played a merely functional role in the realization of observational predictions without corresponding to something in the world was simply implausible. Eckersley, Heising and Hoyt Taylor also embraced such ontological arrogations, but emphasized their operational character for commercial communications. On the other hand, those physicists—above all those with a solid mathematical background—who, while admitting the reality of ionized regions extricated themselves from the radical reification of theoretical terms, all based their stands upon the value of concepts and theories in their ability to generate correct observational predictions. This is the case with Chapman, Pedersen, and Hulburt—*a priori*, epistemological instrumentalists—, for whom theories were mere tools or calculating devices.

But if this social episode can claim to be explained in terms of a triadic communion between academia, engineering, and the military, then one should look for affiliations between a scientist's allegiance to the realism issue and their socio-economic profession and dimension. And, effectively, while acknowledging the heterogeneous and polychrome spectrum of these realms, one finds, however, that a fair proportion of the upper atmospheric physicists most willing to embrace a strict realist interpretation had close contact with, or were involved in, military institutions. The British RRB, the government body through which Appleton, Smith-Rose, Watson-Watt, and their respective teams channelled their activities, had been instituted to safeguard the interests associated with the Empire and the Admiralty. The American NRL, associated with the Navy, played an important role in setting up a Washington network for ionospheric research in conjunction with the DTM and NBS for similar purposes (Hevly, 1994, pp. 143–8). In contrast, those few physicists—remarkably few, we believe—who were most vehemently opposed to the realist reading, belonged to academia: Hollingworth to the College of Technology,

Manchester, and Murray and Barton Hoag to the University of Chicago. On the other hand, with the exception of a very few cases such as A. Meissner and J.E. Taylor in the mid-1920s,[114] practically all radio engineers and physicists working in private industry viewed the layers in strongly realist terms. Such was the case for Eckersley, Millington, and Tremellen at Marconi's—whose stratified view was immaculately propagated in *The Marconi Review* ,[115] Bell Lab's engineers Heising, G.C. Southworth, and W.M. Goodall, and their colleagues at AWA, RCA, and Telefunken. The sort of epistemological approach adopted by these radio engineers and physicists derived from a loose form of operationalism that attributed cognitive meaning to unverified propositions. This approach made the realist interpretation a useful tool for the betterment and prediction of radio communications, but at the cost of turning ionospheric physics into phenomenological knowledge.[116]

Scientists' degree of affinity with the realism issue correlates, therefore, with the socio-economic milieu in which they carried out their research. This by no means fortuitous concurrence suggests that the explanation anticipated by ionospheric physicists for the literal interpretation of layers during the postwar years is plainly unsatisfactory for the interwar years. In light of the evidence displayed, it seems reasonable to conclude that the reification of concepts and the elaboration of the parabolic layer model were in reality accommodations to a very particular environment. In this respect, circumstances of time and space as peculiar as those with which we are dealing here—a quarter of a century of conjectures, a burgeoning industry, quasi-miraculous technical deeds, and Empires searching for communication hegemonies—must in all probability have left their mark on the scientific doctrines. And it is more than likely that

[114]As late as 1926, J.E. Taylor, a Post Office electrical engineer, complained that 'a good deal of time, money, and energy is being wasted on the pursuit of this academic myth, of a useful ionized layer'. Clark (1971, p. 44).

[115]Eckersley (1931, p. 1): 'The evidence for existence of two reflecting and refracting layers has become practically overwhelming.'; Tremellen (1939, p. 12): 'Three main reflecting regions exist in the ionosphere... [their] height and density vary with the time of day and year, sunspot cycle and latitude, and these conditions in turn decide which layers *control* [italics added] the communications'.

[116]Illustrative for its testimony of the picture in strongly realist terms that was transmitted in a technical journal associated with industrial firms is Goodall (1935, pp. 194–9). In it the author puts forward statements such as: 'Today no one questions its existence [that of a reflecting layer]. The evidence admits of no other interpretation.... Without such a reflecting region, long-distance radio communication by short waves would be impossible.... As a result of studies made by the Laboratories, however, it is now known that the ionosphere is composed of at least five, and possibly more, reflecting regions.'

they also affected the way in which scientists conceived the phenomenon, the realism, the experience, in short, the interaction between *observer* and *observance.* However, regardless of the degree of realism that instrumentation and data presentation methods could instil, of the coloration that the magneto-ionic theory could infuse, of the effect that the surrounding milieu could provoke, there is *one* feature that, we believe, has unequivocally characterized the community of upper atmospheric physicists: *a realist interpretation of the ionosphere.*

BIBLIOGRAPHY

A.W.A., 1931. *Marconi School of Wireless: Radio, the Industry of the Future.* Sydney: A.W.A.

——, 1932. *Facts Regarding the Wireless Industry in Australia.* Sydney: A.W.A.

Affronti, F., 1977. *Atmosfera e meteorologia.* Modena: STEM.

Ahlstrom, G., 1982. *Engineers and Industrial Growth, Higher Technical Education and the Engineering Profession during the 19th and Early 20th Century: France, Germany, Sweden, and England.* London.

Aitken, H.G.J., 1985. *The Continuous Wave: Technology and American Radio, 1900–1932.* Princeton, N.J.: Princeton University Press.

——, 1976. *Syntony and Spark: The Origins of Radio.* New York: Wiley Inter-science.

Akasofu, S.I., 1970. 'In Memoriam Sydney Chapman'. *Space Science Reviews*, 11, 599–606.

Akasofu, S.I., Fogle, B., Haurwitz, B. eds., 1968. *Sydney Chapman, Eighty from his Friends.* Boulder, Colorado.

Al'pert, Y.L., 1960. *Radio Wave Propagation and the Ionosphere.* New York: Consultants Bureau.

Albu, A., 1980. 'British Attitudes to Engineering Education: A Historical Perspective'. In Keith Pavitt, ed. *Technical Innovation and British Economic Performance.* London: MacMillan, 19–87.

Aldcroft, D.H., 1966. 'Economic Progress in Britain in the 1920s'. *Scottish Journal of Political Economy*, 13, 298–316.

Allen, C.W., 1978. 'The Beginnings of the Commonwealth Solar Observatory'. *RAAS*, 4, 27–49.

Allibone, T.E., 1984. 'Metropolitan-Vickers Electrical Company and the Cavendish Laboratory'. In John Hendry, ed. *Cambridge Physics in the 30s.* Bristol: Adam Hilger, 151–73.

Alpers, S., 1983. *The Art of Describing: Dutch Art in the Seventeenth Century.* Chicago: The University of Chicago Press.

Alter, Peter, 1987. *The Reluctant Patron. Science and the State in Britain, 1850–1929.* Oxford: BERG.

Amalgamated Wireless. *Radiola Broadcast Receivers: The Supreme achievement in Broadcast Reception*. Melbourne: A.W.A., n.d..

Anthony, I.A., 2000. *Radio Wizard: Edward Samuel Rogers and the Revolution of Communications*. Toronto: Gage Pub. Co.

Apostel, L., 1961. 'Formal Study of Models'. In H. Freudenthal, ed. *The Concept and the Role of the Model in Mathematics and Natural and Social Sciences*. Dordrecht: Reidel Publishing Co., 1–37.

Appleton, E.V., 1919. 'Note on the Effects of Grid Currents in Three-Electrode Ionic Tubes'. *Phil. Mag.*, 37 (217), 129–34.

——, 1924–25. 'Geophysical Influences on the Transmission of Wireless Waves'. *Proc. Phys. Soc. Lond.,* 37, 38D-45D.

——, 1930. 'The Romance of Short Waves'. *World-radio*, 7 Nov 1930, 710.

——, 1932a. 'Cable and Wireless'. *The Times Trade and Engineering Supplement*,
21 May, 1932, 17.

——, 1932b. 'Wireless Studies of the Ionosphere'. *Proc. IEE*, 71, 642–50.

——, 1933a. Co-operative Radio Research in the Empire. *World-Radio*, 19, 669.

——, 1933b. *Empire Communication: The Norman Lockyer Lecture, 1933.* London: British Science Guild.

——, 1935. 'Temperature Changes in the Higher Atmosphere'. *Nature*, 136, 52.

——, 1937a. 'Regularities and Irregularities in the Ionosphere'. *Proc. Roy. Soc. Lon.*, 162, 451–478.

——, 1937b. 'Empire Radio Communications'. *Unit. Emp. RES. J.*, 28, 585.

——, 1955. 'Thermionic Devices from the Development of the Triode up to 1939'. In Institution of Electrical Engineers ed., *Thermionic Valves, 1904–1945: The First Fifty Years* London: IEE, 17–25.

——, 1963. 'Radio and the Ionosphere'. In C. Domb, ed. *Clerk Maxwell and Modern Science*. London: Athlone Press, 70–88.

——, 1964. 'The Ionosphere'. In Nobelphysics, *Nobel Lectures: Physics, 1942–1962.* Amsterdam: Elsevier Publ. Co., 1–8.

Appleton, E.V. & Barnett, M.A.F., 1925a. 'On Some Direct Evidence for Downward Atmospheric Reflection of Electric Rays'. *Proc. Roy. Soc. Lond.,* 109, 621–41.

——, 1925b. 'Local Reflections of Wireless Waves from the Upper Atmosphere'. *Nature*, 115, 333–4.

Appleton, E.V. & Beynon, W.J.G., 1940. 'The Application of Ionospheric data to Radio-Communication Problems: Part I'. *Proc. Phys. Soc. Lond.,* 52, 518–33.

Appleton, E.V. & Builder, G., 1932. 'Wireless Echoes of Short Delay'. *Proc. Phys. Soc. Lon.*, 44, 76–87.

Appleton, E.V. & Green, A.L., 1930. 'On Some Short-Wave Equivalent Height Measurements on the Ionized Regions of the Upper Atmosphere'. *Proc. Roy. Soc. Lond.*, 128, 159–78.

Appleton, E.V., Naismith, R., 1932. 'Some Measurements of Upper Atmosphere Ionization'. *Proc. Roy. Soc. Lon.*, 137, 36–54.

——, 1935. 'Some Further Measurements of Upper Atmospheric Ionization'. *Proc. Roy. Soc. Lon.*, 150, 685–708.

Appleton, E.V. & Ratcliffe, J.A., 1928. 'On a Method of Determining the State of Polarization of Downcoming Wireless Waves'. *Proc. Roy. Soc. Lon.*, A 117: 576–88.

Appleton, E.V. & Van der Pol, B., 1921. ‚On the Form of Free Triode Vibrations'. *Phil. Mag.*, 42, 248: 202–20.

——, 1922. ‚On a Type of Oscillation: Hysteresis in a Simple Triode Generator'. *Phil. Mag.*, 43, 253. 177–93.

Archer, G.L., 1938. *History of Radio to 1926.* New York: American Historical Society Inc.

Armstrong, E.H., 1951. 'Wrong Roads and Missed Chances: Some Ancient Radio History'. *Marconi Review,* 4, Sup., 21–8.

Ashbridge, N., 1933. 'Six Months of Empire Broadcasting'. *World-Radio,* 16, 677–80.

Atkinson, J.T., 1976. *DSIR s First Fifty Years.* Wellington: Department of Scientific and Industrial Research.

Attard, B., 1999. *Australia as a Dependent Dominion, 1901–1939. Working Papers in Australian Studies, 115, Sir Robert Menzies Centre for Australian Studies.* London: University of London, Institute of Commonwealth Studies.

Audretsch, J., 1995. 'Vorläufige Physik und Andere Pragmatische Elemente Physikalischer Naturerkenntnis'. In H. Stachowiak, ed. *Pragmatik, Band III* Hamburg, 1989, 373–92.

Auerbach, F., 1925. *Die Methoden der Theoretischen Physik.* Leipzig: Akad. Verlagsanstalt.

Australian Academy of Technological Sciences and Engineering Comp., 1988. *Technology in Australia, 1788–1988: A Condensed History of Australian Technological Innovation and Adaptation during the First Two Hundred Years.* Melbourne.

Babaian, S.A., 1992. *Radio Communication in Canada: A Historical and Technological survey.* Otawa: National Museum of Science and Technology.

Babe, R.E., 1988. 'Emergence and Development of Canadian Communication: Dispelling the Myths'. In R.M. Lorimer & D.C. Wilson, eds. *Communication*

Canada: Issues in Broadcasting and New Technologies. Toronto: Kagan & Woo Ltd., 58–79.

——, 1990. *Telecommunications in Canada*. Toronto: Toronto University Press.

Baeumler, M., 1925. 'Investigations on the Propagation of Electromagnetic Waves'. *Proc. IRE,* 13, 3–27.

Bailey, E.G. & Healey, R.H., 1939. 'Some Aspects of Valve Manufacture in Australia'. *Proc. Wor. Rad. Con.*, 1–7.

Bailey, V.A., 1934. 'The Influence of Electric Waves on the Ionosphere'. *Philosophical Magazine*, S.7, 18, 369–86; *CSIR Bulletin*, 1935, 88; *RRB Rept*, 7, 40–52.

——, 1960. 'Obituary: Geoffrey Builder'. *Australian Journal of Science*, 23 (5), 155–56.

Bailey, V.A. & Martyn, D.F., 1934a. 'Interaction of Radio Waves'. *Nature*, 133, 218.

——, 1934b. 'The Influence of Electric Waves on the Ionosphere'. *Philosophical Magazine*, S.7, 18, 369–86; *CSIR Bulletin*, 1935, 88; *RRB Rept*, 7, 40–52.

Baker, P. & Hance, B., 1981. 'Round, Henry Joseph, 1881–1966'. In E.T. Williams & Nicholls, C.S., eds. *The Dictionary of National Biography, 1961–1970*. Oxford: Oxford University Press, 897–8.

Baker, W.G., 1926. 'Refraction of Short Waves in the Upper Atmosphere'. *Tran. Am. Ins. Ele. Eng.*, 45, 302–33.

——, 1938. 'Studies in the Propagation of Radio Waves in an Isotropic Ionosphere'. *AWA Tech. Rev.*, 36, 297–320.

Baker, W.J., 1970. A *History of the Marconi Company, 1874–1965*. London: Methuen.

Bannon, J., Higgs, A.J., Martyn, D.F. & Munro, G.H., 1940. 'The Association of Meteorological Changes with Variations of Ionization in the F_2 Region of the Ionosphere'. *Proc. Roy. Soc. Lond.*, A 174, 298–309.

Barfield, R.H. & Ross, W., 1937. 'A Short Wave Adcock Direction-Finder'. *Jour. IEE,* 81, 682–90.

Barnett, C., 1986. *The Audit of War: The Illusion and Reality of Britain as a Great Nation*. London: Macmillan.

Barnett, M.A.F., 1939–40. 'Obituary of E. Kidson, 1882–1939'. *Transactions and Proceedings of the Royal Society of New Zealand*, 69, 186–7.

Barnouw, E., 1978. *The Sponsor: Notes on a Modern Potentate*. New York: Oxford University Press.

Barraclough, D.R., 1989. 'Geomagnetism: Historical introduction'. In D.E. James, ed. *The Encyclopedia of Solid Earth Geophysics*. New York: Van Nostrand Reinholt, 584–92.

Barrell, H., 1932. 'Kurzer Überblick über die Physik der Hohen Atmosphäre'. *Zeitschrift für technische Physik,* 13, 611–616.

——, 1939. 'Some Problems of Terrestrial Magnetism and Electricity'. In J.A. Fleming, ed. *Terrestrial magnetism and electricity.* New York: Dover, 385–433.

——, 1969. 'Kew Observatory and the National Physical Laboratory'. *Met. Mag.*, 98, 171–80.

Bartlett, M., 1995. *Education for Industry: Attitudes and Policies Affecting the Provision of Technical Education in Britain, 1916–1929.* Oxford University, PhD. diss.

Barton Hoag, J., 1942. *Basic Radio: The Essentials of Electron Tubes and their Circuits.* London: Chapman & Hall.

Barty-King, H., 1979. *Girdle Round the Earth: The Story of Cable and Wireless and its Predecessors to Mark the Group s Jubilee, 1929–1979.* London: Heinemann.

Bates, D.R., 1973. 'The Normal E- and F-layers', *Jour. Atm. Terr. Phys.*, 35, 1,935–52.

Beaglehole, J.C., 1949. *Victoria University College: An Essay Towards a History.* Wellington: New Zealand University Press.

Beer, G., 1996. 'Wireless: Popular Physics, Radio and Modernism'. In F. Spufford & J. Uglow, eds. *Cultural Babbage: Technology, Time and Invention.* London: Faber & Faber, 149–66.

Belfort, R., 1929. 'Outlook for Cable-Radio'. *Unit. Emp. RES. J.*, 20, 19–23.

Bellini, E., 1921. The Errors of Direction-Finders. *The Electrician,* 86, 220–2.

Benedetto, G. di, ed., 1974. *Bibliografia Marconiana.* Genova: Giunti.

Berg, J.S., 1999. *On the Short Waves, 1923–1945: Broadcast Listening in the Pioneer Days of Radio.* London: McFarland &Co.

Berger, C., 1976. *The Writing of Canadian History: Aspects of English-Canadian Historical Writing. Since 1900.* Toronto: Toronto University Press.

Berkner, L.V., 1941. 'Contributions of Ionospheric Research to Geomagnetism'. *Proceedings of the American Philosophical Society*, 84, 309–21.

Beyerchen, A., 1996. 'From Radio to Radar. Interwar Military Adaptation to Technological Change in Grmany, the United Kingdom, and the United States'. In W. Murray & A.R. Millett, eds. *Military Innovation in the Interwar Period.* Cambridge: Cambridge University Press, 265–99.

Beynon, W.J.G., 1967. 'Preface to Special Issue on the Analysis of Ionograms for Electron Density Profiles'. *Radio science,* 2, 1118.

——, 1975a. 'Marconi, Radio Waves, and the Ionosphere'. *Radio Science*, 10, 657–64.

Beynon, W.J.G., 1975b. 'U.R.S.I. and the Early History of the Ionosphere'. *Phil. Trans. Roy. Soc. Lond.*, 280, 47–54.

Blackett, P.M.S., 1933. 'The Craft of Experimental Physics'. In H. Wright, ed. *Cambridge University Studies.* London: Ivor Nicholson & Watson, 67–96.

——, 1960. Charles Thomson Rees Wilson, 1869–1959. *Biog. M. Fell. Roy. Soc.*, 6, 269–95.

Blackwell, M.J., 1958. Eskdalemuir Observatory: The First Fifty Years. *Met. Mag.*, 87, 129–32.

Bleaney, B., 1994. 'The Physical Sciences in Oxford, 1918–1939 and Earlier'. *Notes and Records of the Royal Society*, 48, 247–61.

Blondel, A., 1903. 'Quelques remarques sur les effets des antennes de transmission'. *Association française pour l'avancement des sciences*, 407.

Bolton, H.C., 1990. 'Optical Instruments in Australia in the 1939–45 War: Successes and Lost Opportunities'. *Australian Physicist*, 27 (3), 31.

Born, M., 1953. 'Physical reality'. *Philosophical Quarterly*, 3, 140.

Boswell, R.W., 1936. 'Sources of Atmospherics over the Tasman Sea'. *RRB Rept.*, 10, 9–17.

Boswell, R.W., Wark, W.J., & Webster, H.C., 1936. 'A Directional Recorder for Atmospherics'. *CSIR Bulletin*, 100, *RRB Rep.*, 10, 9–17.

Bourassa, H., 1926. *Le Canada, nation libre? Discours prononcé à Papineauville le 18 juillet 1926.* Montreal: Devoir.

Bowen, E.G., 1984. 'The Origins of Radio Astronomy in Australia'. In W.T. Sullivan, ed. *The Early Years of Radioastronomy.* Cambridge: Cambridge University Press, 85–111.

Bowhill, S.A. & Schmerling, E.R., 1961. 'The Distribution of Electrons in the Ionosphere'. In L. Marton, ed. *Advances in Electronics and Electron Physics.* New York: Academic Press, 265–326.

Bown, R., 1937. 'Transoceanic Radio Telephone Development'. *Proc. IRE*, 25, 1,124–35.

Boyd, M., 1979. 'Australian-New Zealand Relations'. In W.S. Livingston & W.R. Louis, ed. *Australia, New Zealand and the Pacific Islands since the First World War.* Texas, 47–61.

Bradie, M., 1980. 'Models, Metaphors, and Scientific Realism'. *Nature and System*, 2, 3–20.

Bradley, F.R., 1934. 'History of the Electric Telegraph in Australia'. *Journal of the Royal Australian Historical Society*, 20.

Bradley, J., Excell, P., & Rowlands, P., 2000–01. 'From Triode Oscillations to Modulated Radar: Sir Edward Appleton's Involvement'. *Trans. Newcomen Soc.,* 72, 115–25.

Bragg, M., 2002. *RDF 1: The Location of Aircraft by Radio Methods 1935–1945.* Paisley: Hawkhead.

Brannigan, A., 1981. *The Social Basis of Scientific Discoveries.* Cambridge: Cambridge University Press.

Breit, G. & Tuve, M.A., 1925. 'A Radio Method of Estimating the Height of the Conducting Layer'. *Nature,* 116, 357.

——, 1926. 'A Test of the Existence of the Conducting Layer'. *Physical Review,* 28, 554–76.

Bridge, C. & Attard, B., eds., 2000. *Between Empire and Nation: Australia's External Relations from Federation to the Second World War.* Melbourne: Australian Scholarly Publishing.

Briggs, A., 1961. *The Birth of Broadcasting: The History of Broadcasting in the United Kingdom.* London: Oxford University Press.

Bright, A.A., 1949. *The Electric Lamp Industry.* New York: MacMillan.

Bright, C., 1923. 'The Empire's Telegraph and Trade'. *Fortnight Review,* 113, 457–74.

British Information Services, 1963. *Britain and Commonwealth Telecommunications.* London.

Brock, G.W., 1981. *The Telecommunications Industry: The Dynamics of Market Industry.* Cambridge, Mass.: Harvard University Press.

Brooking, T., 1981. 'Economic Transformation'. In W.H. Oliver & B.R. Williams, eds. *The Oxford History of New Zealand.* Wellington: Oxford University Press, 226–49.

Brown, A.C., 1977. *A History of Scientific Endeavour in South Africa.* Cape Town: Royal Soc. South Africa.

Brown, F.J., 1927. *The Cable and Wireless Communications of the World: A Survey of Present Day Means of International Communication by Cable and Wireless, containing Chapters on Cable and Wireless Finance.* London.

Brown, H.P., 1938. 'Broadcasting in Australia'. *Proc. Wor. Rad. Con.,* 1–15.

Brown, O.F., 1924. 'The Heaviside Layer and How It may be produced'. *EW&WE,* 1, 595–597.

——, 1927. *The Elements of Radio-Communication.* London: Oxford University Press.

Bucher, E.E. et al., 1928. *The Radio Industry: The Story of its Development.* Chicago.

Buchwald, J., 1985. *From Maxwell to Microphysics: Aspects of Electromagnetic Theory in the last Quarter of the Nineteenth Century.* Chicago: University of Chicago Press.

Budden, K.G., 1961. *Radio Waves in the Ionosphere* Cambridge: Cambridge University Press.

——, 1988. 'John Ashworth Ratcliffe, 12 December 1902–25 October 1987'. *Biog. M. Fell. Roy. Soc.*, 34, 671–711.

Builder, G., 1932. 'The Existence of More than One Ionized Layer in the Upper atmosphere'. *EW & WE,* Dec 1932, 667–72.

——, 1933. 'Wireless Apparatus for the Study of the Ionosphere'. *J. Ins. Elec. Eng.*, 73, 419–436.

Builder, G., Benson, J.E., 1936. 'Noise Interference in Radio Receivers'. *AWA Tech. Rev.*, 2, 1, 23–32.

——, 1938. 'Precision Frequency-Control Equipment Using Quartz Crystals'. *AWA Tech. Rev.*, 34, 157–214.

Bunge, M., 1968. 'Les Concepts de Modèle'. *L Âge de la Science*, July, 165–80.

Burch, L.L.R., 1980. *The Flowerdown Link: A Story of Telecommunications and Radar Throughout the Royal Flying Corps and Royal Air Force.* Weymouth: Sherren.

Burdon, R.M., 1965. *The New Dominion and Social and Political History of New Zealand, 1918–39.* Wellington: Reed.

Bureau of Standards, 1931. 'Bibliography on Radio Wave Phenomena and Measurement of Radio Field Intensity'. *Proc. IRE,* 19, 1,034–89.

Bureau, M.R., 1926. 'Les Atmosphériques', *Onde Élec.*, 5, 301–44.

Bureau, R., 1927. 'Note sur Certaines Anomalies dans la Propagation des Ondes Courtes'. *L Onde Électrique*, 661, 52–5.

Burnett, A.A. & Burnett, R., 1980. *The Australian and New Zealand Nexus: Annotated Documents.* Canberra: Australian Inst. of International Affairs.

Burton, P., 1965. *The New Zealand Geological Survey, 1865–1965.* Wellington: DSIR.

Bush, S.G. & Gillmor, C.S., 1995. Geophysics. In L.M. Brown, A. Pais, & B. Pippard, eds. *Twentieth-Century Physics.* Bristol: Institute of Physics Publ., 3, 1,943–2,016.

Bussey, G., 1976. *Vintage Crystal Sets, 1922–1927.* London: IPC Press.

——, 1990. *Wireless the Crucial Decade: History of the British Wireless Industry, 1924–34.* London: Peter Peregrinus Ltd.

Butcher, B.W., 1984. 'Science and the Imperial Vision: The Imperial Geophysical Experimental Survey, 1928–1930'. *HRAS*, 6, 1. 31–43.

Butlin, N.G., 1970. 'Some Perspectives of Australian Economic Development, 1890–1965'. In C. Forster, ed. *Australian Economic Growth in the Twentieth Century*. London: G. Allen & Unwin.

Buttner, H.H., 1931. 'The Role of Radio in the Growth of International Communication'. *Electrical Communication*, 9, 249–54.

Buxton, N.K. & Aldcroft, D.H., eds., 1979. *British Industry Between the Wars: Instability and Industrial Development, 1919–1939*. London: Scolar Press.

Caccialupi, P., 1939. *Il Dominatore dell Infinito Guglielmo Marconi.* Milano: La Prora.

Canuel, A., 1986. 'La Présence de l'Impérialisme dans les Débuts de la Radiophonie au Canada, 1900–1928'. *Journal of Canadian Studies,* 20, 45–76.

——, 1988. *Les Rapports Entre la Radiophonie et l'Impérialisme dans le Contexte Socio-Politique Canadien de 1901 à 1928*. Montréal: Université de Montréal, PhD diss.

Cave, C.J.P. & Watson, R.A., 1923. 'Study of Radiotelegraphic Atmospherics in Relation to Meteorology'. *Quart. J. Roy. Met. Soc.*, 49, 35–42.

Cawood, J., 1979. 'The Magnetic Crusade: Science and Politics in Early Victorian Britain'. *Isis*, 70, 493–518.

Chapman, R.M. & Malone, E.P., 1969. *New Zealand in the Twenties: Social Change and Material Progress*. Auckland: Heinemann, 1969.

Chapman, S., 1913. 'On the Diurnal Variations of the Earth's Magnetism Produced by the Moon and Sun'. *Phil. Trans. Roy. Soc. Lond.*, 213, 279–321.

——, 1914. 'On the Lunar Variation of the Earth s Magnetism at Pavlovsk and Pola, 1897–1903'. *Phil. Trans. Roy. Soc. Lond.,* 1914, 214, 295–317.

——, 1915. 'Lunar Diurnal Magnetic Variation and its Change with Lunar Distance'. *Phil. Trans. Roy. Soc. Lond.,* 215, 161–76.

——, 1919. 'Solar and Lunar Diurnal Variations of Terrestrial Magnetism'. *Phil. Trans. Roy. Soc. Lond.,* 218, 1–118.

——, 1924–25. 'The Evidence of Terrestrial Magnetism for the Existence of Highly Ionized Regions in the Upper Atmosphere', *Proc. Phys. Soc. Lond.,* 37, 38D-45D.

——, 1928. 'On the Theory of Solar Diurnal Variation of the Earth's Magnetism'. *Proc. Roy. Soc. Lond.,* 122, 369–86.

——, 1930a, 'A Theory of Upper Atmospheric Ozone'. *Mem. Roy. Met. Soc.*, 3, 103–125;

Chapman, S., 1930b. 'On the Annual Variation of Upper-Atmospheric Ozone'. *Phil. Mag.,* 10, 345–52.

——, 1930c. 'On Ozone and Atomic Oxygen in the Upper Atmosphere'. *Phil. Mag.,* 10, 369–83.

——, 1931a. 'The Absorption and Dissociation or Ionizing Effect of Monochromatic Radiation in an Atmosphere on a Rotating Earth'. *Proc. Phys. Soc.*, 43, 26–45, 483–501.

——, 1931b. 'Some Phenomena of the Upper Atmosphere'. *Proc. Roy. Soc. Lon.,* 132, 353–74.

——, 1931c. 'Some Phenomena of the Upper Atmosphere'. *Nature*, 128, 415, 464–5.

——, 1934. 'Radio Exploration of the Ionosphere'. *Nature,* 133, 908.

——, 1941. 'Charles Chree and his Work on Geomagnetism'. *Proc. Phys. Soc.,* 53, 629–634.

——, 1946a. 'A Plea for the Abolition of "Meteorology"'. *Weather*, 1, 146–47.

——, 1946b. 'Some Thoughts on Nomenclature'. *Nature*, 157, 405.

——, 1950. 'Upper Atmosphere Nomenclature'. *Jour. Geo. Res.*, 55, 395–99 [also in *Jour. Atm. Terr. Phys.*, 1, 121–4, 201; *Bull. Am. Met. Soc.*, 31, 288–90]

——, 1956. 'The Electrical Conductivity of the Ionosphere: A Review'. *Il nuovo cimento,* 4, 1,385–1,412.

——, 1967a. 'Historical Introduction to Aurora and Magnetic Storms'. *Annals of geophysics,* 24, 497–505.

——, 1967b. 'Perspective in Physics of Geomagnetic Phenomena'. In S. Matsushita & W.H. Campbell, eds. *Physics of Geomagnetic Phenomena*. London: Academic Press, 3–28.

Chapman S. & Bartels, J., 1940. *Geomagnetism.* Oxford: Oxford University Press.

Chapman, S. & Milne, E.A., 1920. 'The Composition, Ionization and Viscosity of the Atmosphere at Great Heights'. *Quart. Jour. Roy. Met. Soc.,* 46, 357–98.

Childs, W.W., 1924. 'Problems in the Radio Industry'. *The American Economic Review,* 14, 520–3.

Chree, C., 1915. 'Atmospheric Electricity Potential Gradient at Kew Observatory, 1898 to 1912'. *Phil. Trans. Roy. Soc. Lond.*, 215, 133–59.

——, 1926. 'Atmospheric Ozone and Terrestrial Magnetism'. *Proc. Roy. Soc. Lon.*, 110, 693–9.

——, 1927. 'Wireless Communication and Terrestrial Magnetism'. *Nature,* 119, 82–3.

Chrishop, I.F., 1986. 'A Short History of Electrical Education and Training in the Royal Navy'. In *Papers Presented at the 13th IEE Weekend Meeting on the History of Electrical Engineering*. London, 3/1–3/10.

Christie, M., 2000. *The Ozone Layer: A Philosophy of Science Perspective.* Cambridge: Cambridge University Press.

Churchland, P.M. & Hooker, C.W., 1985. *The Image of Science.* Chicago: University of Chicago Press.

City and Guilds of London Institute, 1993. *A Short History, 1878–1992.* London.

Clark, K., 1931. *International Communications: The American Attitude.* New York: Columbia University Press.

Clark, M., 1993. *A History of Australia.* Melbourne: Melbourne University Press.

Clark, R., 1971. *Sir Edward Appleton.* Oxford, New York: Pergamon Press.

Clark, R.W., 1965. *Tizard.* Cambridge, Mass.: MIT Press.

Clarricoats, J., 1967. *World at Their Fingertips.* London: Radio Society of Great Britain.

Clayton, R. & Algar, J., 1989. *The GEC Research Laboratories: 1919–1984.* London: Peter Peregrinus Ltd.

Clifford, G.D. & Sharp, F.W., 1989. *A 20th Century Professional Institution. The Story of the I.E.R.E., 1925–1988.* London: Peter Peregrinus.

Close, C., 1983. 'Thomas Howell Laby, 1880–1946. Physicist'. *ADB*, 9, 640–1.

Coates, V.T., Finn, B., eds., 1979. *A Retrospective Technology Assessment: Submarine Telegraphy.* San Francisco: San Francisco Press.

Codel, M., ed. 1930. *Radio and Its Future.* New York: Harper & Brothers.

Cohen, I.B., 1949. *Science, Servant of Man: A Layman's Primer for the Age of Science.* London: Sigma Books Ltd.

Collin, H.M., 1985. *Changing Order: Replication and Induction in Scientific Practice.* London: Sage.

Collins, R., 1977. *A Voice from Afar. The History of Telecommunications in Canada.* Toronto: McGraw Hill.

Copland, D.B. & Janes, C.V., 1937. *Australian Trade Policy: A Book of Documents, 1932–1937.* Sydney: Angus & Robertson.

Corbett, A.H., 1961. 'The First Hundred Years of Australian Engineering Education, 1861–1961'. *Jour. Inst. Eng. Aus.*, 33, 147–58.

Counihan, M., 1982. 'The Formation of a Broadcasting Audience: Australian Radio in the Twenties'. *Meanjin*, 41, 196–209.

Cowling, T.G., 1971. 'Sydney Chapman, 1888–1970'. *Biog. M. Fell. Roy. Soc.*, 17, 53–89.

Crawford, M., 1931. 'Communication Engineering in Australia'. *Jour. Inst. Eng. Aus.*, 3, 41–53.

Crean, S.M., 1976. *Who's Afraid of Canadian Culture?*. Don Mills, Ont.: General Publications.

Crichton, J., 1950. 'Eskdalemuir Observatory'. *Met. Mag.*, 79, 337–40.

Cromer, H.W., 1933. 'Industry and the Universities I: Chemistry, Chemical Engineering, Physics, Geology and Geography at King's College, University of London'. *JC*, 12, 50–4.

Crowther, J.A., 1926. 'Research Work in the Cavendish Laboratory in 1900–1918'. *Nature* Supplement, 118, 58–60.

——, 1974. *The Cavendish Laboratory, 1874–1974.* London: MacMillan.

Cumming, N.D., 1931. 'Radio Communication: What it Means to the Empire and the World'. *Unit. Emp. RES. J.*, 22, 506–7.

Curnow, R., 1963a. 'Communications and Political Power'. In I. Bedford & R. Curnow, *Initiative and Organization.* Melbourne: Cheshire.

——, 1963b. 'The Origins of Australian Broadcasting, 1900–1923'. In I. Bedford & R. Curnow, *Initiative and Organization.* Melbourne: Cheshire.

Currie, G. & Graham, J., 1966. *The Origins of CSIRO: Science and the Commonwealth Government, 1901–1926.* Melbourne: CSIRO.

——, 1968. 'Growth of Scientific Research in Australia: The Council for Scientific and Industrial Research and the Empire Marketing Board'. *RAAS*, 13, 25–35.

——, 1971. 'G.A. Julius and Research for Secondary Industry'. *RAAS*, 2 (1), 10–28.

——, 1974. 'CSIR 1926–1939'. *Public Administration*, 33(3), 230–52.

Cushing, J.T., 1982. 'Models and Methodologies in Current Theoretical High-Energy Physics'. *Synthese*, 50, 5–101.

Czarnocka, M., 1995. 'Models and the Symbolic Nature of Science'. *Poznan Stud. Phil. Sc. Hum.*, 44, 27–36.

Darrow, K.K., 1940a. 'Analysis of the Ionosphere'. *Bell System Technical Review*, 19, 455–488.

——, 1940b. 'Introduction to the Ionosphere'. *Proceedings of the American Philosophical Society*, 83, 429–45.

Davies, K., 1965. *Ionospheric Radio Propagation.* Washington, D.C.

Davies, Susan, 1983. 'Rutherford and Physics in Australia'. *Aus. Phys.*, 20, 219–23.

Davis, N.E., 1930. 'The Marconi-Adcock Direction Finder'. *The Marconi Review,* 21, 1–8.

Day, Alan A., 1966. 'The Development of Geophysics in Australia'. *Journal and Proceedings of the Royal Society of New South Wales*, 100, 33–60.

Day, P., 1994. *The Radio Years: A History of Broadcasting in New Zealand.* Auckland: Auckland University Press.

De Soto, C.B., 1936. *Two Hundred Meters and Down: The Story of Amateur Radio.* West Hartford, Conn.

Dean, K., 2003. 'Inscribing Settler Science: Ernest Rutherford, Thomas Laby and the Making of Careers in Physics'. *History of Science*, 41, 217–40.

Decibel, 1937. *Wireless Terms Explained.* London: Isaac Pitman & Sons.

Dellinger, J., 1947. 'The Ionosphere'. *Scientific Monthly,* 65, 115–26.

Dellinger, J.H., 1939. 'The Role of the Ionosphere in Radio Wave Propagation'. *Transactions of the American Institute of Electrical Engineers*, 58, 803–21.

Deloraine, E.M., 1930. 'The Use of Short Waves in Radio Communication'. *Electrical Communication*, 8, 213–15.

Deloy, L., 1924. 'Communications Transatlantiques sur Ondes de 100 Mètres'. *L'onde électrique,* 3, 38–42.

Denny, L., 1930. *America Conquers Britain: A Record of Economic War.* New York, London: A.A. Knopf.

Department of Marine and Fisheries, 1924. *Annual Report for 1922–23.* Ottawa.

Department of the Naval Service, 1920. *Annual Report for 1918–19.* Ottawa.

DeVorkin, D.H., 1998. 'Ozone'. In G.A. Good, ed., *Sciences of the Earth. An Encyclopaedia of Events, People and Phenomena.* New York & London: Garland Publishing, 641–46.

Dewar, J., 1902. 'Problems of the Atmosphere'. *Proc. Roy. Soc. Lond.,* 17, 223–30.

Dieminger, W., 1948. *FIAT, Review of German Science*, 17, 93–163.

——, 1974. 'Early Ionospheric Research in Germany'. *Jour. Atm. Terr. Phys.*, 36, 2,085–93.

——, 1975. 'Trends in Early Ionospheric Research in Germany'. *Phil. Trans. Roy. Soc. Lond.*, 280, 27–34.

Dilworth, C., 1990. 'Empiricism Vs. Realism: High Points in the Debate During the Past 150 Years'. *SHPS*, 21(3), 431–62.

Dingle, H., 1941. 'Alfred Fowler, 1868–1940'. *Ob. Not. Fell. Roy. Soc.,* 3, 483–97.

Divall, C., 1990. 'A Measure of Agreement: Employers and Engineering Studies in the Universities of England and Wales, 1897–1939'. *Soc. Stud. Sc.*, 20, 65–112.

Dixon, F., 1975. *Inside the ABC: A Piece of Australian History.* Melbourne: Hawthorn Press.

Dobson, G.M.B., (1926), *The Uppermost Regious of the Earth's Atmosphere.* Oxford: Clarendon Press.

——, 1930. 'Observations of the Amount of Ozone in the Earth's Atmosphere and its Relation to Other Geophysical Conditions'. *Proc. Roy. Soc.,* 129, 411–33.

——, 1966. *Forty Years Research on Atmospheric Ozone at Oxford –a History.* Oxford: Clarendon Laboratory, reprinted in *Applied Optics*, 1968, 7, 387–405.

Dobson, G.M.B., Harrison, D.N., & Lawrence, J., 1926. 'Measurements of the Amount of Ozone in the Earth's Atmosphere and its Relation to Other Geophysical Conditions'. *Proc. Roy. Soc. Lond.,*110, 660–93; 1927, 114, 521–41; 1929, 122, 456–86.

Doel, R.E., 1997. 'The Earth Sciences and Geophysics'. In J. Krige & D. Pestre, eds., *Science in the 20th Century.* Amsterdam: Harwood Academy, 391–416.

Donald, R., 1928. 'Empire Cable and Wireless: A Bold Scheme Needed'. *The Times*, 24th March, 1928.

Donaldson, F.L., 1962. *The Marconi Scandal.* London: R. Hart-Davis.

Dougherty, I., 1993. 'Marconi a Phoney? Change and Expansion in New Zealand Telecommunications in the Interwar Years'. *Stout Centre Review*, March, 9–16.

——, 1997. *Ham Shacks, Brass Pounders and Rag Chewers: A History of Amateur Radio in New Zealand.* Upper Hut: New Zealand Association of Radio Transmitters.

Douglas, S.J., 1987. *Inventing American Broadcasting, 1899–1922.* Baltimore: Johns Hopkins University Press.

Dowsett, H.M., 1915. The Physical and Electrical State of the Atmosphere. *The Wireless World*, 3, 278–82.

——, 1929. 'Commercial Short Wave Wireless Communication'. *The Marconi Review*, 1 (13), 14–30; 1 (14), 1–15; 1 (15), 1–17.

——, 1934. 'Marconi s Place in History'. *The Marconi Review*, 51, 1–3.

——, 1937. 'The Marconi School of Wireless Communication'. *The Marconi Review*, 66, 1–14.

Drummond, I.M., 1974. *Imperial Economic Policy 1917–1939. Studies in Expansion and Protection.* London: Allen and Unwin.

Dubois, J.L., Multhauf, R.P., & Ziegler, Ch.A., 2002. *The Invention and Development of the Radiosonde.* Washington, D.C.: Smithsonian Institution Press.

Dummelow, J., 1949. *A History of the Metropolitan-Vickers Electrical Company Limited.* Manchester: Metropolitan Vickers Electrical Company Ltd.

Duncan, R.L. & Drew, C.E., 1929. *Radio Telegraphy and Telephony.* New York: John Wiley & Sons.

Dunlap, O.E., 1937. *Marconi: The Man and his Wireless.* New York: MacMillan Co.

Duval, R., 1980. *Histoire de la Radio en France.* Paris: Alain Moreau.

Dyster, B. & Meredith, D., 1990. *Australia in the International Economy in the Twentieth Century.* Melbourne: Cambridge University Press.

E.H.S., 1926. 'The Post Office Wireless Services'. *Post Office Electrical Engineers Journal*, 19, 58–66.

Eccles, W.H., 1912. 'On the Diurnal Variations of the Electric Waves Occurring in Nature, and on the Propagation of Electric Waves Round the Bend of the Earth'. *Proc. Roy. Soc. Lond.*, 87, 79–99.

——, 1913a. 'On Certain Phenomena Accompanying the Propagation of Electric Waves over the Surface of the Globe'. *The Electrician*, 69, 1,015–19.

——, 1913b. Atmospheric Refraction in Wireless Telegraphy. *The Electrician,* 69, 969–70.

——, 1922. 'Imperial Wireless Communication'. *Jour. Roy. Soc. Arts*, 70, 509–22.

——, 1923. 'The Amateur's Part in Wireless Development'. *The Wireless World*, 13, 50–4.

——, 1924. 'The Importance of the Amateur'. *The Wireless World*, 13, 620–23.

——, 1925. The Work of a Wireless Engineer. *The Wireless World*, 16, 649–51.

——, 1926a, 'Wireless Development since the War'. *EW & WE*, 3, 740–42.

——, 1926b, 'Radio Communication and Imperial Development'. *Nature*, 1926, 117, 659–62.

——, 1927. 'Wireless Communication and Terrestrial Magnetism'. *Nature,* 119, 157.

——, 1930. *The Influence of Physical Research on the Development of Wireless.* London: Institute of Physics.

——, 1945. 'John Ambrose Fleming, 1849–1945'. *Obituary Notices of Fellows of the Royal Society*, 5, 231–42.

Eckersley, P.P., 1932. *A Planned System for Australia Broadcasting. A Report to Amalgamated Wireless Limited.* London: Institution of Radio Engineers.

——, 1933. 'The Future of Wireless Engineering. Considerable Expansion Inevitable'. *JC*, 12, 7–10.

Eckersley, T.L., 1921. 'The Effect of the Heaviside Layer on the Apparent Direction of Electromagnetic Waves'. *The Radio Review,* 2, 60–5, 231–48.

——, 1925. 'A Note on Musical Atmospheric Disturbances'. *Philosophical Magazine*, 49, 1,250–60.

——, 1929. 'An Investigation of Short-Waves'. *Jour. IEE*, 67, 992–1,029.

——, 1930. 'Multiple Signals in Short-Wave Transmission'. *Proc. IRE*, 18, 106–22.

——, 1931a. 'On the Connexion Between the Ray Theory of Electric Waves and Dynamics'. *Proc. Roy. Soc. Lond.*, A 132, 83–98.

——, 1931b. '1929–1930 Developments in the Study of Radio Wave Propagation'. *The Marconi review,* 31, 1–8.

——, 1932, 'Studies in Radio Transmission'. *Jour. IEE*, 71, 405–54.

——, 1937. 'Ultra-short Wave Refraction and Diffraction'. *Jour. IEE*, 80, 286–304.

Eckersley, T.L. & Millington, G., 1939. 'The Application of the Phase Integral Method to the Analysis of Diffraction and Reflection of Wireless Waves Round the Earth'. *Philosophical Transactions*, A 237, 273–309.

Eckersley, T.L. & Tremellen, K.W., 1929. 'World-wide Communications with Short Wireless Waves'. In *Proceedings of the world engineering Congress, Tokyo*, 20, 177–212.

Edelstein, M., 1987. *Professional Engineers in the Australian Economy: Some Quantitative Dimensions, 1866–1980.* Canberra: Australian National University, Working Papers in Economic History, 93.

Edgeloe, V.A., 1989. *Engineering Education in The University of Adelaide 1889–1980: A Registrar's Retrospect.* Adelaide: Kinhill Engineers Pty Ltd.

Edgerton, D.E.H., 1994. 'British Industrial R&D, 1900–1970'. *Journal of European Economic History*, 23, 49–68.

——, 1996. *Science, Technology and the British Industrial Decline, 1870–1970.* Cambridge: Cambridge University Press.

Edgerton, D.E.H. & Horrocks, S.M., 1994. 'British Industrial Research and Development before 1945'. *Economic Historical Review*, 47, 213–38.

Egan, M.B., 1920. 'The Training of a R.A.F. Wireless Operator'. *The Wireless World*, 7, 621–37.

Egerton, A.C., 1949. 'Lord Rayleigh, 1875–1947'. *Ob. Not. Fell. Roy. Soc.*, 6, 503–38.

Eggleston, W., 1978. *National Research in Canada*. Toronto: Clarke, Irwin & Co.

Eisemon, T.O., 1984. 'The Development of Science in New Zealand'. *Pacific Viewpoint*, 25 (1), 45–56.

Engel, A. von, 1957. 'John Sealy Edward Townsend'. *Biog. M. Fell. Roy. Soc.*, 3, 257–272.

Evans, W.F., 1970. *History of the Radiophysics Advisory Board, 1939–45*. Melbourne: CSIRO.

——, 1973. *History of Radio Research Board, 1926–1945*. Melbourne.

Excell, P.S., 1994. 'Sir Edward Appleton and Joseph Priestley: Two Giants of Electrical Science', *Jour. Atm. Terr. Phys.*, 56 (6), 693–704.

Falciasecca, G. & Valotti, B., eds., 2003. *Guglielmo Marconi: Genio, Storia e Modernità*. Milano: G. Mondadori.

Ferguson, R.W., ed., 1935. *Training in Industry: A Report Embodying the Results of Enquiries Conducted between 1931 and 1934 by the Association for Education in Industry and Commerce*. London.

Ferraro, V.C.A., 1968. 'Sydney Chapman: An Appreciation on the Occasion of his 80th Birthday'. *Geo. Jour. Roy. Astr. Soc.,* 15, 1–5.

——, 1971. Sydney Chapman. *Bull. Lon. Math. Soc.*, 3, 221–50.

Ferris, J.R., 1989. *The Evolution of British Strategic Policy 1919–1926*. London: Macmillan.

Fessenden, R., 1908. 'Wireless Telegraphy'. *Proceedings of the American Institute of Electrical Engineers,* 27, 553–629.

Fisk, D., 1934. *Exploring the Upper Atmosphere*. London: Faber & Faber Ltd.

Fisk, E.K., 1995. *Hardly Ever a Dull Moment*. Canberra: Australian National University.

Fisk, E.T., 1923a, 'Wireless Service with Australia'. *United Empire. R.C.I. Journal,* 14, 89–98.

——, 1923b, 'The Application and Development of Wireless'. *Proceedings of the 2nd Pan-Pacific Science Congress, Australia*, 1, 619–31.

——, 1935. 'Foreword'. *AWA Tech. Rev.*, 1, 2.

Fitzgerald, G.F., 1893. 'On the Period of Vibration of Disturbances of Electrification of the Earth'. *Nature,* 48, 526.

F.J.W.W., 1929. 'The Conductivity of the Atmosphere'. *Nature,* 123, 155–6.

Fleming, C.A., 1971. 'Ernest Marsden, 1889–1970'. *Biog. M. Fell. Roy. Soc.*, 17, 463–96.

Fleming, D., 1962. 'Science in Australia, Canada, and the United States: Some Comparative Remarks'. In *Proceedings of the Tenth International Congress of the History of Science, Ithaca, 1962*. Paris: Hermann, 179–96.

Fleming, J.A., 1914. 'On Atmosphere Refraction and its Bearing on the Transmission of Electromagnetic Waves round the Earth's Surface'. *Proc. Phys. Soc. Lond.*, 26, 318–33.

——, 1915. 'On the Causes of Ionization of the Atmosphere'. *The Electrician,* 75, 348–50.

——, 1921. 'The Coming of Age of Long Distance Radiotelegraphy and Some of its Scientific Problems'. *Jour. Roy. Soc. Arts,* 70, 66–78, 82–97.

——, 1925. 'The Propagation of Wireless Waves of Short Wave-Length round the World'. *Nature*, 115, 123–124.

——, 1927. 'Electrical Engineering as a Career'. *JC*, 6, 19–23.

——, 1929. 'Engineering and Other Openings in the Wireless Industry'. *JC*, 8, 9–12.

——, 1930. 'Electro-Communication Engineering as a Career'. *JC*, 9, 7–12.

——, 1932. 'Looking Ahead in Electrical Engineering. Future Openings for Trained Engineers'. *JC*, 11, 7–14.

——, 1937. 'Guglielmo Marconi and the Development of Radio Communications'. *Jour. Roy. Soc. Arts*, 86, 42–63.

——, 1939. 'Physics and the Physicists of the Eighteen Seventies'. *Nature*, 143, 99–102.

Flohn, H. & Penndorf, R., 1950. 'The Stratification of the Atmosphere'. *Bull. Am. Met. Soc.*, 31, 71–8, 126–30.

Forman, P., 1971. 'Weimar Culture, Causality, and Quantum Theory, 1918–1927: Adaptation by German Physicists and Mathematicians to a Hostile Intellectual Environment'. *HSPS*, 3, 1–115.

Forster, C., 1964. *Industrial Development in Australia, 1920–1930.* Canberra: Australian National University.

Fortner, R.S., 2005. *Radio, Morality, and Culture: Britain, Canada, and the United States, 1919–1945.* Carbondale, I.L: Southern Illinois University Press.

Foster, L., 1986. *High Hopes: The Men and Motives of the Australian Round Table.* Melbourne: Melbourne University Press.

Fowler, A. & Strutt, R.J., 1917. 'Absorption Bands of Atmospheric Ozone in the Spectra of Sun and Stars'. *Proc. Roy. Soc. Lond.,* 93, 577–86.

Frame, T. & Faulkner, D., 2003. *Stromlo. An Australian Observatory.* New South Wales: Allen Unwin.

Franklin, C.S., 1922. 'Short-Wave Directional Wireless Telegraphy'. *Wireless World and Radio Review*, 10, 219–25.

Friedman, R.M., 1995. 'Civilization and National Honour: The Rise of Norwegian Geophysical and Cosmic science'. In J.P. Collet, ed. *Making Sense of Space. The History of Norwegian Space Activities.* Oslo: Scandinavian University Press, 3–39.

Frigon, A., 1929. 'The Organization of Radio Broadcasting in Canada'. *Revue Trimestrielle Canadienne,* 60, 395–411.

Galbreath, R., 1998a, *DSRI: Making Science Work for New Zealand: Themes from the History of the Department of Scientific and Industrial Research, 1926–1992.* Wellington: Victoria University.

——, 1998b. 'Marsden, Ernest, 1889–1970'. *DNZB*, 4.

——, 1999. 'New Zealand Scientists in Action: The Radio Development Laboratory and the Pacific War'. In R. MacLeod, ed. *Science and the Pacific War, Science and Survival in the Pacific, 1939–45.* Dordrecht: Kluwer Academic Pub., 211–27.

Gardiner, G.W., 1962a, 'Radio Research at Ditton Park. Some Historical Notes'. *R.R.O. Newsletter*, 15th Jan., 9.

——, 1962b. 'Radio Research at Ditton Park II, 1922–27'. *R.R.O. Newsletter*, 15th Feb 1962, 10.

——, 1962c. 'Radio Research at Ditton Park III, 1927–33'. *R.R.O. Newsletter*, 15th March 1962, 11.

——, 1962d. 'Radio Research at Ditton Park IV, Radiolocation'. *R.R.O. Newsletter*, 15th April 1962, 12.

——, 1962e. 'Radio Research at Ditton Park V, 1935–45'. *R.R.O. Newsletter*, 10th May 1962, 13.

Gardiner, G.W., Lane, J.A., & Rishbeth, H., 1982. 'Radio and Space Research at Slough, 1920–1981'. *The Radio and Electronic Engineer*, 52 (3), 111–21.

Gardner, J.E., 1997. *Stormy Weather: A History of Research in the Bureau of Meteorology.* Melbourne: Bureau of Meteorology.

Garratt, G.R.M., 1950. *One Hundred Years of Submarine Cables.* London: Science Museum.

Geddes, K., 1974. *Guglielmo Marconi, 1874–1937.* London: Science Museum.

Geddes, K. & Bussey, G., 1991. *The Setmakers: A History of the Radio and Television Industry.* London: Brema.

Geddes, M., 1939. 'Photographic Determination of Height and Position of Aurorae Determined in New Zealand in 1937'. *Radio Research Bulletin*, 3.

Geeves, P., 1993. *The Dawn of Australia Radio Broadcasting*. Sydney: Electronics Australia.

Gibbs, A., 1921. 'Effects of the Recent Aurora on Telegraph-Lines, Telephone-Lines, and Wireless Stations'. *N. Z. Jour. Sc. Tech.*, 4, 183–8.

Gilliland, T.R., 1933. 'Note on a Multi-Frequency Automatic Recorder of Ionosphere Heights'. *Jour. Res. NBS*, 11, 561–6.

——, 1935. 'Multifrequency Ionosphere Recording and its Significance'. *Proc. IRE,* 23, 1,076–101.

Gilliland, T.R., Kirby, S.S., Smith, N., & Reymer, S.E., 1937. Characteristics of the Ionosphere and their Application to Radio Transmission. *Jour. Res. NBS,* 18, 645–67.

——, 1938. 'Maximum Usable Frequencies for Radio Sky-Wave Transmission, 1933 to 1937'. *Jour. Res. NBS,* 20, 627–39.

Gillmor, C. S., 1976. 'The History of the Term *Ionosphere*'. *Nature*, 262, 347–8.

——, 1981. 'Threshold to Space: Early Studies of the Ionosphere'. In P. Hanle & V. del Chamberlain, eds. *Space Science Comes of Age: Perspectives in the History of the Space Sciences.* Washington D.C.: Smithsonian Institution Press, 101–14.

——, 1982. 'Wilhelm Altar, Edward Appleton, and the Magneto-Ionic Theory'. *Proc. Am. Phil. Soc.*, 126, 395–423.

——, 1986. 'Federal Funding and Knowledge Growth in Ionospheric Physics, 1945–81'. *Soc. Stud. Sc.*, 16, 105–33.

——, 1991. 'Ionospheric and Radio Physics in Australian Science Since the Early Days'. In Home, R.W. & Kohlstedt, S.G., eds., 1991. *International Science and National Scientific Identity.* London: Kluwer Academic Publishers, 181–204.

——, 1994. 'The Big Story: Tuve, Breit, and Ionospheric Sounding, 1923–1928'. In G.A. Good, ed., *History of Geophysics.* Washington, D.C.: American Geophysical Union, 133–41.

——, 1997. 'The Formation and Early Evolution of Studies of the Magnetosphere'. In *Discovery of the Magnetosphere.* Washington, DC: American Geophysical Union, 1–12.

Gilmour, J., 1928. *Report of the Imperial Wireless and Cable Conference.* London.

Gingras, Y., 1991. *Physics and the Rise of Scientific Research in Canada.* Montreal: McGill-Queen s University Press.

Girardeau, E., 1951. *Comment Furent Créées et Organisées les Radio-Communications Internationales.* Paris.

Glasgow, R.S., 1936. *Principles of Radio Engineering.* New York, London: McGraw-Hill.

Godfrey, D., 1982. 'Canadian Marconi: CFCF, the Forgotten Case'. *Canadian Journal of Communications*, 8, 56–71.

Godfrey, G.H. & Price, W.L., 1937. 'Thermal Radiation and Absorption in the Upper Atmosphere'. *Proc. Roy. Soc. Lond.*, A 163, 228–49.

Going, T.C.H., 1995. 'The Growth of the Electron Tube Industry'. In Newcomen Society, *History of Thermionic Devices*, Conference Proceedings. London, 41–56.

Gold, E., 1909. 'The Isothermal Layer of the Atmosphere and Atmospheric Radiation'. *Proc. Roy. Soc, Lond.,* 82, 43–70.

Gold, E., 1965. 'Simpson, George Clark'. *Biog. M. Fell. Roy. Soc.*, 11, 157–75.

Good, G.A., 1988. 'The Study of Geomagnetism in the late 19th Century'. *EOS: Tr. Am. Geo. Un.,* 69, 218–28.

——, 1994. 'Vision of a Global Physics: The Carnegie Institution and the First World Magnetic Survey'. In G.A. Good, ed. *The Earth, the Heavens, and the Carnegie Institution of Washington.* Washington, DC: American Geophysical Union, 29–36.

——, 1998. 'Geomagnetism. Theories between 1800 and 1900'. In G.A. Good, ed. *Sciences of the Earth. An Encyclopedia of Events, People and Phenomena.* New York: Garland Publishing, 350–7.

——, 1999. 'Sydney Chapman'. In *Biographies in American National Biography.* Oxford: Oxford University Press, 4, 715–17.

——, 2000. 'The Assembly of Geophysics: Scientific Disciplines as Frameworks of Consensus'. *SHPMP*, 31: 259–92.

——, 2002. 'From Terrestrial Magnetism to Geomagnetism: Disciplinary Transformation in the Twentieth Century'. In D.R. Oldroyd, ed. *The Earth Inside and Out: Some Major Contributions to Geology in the Twentieth Century.* London: Geological Society, 229–39.

Goodall, W.M., 1935. 'The Ionosphere'. *Bell Laboratories Record,* 13, 194–9.

Goot, M., 1981. 'Fisk, Sir Ernest Thomas, 1886–1965'. In B. Nairn & G. Serle eds., *Australian Dictionary of Biography.* Melbourne: Melbourne University Press, 8, 508–10.

Gossling, B.S., 1920. 'The Development of Thermionic Valves for Naval Uses'. *Jour. IEE*, 58, 670–703.

Gough, J., 1993. *Watching the Skies: The History of Ground Radar in the Air Defence of the United Kingdom.* London: HMSO.

Graves, C.G., 1931. 'Empire Broadcasting'. *Unit. Emp. RES. J.*, 24, 479–80.

Green, A.L., 1932. 'The State of Polarization of Sky Waves and Height Measurements of the Heaviside Layer in the Early Morning'. *CSIR Bulletin*, 59; *RRB. Rep.* 2.

Green, A.L., 1936. 'Non-Fading Noise-Free Broadcasting Stations'. *Journal of the Institution of Engineers, Australia*, 8, 203–15.

——, 1937. 'The Service Area of a Long-Wave Telegraphic Transmitter'. *AWA Tech. Rev.*, 3 (1), 7–29.

——, 1946. 'Early History of the Ionosphere'. *AWA. Tech. Rev.*, 7, 177–228.

Green, A.L. & Martyn, D.F., 1935. 'The Characteristics of Downcoming Radio Waves'. *Proc. Roy. Soc. Lond.*, A 148, 104–20.

Grimal, H., 1980. 'L'évolution du concept d'empire en Grande-Bretagne'. In M. Duverger, ed. *Le Concept d'Empire*. Paris: Presses Universitaires de France, 337–64.

Griset, P., 1989. 'Câbles et radio dans le premier tiers du XXe siècle: Service public, initiative privée et mutations technologiques dans le domaine des télécommunications internationales'. *Revue de la S.A.G.M. et de l'A.S.I.T.A.: Techniques Avancées*, 9, 24–30.

Groenewold, H.J., 1961. 'The Model in Physics'. In H. Freudenthal, ed. *The Concept and the Role of the Model in Mathematics and Natural and Social Sciences.* Dordrecht: Reidel Publishing Co., 98–103.

G.T., 1937. 'Italianità di Marconi'. *Bologna. Rivista Mensile del Comune*, 24 (7), 28–9.

Guagnini, A., 1994. 'Guglielmo Marconi e la formazione della Marconi Wireless Company'. *Museoscienza*, 7, 22–7.

Hacking, I., 1983. *Representing and Intervening.* Cambridge: Cambridge University Press.

Hales, A.L., 1992. 'Lloyd Viel Berkner, February 1, 1905–June 4, 1967'. *Bibliographical Memoirs of the National Academy of Sciences*, 61, 3–24.

Hall, J.H., 1980. *The History of Broadcasting in New Zealand, 1920–1954.* Wellington: Broadcasting Corporation of New Zealand.

Hallborg, H.E., 1927. 'Some Practical Aspects of Short-Wave Operation at High Power'. *Proc. IRE*, 15, 501–18.

Hallborg, H.E., Briggs, L.A., & Hansell, C.W., 1927. 'Short-Wave Commercial Long-Distance Communication'. *Proc. IRE*, 15, 467–500.

Hancock, H.E., 1950. *Wireless at Sea: The First Fifty Years.* Chelmsford: Marconi International Marine Communication Co.

Harbord, James, 1926. 'America's Position in Radio Communication'. *Foreign Affairs*, 4, 465–74.

——, 1930. 'Radio in World Communications'. In M. Codel, ed. *Radio and its Future.* London: Harper & Brothers, 95–105.

Harcourt, E., 1987. *Taming the Tyrant: The First 100 years of Australia s International Telecommunication Service.* Sydney: Allen & Unwin.

Harcourt, P. & Downes, P., 1976. *Voices in the Air.* Wellington: Methuen.

Harding, G., 1984. '75 Years of Weather Forecasting in Australia'. *Australian Science Magazine*, 1, 18–46.

Hargreaves, J.K. & Hunsucker, R.D., 2000. *The High-Latitude Ionosphere and its Effects on Radio Propagation.* Cambridge: Cambridge University Press.

Harper, W.G., 1950. 'Lerwick Observatory'. *Met. Mag.*, 79, 309–14.

Harrison, A.P., 1979. 'Single-Control Tuning: An Analysis of an Innovation. *Technology and Culture*', 20, 296–321.

Harrison, D.N., 1969. 'The British Radiosonde: Its Debt to Kew'. *Met. Mag.*, 98, 186–90.

Harrison, J.A., 1958. *The Story of the Ionosphere or Exploring with Wireless Waves.* London: Hulton Ed. Pub.

Harrison, R.G., 2003. 'Twentieth-Century Atmospheric Electrical Measurements at the Observatories of Kew, Eskdalemuir and Lerwick'. *Weather*, 58, 11–19.

Hartmann, S., 1995. 'Models as a Tool for Theory Construction: Some Strategies of Preliminary Physics'. *Poznan Stud. Phil. Sc. Hum.*, 44, 49–67.

Hartree, D.R., 1931. 'The Propagation of Electromagnetic Waves in a Refractive Medium in a Magnetic Field'. *Proc. Cam. Phil. Soc.*, 27, 143–62.

Harvard University, 1928. *The Radio Industry.* New York: A.W. Shaw Co.

Hatton, T.J. & Chapman, B.J., 1987. *Post-School Training in Australia in 1900–1980.* Canberra: Australian National University, Working Papers in Economic History, 94, 9–13.

Headrick, D.R., 1988. *The Tentacles of Progress. Technology Transfer in the Age of Imperialism, 1850–1940.* Oxford: Oxford University Press.

——, 1991. *The Invisible Weapon: Telecommunications and International Politics, 1851–1945.* New York: Oxford University Press.

——, 1994. Shortwave Radio and its Impact on International Telecommunications Between the Wars. *History and Technology*, 11, 21–32.

Healey, R.H., 1938. 'The Effect of a Thunderstorm on the Upper Atmosphere'. *AWA Tech. Rev.*, 3 (4), 215–224.

Heath, F., 1938. 'Report on the Organization of Scientific and Industrial Research in the Dominion of New Zealand'. *App. J. Ho. Rep. of N.Z.*, 2, H-27, 12 March.

Heaviside, O., 1902. 'The Theory of Electric Telegraphy'. In *Encyclopaedia Britannica.* London, 33, 215.

Heilbron, J.L., 1985. 'Bohr s First Theories of the Atom'. In A.P. French & P.J. Kennedy, eds. *Niels Bohr. A Centenary Volume.* Cambridge Mass.: Harvard University Press, 33–49.

Heising, R.A., 1928. 'Experiments and Observations Concerning the Ionized Regions of the Atmosphere'. *Proc. IRE,* 16, 75–99.

Henderson, J.T., 1933. 'Measurements of Ionisation in the Kennelly-Heaviside Layer During the Eclipse of 1932'. *Can. Jour. Res.*, 8, 1–14.

Henderson, J.T. & Rose, D.C., 1933. 'Fading and Signal-Strength Measurements Taken during the Solar Eclipse of August 31, 1932'. *Can. Jour. Res.*, 8, 29–36.

Hess, V.F., 1928. *The Conductivity of the Atmosphere: The Electrical Conductivity of the Atmosphere and its Causes,* (trans. L.W. Cold). London: Constable.

Hesse, Mary B., 1963. *Models and Analogies in Science.* Paris: Université de Notre Dame.

Hevly, B., 1987. *Basic Research within a Military Context: The Naval Research Laboratory and the Foundations of Extreme Ultraviolet and X-ray Astronomy, 1923–1960.* Baltimore: Johns Hopkins University, PhD diss.

Hevly, B., 1994. 'Building a Washington Network for Atmospheric Research'. In G.A. Good, ed. *The Earth, the Heavens, and the Carnegie Institution of Washington.* Washington, D.C.: American Geophysical Union.

Hezlet, A.R., 1975. *The Electron and Sea Power.* London: P. Davies.

Higgs, A.J., 1934. 'Measurements of Atmospheric Ozone made at the Commonwealth Solar Observatory, Mount Stromlo, Canberra, during the years 1929 to 1932. *Memoirs of the Commonwealth Solar Observatory*, 3.

Higgs, A.J., Martyn, D.F., Munro, G.H., & Williams, S.E., 1937. 'Ionospheric Disturbances, Fadeouts and Bright Hydrogen Solar Eruptions'. *Nature*, 140, 603–5.

Higgs, A.J., Munro, G.H., & Webster, H.C., 1935. 'Simultaneous Observations of Atmospherics with Cathode Ray Direction-Finders at Toowoomba and Canberra'. *CSIR Bulletin*, 89, *RRB. Rep.,* 8, 9–42.

——, 1936. 'Thunderstorms in the Tasman and Coral Sea, March 1934'. *Meteorological Magazine*, 71, 59–62.

Hoare, M.E., 1976. 'The Relationship between Government and Science in Australia and New Zealand'. *Journal of the Royal Society of New Zealand*, 63. 381–94.

Hobsbawm, E.J., 1987. *Industry and Empire. An Economic History of Britain since 1750.* London: Weidenfeld and Nicolson.

Holland, R.F., 1981. *Britain and the Commonwealth Alliance, 1918–1939.* London: Macmillan.

Hollingworth, J., 1933. 'Some Characteristics of Short-Wave Propagation'. *Jour. IEE,* 72, 229–51.

——, 1934. 'Structure of the Ionosphere'. *Nature,* 134, 462.

——, 1935. 'The Structure of the Ionosphere'. *Proc. Phys. Soc. Lond.,* 47, 843–51.

Hollingworth, J. & Naismith, R., 1929. 'A Portable Radio Intensity Measuring Apparatus for High Frequencies'. *Jour. IEE*, 67, 1,033–44.

Home, R.W., 1981. 'W.H. Bragg and J.P.V. Madsen: Collaboration and Correspondence, 1905–1911'. *HRAS*, 5 (2), 1–29.

——, 1987. 'The Beginnings of an Australian Physics Community'. In Nathan Reingold and Marc Rothenberg eds. *Scientific Colonialism: A Cross Cultural Comparison.* Washington, D.C.: Smithsonian Institution Press, 3–34.

——, 1988a. 'The Physical Sciences: String, Sealing Wax and Self-Sufficiency'. In R. Macleod, ed. *The Commomwealth of Science. ANZAAS and the Scientific Enterprise in Australasia, 1888–1988.* New York: Oxford University Press, 147–65.

——, 1988b. *Australian Science in the Making.* Cambridge: Cambridge University Press.

——, 1991. 'Georg Neumayer and the Flagstaff Observatory, Melbourne'. In D. Walker & J. Tampke eds. *From Berlin to the BurdekIn: The German Contribution to the Development of Australian Science, Exploration and the Arts.* Sydney: New South Wales University Press, 40–53.

——, 1993a. 'Builder, Geoffrey, 1906–1976. Physicist and Radio Engineer'. *ADB*, 13, 290–91.

——, 1993b. 'Victor Albert Bailey, 1895–1964'. *ADB*, 13, 90–1.

——, 1994a. 'Defining the Boundaries of the Field: Early Stages of the Physics Discipline in Australia'. In R. McLeod & R. Jarrell, *Dominions Apart: Reflections on the Culture of Science and Technology in Canada and Australia, 1850–1945.* Special Issue of *Scientia Canadensis*, 17, 53–70.

——, 1994b. 'To Watheroo and Back: The DTM in Australia, 1911–1947'. In G. Good ed. *The Earth, the Heavens and the Carnegie Institution of Washington.* Washington: American Geophysical Union, 149–160.

——, 2000. 'David Forbes Martyn, 1906–1970. Physicist'. *ADB*, 15, 320–2.

Home, R.W. & Kohlstedt, S.G., eds., 1991. *International Science and National Scientific Identity.* London: Kluwer Academic Publishers.

Home, R.W. & Needham, P.J., 1990. *Physics in Australia to 1945: Bibliography and Biographical Register.* Melbourne: University of Melbourne/National Centre for Research.

Hong, S., 1996. 'Styles and Credit in Early Radio Engineering: Fleming and Marconi on the First Transatlantic Wireless Telegraphy'. *Annals of Science*, 53, 431–65.

——, 2001. *Wireless: from Marconi's Black-Box to the Audion.* Cambridge, Mass.: The M.I.T. Press.

Hooke, L.A., 1938. 'Australian Radio Communication Services'. *AWA Tech. Rev.*, 3, 229–51.

Hopkins, J.C., 1926. 'Relations with the Empire'. *The Canadian Annual Review of Public Affairs,* 52–3.

Hopkins, W. *History of Canadian Marconi Company* Montreal, s.d.

Houghton, J.T. & Walshaw, C.D., 1977. 'Gordon Miller Bourne Dobson, 1889–1976.' *Biog. M. Fell. Roy. Soc.*, 23, 41–57.

Howden, S., 1997. 'Australian Radio and Cultural Formation'. *Access: History*, 1 (1), 29–47.

Howe, G.W.O., 1922. 'Notes on Wireless Matters'. *The Electrician*, 89, 260–1.

——, 1924. 'A New Theory of Long-Distance Radio-Communication. *The Electrician,* 93, 282–3, 548.

Howeth, L.S., 1963. *History of Communications-Electronics in the United States Navy* Washington, D.C.: *Bureau of Ships and Office of Naval History.*

Hoyt Taylor, A., 1924. 'The Navy's Work on Short Waves'. *QST,* 8, 9–14.

——, 1926. 'Relation between the Height of the Kennelly-Heaviside Layer and High-Frequency Radio Transmission Phenomena'. *Proc. IRE,* 14, 521–40.

Hughes, D.W., 1990. 'Meteors and Meteor Showers: An Historical Perspective, 1869–1950'. In J.J. Roche, ed., *Physicists Look Back: Studies in the History of Physics.* Bristol, New York: Adam Hilger, 261–305.

Hughes, J., 1998. 'Plasticine and Valves: Industry, Instrumentation and the Emergence of Nuclear Physics'. In J.P. Gaudillière & I. Löwy, *Invisible Industrialist: Manufactures and the Production of Scientific Knowledge.* London: MacMillan Press, 58–101.

Hughes, P., 1998. *AWA Radiolettes, 1932–1949.* Victoria.

Hughes, T.P., 1988. 'Model Builders and Instrument Makers'. *Science in Context*, 2, 59–75.

Hugill, P.J., 1999. *Global Communications since 1844. Geopolitics and Technology.* Baltimore: The John Hopkins University Press.

Hulburt, E.O., 1928. 'Ionization in the Upper Atmosphere of the Earth'. *Phys. Rev.*, 31, 1,018–37.

——, 1974. 'Early Theory of the Ionosphere'. *Jour. Atm. Terr. Phys.*, 36, 2,137–40.

Hull, A., 1999. 'War of Words: The Public Science of the British Scientific Community and the Origins of the Department of Scientific and Industrial Research, 1914–16.' *British Journal for the History of Science*, 32, 461–81.

Hulthén, E., 1964. 'Physics 1947'. In *Nobel Lectures Physics: 1942–62.* London: Elsevier Publ. Co., 75–8.

Hutchinson, E., 1970. 'Scientists as an Inferior Class: The Early Years of the DSIR'. *Minerva*, 8, 396–411.

Hyde, J., 1982. 'The Development of Australian Tertiary Education to 1939'. In S. Murray-Smith, ed. *Melbourne Studies in Education.* Melbourne: Melbourne University Press, 105–40.

Ihde, D., 1991. *Instrumental Realism: The Interface between Philosophy of Science and Philosophy of Technology.* Indianapolis: Indiana University Press.

Iles, F.A., 1927. 'Imperial Wireless'. *Journal of the Royal Engineers*, 41, 259–85.

Ionescu, T. & Mihul, C., 1934. 'Sur la Structure de la Couche Ionisée de l'Atmosphère (Ionosphère)'. *Comptes Rendus Hebdomadaires des Séances*, 199, 1,301–3.

Isaacs, G.C., 1923. *The Past, Present and Future of Wireless Telegraphy and its Relation to Empire Communications. An Address to Members of the Industrial Group of the House of Commons on May 8th, 1923.* London.

Isbell, A.A., 1930. 'The RCA World-wide Radio Network'. *Proc. IRE*, 18, 1,732–42.

Isted, G.A., 1974. 'Marconi: A Turning Point in Radio Communication'. *Electronics & Power*, 20, 315–19.

Jacobs, L., 1969. 'The Two Hundred Years Story of Kew Observatory'. *Met. Mag.*, 98, 162–71.

Jacot, B.L. & Collier, D.M.B., 1935. *Marconi: Master of Space.* London: Hutchison.

Jansky, C., 1926. 'Collegiate Training for the Radio Engineering Field'. *Proc. IRE*, 14, 431–45.

Jarrell, R.A., 1988. *The Cold Light of Dawn: the History of Canadian Astronomy.* Toronto: Toronto University Press.

Jenkin, J.G., 1983. 'The Cavendish Tradition in Australian Physics: Time for Change'. *Aus. Phys.*, 20, 46–50.

——, 1990. 'British Influence on Australian Physics, 1788–1988'. *Berichte zur wissenschaftsgeschichte*, 13, 93–100.

Jenkinson, S.H., 1940. *New Zealanders and Science.* Wellington: Department of Internal Affairs.

Johnson, L., 1981. 'Radio and Everyday Life. The Early Years of Broadcasting in Australia, 1922–1945'. *Media Culture and Society*, 3, 167–78.

——, 1988. *The Unseen Voice: A Cultural Study of Early Australian Radio.* London: Routledge.

Johnston, H.F., 1926. 'Determination of the Atmospheric Potential-Gradient Reduction-Factor at the Watheroo Magnetic Observatory, Western Australia'. *Report of the 18th Meeting of the Australian Association for the Advancement of Science*, 128.

Jolly, W.P., 1972. *Marconi.* London: Constable.

Jome, H.L., 1925. *Economics of the Radio Industry.* New York: A.W. Shaw Co.

Jones, S., 1972. 'The Development of the Australian Telecommunications Manufacturing Industry'. In *The Centenary of the Adelaide-Darwin Overland Telegraph Line.* Adelaide: Institution of Engineers and Australian Post Office.

Jones-Imhotep, E.C., 2000. 'Disciplining Technology: Electronic Reliability, Cold-War Military Culture and the Topside Ionogram'. *History and technology,* 17, 125–75.

——, 2001. *Communicating the Nation: Northern Radio, National Identity and the Ionospheric Laboratory in Cold War Canada.* Harvard: Harvard University, PhD diss.

Jouast, R., 1936. 'La Constitution de l'Ionosphère'. *Journal de Physique*, Jul 1936, 286–96.

Kaiser, D., 2000. 'Stick-Figure Realism: Conventions, Reification, and the Persistence of Feynman Diagrams, 1948–1964'. *Representations,* 70, 49–86.

Kargon, Robert H., 1977. *Science in Victorian Manchester. Enterprise and Expertise.* London: Johns Hopkins University Press.

Kaur, P., Srivastava, M.P., Nath, N., & Setty, C.S.G.K., 1973. 'Phase Integral Corrections to Radio Wave Absorption and Virtual Height for Model Ionospheric Layers'. *Jour. Atm. Terr. Phys.,* 35, 1,745–54.

Keentok, M., 2004. 'AWA in Ashfield'. *Journal of the Ashfield and District Historical Society*, 15, 22–41.

Kendall, P.C., 1970. 'Obituary: Professor Sydney Chapman'. *Plan. Sp. Sc.*, 18, 1,871–72.

Kendle, J., 1975. *The Round Table Movement and Imperial Union.* Toronto: University of Toronto Press.

Kennedy, J. de N., 1950. *History of the Department of Munitions and Supply, Canada, in the Second World War.* Ottawa: King's Printer.

Kennedy, P., 1987. *The Rise and Fall of the Great Powers: Economic Change and Military Conflict from 1500 to 2000.* New York: Random.

Kennelly, A.E., 1902. 'On the Elevation of the Electrically-Conducting Strata of the Earth's Atmosphere'. *Electrical World and Engineer,* 39, 473.

Kenrick, G.W. & Pickard, G.W., 1930. 'Summary of Progress in the Study of Radio Wave Propagation Phenomena'. *Proc. IRE,* 18, 649–68.

Kerr, F.J., 1984. 'Early Days in Radio and Radar Astronomy in Australia'. In W.T. Sullivan, ed., *The Early Years of Radioastronomy* Cambridge: Cambridge University Press, 133–45.

Khrgian, A.Kh., 1973. *Fizika atmosfernogo ozona.* Leningrad, trans. from Russian into English by D. Lederman, 1975. *The Physics of Atmospheric Ozone.* Jerusalem: Israel Program for Scientific Translations.

Kidson, I.M., 1941. *Edward Kidson.* Christchurch: Whitcombe & Tombs Ltd.

King, R.W.P., Mimno, H.R., & Wing, A.H., 1945. *Transmission Lines Antennas and Wave Guides.* London: McGraw-Hill.

Kintner, S.M., 1925. 'History and Future of Radio'. *Iron and Steel Engineer*, October 1925, 423–4.

Kirby, S.S., Berkner, L.V., & Stuart, D.M., 1934. 'Studies of the Ionosphere and their Application to Radio Transmission'. *Jour. Res. BNS,* 12, 15–51.

Kushner, D., 1993. 'Sir George Darwin and a British School of Geophysics'. *Osiris*, 8, 196–223.

Laby, T.H., 1936. 'Contribution to Discussion on Thunderstorm Researches. *Quart. Jover. Roy. Met. Soc.*, 62, 507–16, 525–7.

Ladner, A.W. & Stoner, C.R., 1932. *Short Wave Wireless Communication.* London: Chapman & Hall.

Laeter, J.R. de, 1989. 'The Influence of American and British Thought on Australian Physics Education'. *International Science Education*, 73 (4), 445–57.

Landry, M.P., 1924. 'Etude sur les Irrégularités de Propagation des Ondes Courtes'. *L'Onde Électrique,* 3, 449–510.

Lang, H.R., 1934. 'Physics in Industry: A Developing Field of Work'. *JC*, 13, 527–31.

Larmor, J., 1924. 'Why Wireless Electric Rays Can Bend Round the Earth'. *Philosophical Magazine,* 48, 1,025–36.

Lassen, H., 1926. 'On the Ionization of the Atmosphere and its Influence on the Propagation of Short Electric Waves of Wireless Telegraphy'. *Jahr. drath. Tel.*, 28, 109–13, 139–47.

Lee, A.G., 1930. 'The Radio Communication Services of the British Post Office'. *Proc. IRE*, 18, 1,690–731.

Lee, N.J., 1992. *Industry, Education and the State: The Development of British Elites, Perceptions of the Question of Trained Manpower, 1918–1944*. Sussex University, PhD diss.

Levine, G. ed., 1983. *Realism and Representation: Essays on the Problem of Realism in Relation to Science, Literature, and Culture.* Madison: University of Wisconsin Press.

Lied, F., 1962. 'Introductory Speech'. In B. Maehlum, ed. *Electron Density Profiles in the Ionosphere and Exosphere. NATO Conference Series, Vol. 2.* New York: Pergamon Press.

Lightman, B., ed., 2004. *The Dictionary of Nineteenth-Century British Scientists.* London: Thoemmes Continuum, 4, 1,780–4.

Lindemann, A.F., Dobson, G.M.B., 1923a. 'A Theory of Meteors, and the Density of Temperatures of the Outer Atmosphere to Which it Leads'. *Proc. Roy. Soc. Lond.*, 102, 411–37.

——, 1923b. 'Note on the Photography of Meteors'. *Mon. Not. Roy. Astr. Soc.*, 83, 163–6.

Livingstone, K., 1996. *The Wired Nation Continent: The Communication Revolution & Federating Australia.* Melbourne: Oxford University Press.

Lloyd, B.E., 1988. *In Search of Identity: Engineering in Australia.* Melbourne: University of Melbourne, PhD diss.

Lodge, J.A., 1987. 'THORN EMI Central Research Laboratories: An Anecdotal History'. *Physics in Technology*, 18 (6), 258–68.

Long, J., 1985. *A First Class Job! Frank Murphy, Radio Pioneer.* Sheringham: Norfolk.

Lorimer, R.M. & Wilson, D.C., eds., 1988. *Communication Canada: Issues in Broadcasting and New Technologies.* Toronto: Kagan & Woo Ltd..

Love, A.E.H., 1915. 'The Transmission of Electric Waves Over the Surface of the Earth'. *Phil. Trans. Roy. Soc. Lond.*, 215, 105–31.

Low, A.M., 1924. *Wireless Possibilities.* London: Kegan Paul, Trubner, & Co.

Lundgreen, P., 1990. 'Engineering Education in Europe and the USA, 1750–1930: The Rise to Dominance of School Culture and the Engineering Professions'. *Annals of Science*, 47, 33–75.

Lynch, M., 1985. 'Discipline and the Material Form of Images: An Analysis of Scientific Visibility'. *Soc. Stud. Sc.,* 15, 37–66.

Lyon, Meter, ed., 1976. *Britain and Canada. Survey of a Changing Relationship.* London: Frank Cass.

Macdonald, H.M., 1903. 'The Bending of Electric Waves Round a Conducting Obstacle'. *Proc. Roy. Soc. Lond.*, 71, 251–8.

Macintyre, S., 1986. *The Oxford History of Australia, 1901–1942: The Succeeding Age.* Melbourne: Oxford University Press.

Mackay, R.R., 1938. 'Education Aspects of the Radio Profession in Australia'. *Proc. Wor. Rad. Con.*, 1–8.

MacLaurin, W.R., 1949. *Invention and Innovation in the Radio Industry.* New York: Macmillan Co.

MacLeod, R.M., 1971. 'Scientific Advice in the War at Sea, 1915–1917: The Board of Invention and Research'. *Journal of Contemporary History*, 6, 3–40.

——, 1982. 'On Visiting the *Moving Metropolis*: Reflections on the Architecture of Imperial Science'. *HRAS,* 5, 1–16

——, 1988a. 'The Practical Man: Myth and Metaphor in Anglo-Australian Science'. *Aus. Cul. Hist.*, 8, 24–49.

——, 1988b. 'From Imperial to National Science'. In Roy Macleod, ed. *The Commomwealth of Science. ANZAAS and the Scientific Enterprise in Australasia, 1888–1988.* New York: Oxford University Press, 40–72.

——, 1993. 'Passages in Imperial Science: From Empire to Commonwealth. *Journal of World History*, 4, 117–50

——, 1995. 'Colonial Engineers and the Cult of Practicality: Themes and Dimensions in the History of Australian Engineering'. *History and Technology*, 12, 147–62.

——, 1999, ed. *The Boffins of Botany Bay: Radar at the University of Sydney, 1939–1945.* Canberra: Australian Academy of Science.

MacLeod, R.M. & Andrews, E.K., 1970. 'The Origins of the D.S.I.R.: Reflections on Ideas and Men, 1915–16'. *Public Administration*, 48, 23–48.

MacRae, A.E., 1938. 'Some Aspects of Patent Protection'. *The Queen's Review,* November 1938.

Maddock, R. & McLean, I.W. eds., 1987. *The Australian Economy in the Long Run.* New York: Cambridge University Press.

Madsen, J.P.V., 1935. *The Ionosphere and its Influence upon the Propagation of Radio Waves* Sydney: Simmons Ltd.

——, 1937. 'Radio Research'. *Report of the Meeting of A.N.Z.A.A.S.*, 23, 14–18.

Maeda, K.I., 1986. 'Fifty Years of the Ionosphere'. *Jour. RRL*, 33, 103–68.

Mahant, E.E. & Mount, G.S., 1984. *An Introduction to Canadian-American Relations.* Ontario: Methuen.

Malin, S.R.C., 1996. 'Geomagnetism at the Royal Observatory'. *Quart. Jour. Roy. Astr. Soc.*, 37, 65–74, 71.

Manning, L.A., 1962. *Bibliography of the Ionosphere.* Stanford: Stanford University Press.

Marconi, D., 1962. *My Father Marconi.* London: F. Muller.

Marconi, G., 1902. 'The Progress of Electric Space Telegraphy'. *Proc. Roy. Soc. Lond.,* 17, 195–210.

——, 1922. 'Radio Telegraphy'. *Proc. IRE*, 10, 215–38.

——, 1924. 'Radio Communications. Presidential Address, 11 Dec 1924'. *Jour. Roy. Soc. Arts*, 62.

——, 1924a. *Results Obtained over very Long Distances by Short Wave Directional Wireless Telegraphy. More Generally Referred to as the Beam System.* Hastings.

——, 1924b. 'Risultati ottenuti su lunghissime distanze mediante la radiotelegrafia direzionale ad onde corte, piu generalmente nota come "il sistema a fascio"'. *Telegrafi e Telefoni*, 5 (4), 179–84.

——, 1928. 'Radio Communication'. *Proc. IRE*, 16, 40–69.

——, 1933a. 'Sulla Propagazione di Micro-onde a Notevole Distanza'. *La Ricerca Scientifica*, 24. 71–2.

——, 1933b. 'Radio Communications by Means of Very Short Electric Waves'. *Nature*, 131, 292–4.

——, 1935. 'Fenomeni Accompagnanti le Radiotrasmissioni. Discorso Pronunziato l'11 Settembre 1930 a Trento'. In B.L. Jacot & D.M.B. Collier, *Marconi: Master of Space.* London: Hutchinson.

Marconi School of Wireless Communication. *The Marconi School of Wireless Communication.* London, n.d.

Marconi's Co., *The Marconi Beam System for Long-Distance Communications. A Revolution in Present-Day Practice.* London: Marconi House n.d.

Maris, H.B., 1928. 'The Upper Atmosphere'. *Terr. Mag. & Atm. Elect.,* 33, 233–55.

Martin, D., 1986. *Thorn EMI: 50 Years of Radar: 50 Years of Company Involvement with Radar Technology 1936–1986.* Hayes: Thorn EMI.

Martin, L.H., 1946. 'Obituary: Professor T.H. Laby, F.R.S.'. *Australian Journal of Science*, 9 (2), 64–5.

Martyn, D.F., 1927. 'On Frequency Variations of the Triode Oscillator'. *Philosophical Magazine*, 4, 922–42; 1928, 6, 223–8; 1929, 7, 1,094–6.

——, 1929–30. 'On a New Method of Measurement of Minute Alternating Currents'. *Proc. Roy. Soc. Edi.*, 50, 166–74.

——, 1934. 'Atmospheric Pressure and the Ionisation of the Kennelly-Heaviside Layer'. *Nature*, 133, 294–5.

Martyn, D.F., Munro, G.H., & Piddington, J.H., 1937. 'The Polarization of Radio Echoes'. *Proc. Roy. Soc. Lond.*, A 158, 536–51; *CSIR Bulletin*, 1937, 110; *RRB Rep.*, 12, 7–18.

Martyn, D.F. & Pulley, O.O., 1936. 'The Temperature and Constituents of the Upper Atmosphere'. *Proc. Roy. Soc. Lond.*, A 154, 455–86.

Martyn, D.F. & Wood, H.B., 1933. 'A Frequency Recorder'. *J. Ins. Elec. Eng.*, 5, 6–13; *CSIR Bulletin*, 1935, 87; *RRB. Rep.*, 1935, 6, 49–58.

Massey, H., 1971. 'David Forbes Martyn, 1906–1970'. *Biog. M. Fell. Roy. Soc.*, 17, 497–510.

——, 1974. 'Theories of the Ionosphere, 1930–1955'. *Jour. Atm. Terr. Phys.*, 36, 2,141–58.

——, 1980. 'T.H. Laby F.R.S.- The Laby Memorial Lecture'. *Aus. Phys.*, 17, 181–7.

Mathews, S.A. et al*.,* 1938. 'A Review of Radio Development as Reflected in Applications in the Australian Patent Office during 1936–37'. *Proc. Wor. Rad. Con.,* 7.

Maynard, F.S., 1936. 'Dummy Aerials for Broadcast Frequencies'. *AWA Tech. Rev.*, 2 (1), 33–41.

McCarthy, J., 1971. 'Australia and Imperial Defence: Co-operation and Conflict, 1918–1939'. *Aus. Jour. Pol. Hist.*, 17, 19–32.

——, 1976. *Australia and Imperial Defence, 1918–39*. Brisbane: Sta. Lucia.

McIlwaine, J., 2004. 'Wartime Activities of AWA-Radar'. *Radio Waves*, 87, 35.

McKercher, B.J.C., ed., 1991. *Anglo-American Relations in the 1920s: The Struggle for Supremacy*. Basingstoke: Macmillan.

McLeish, C.W., 1940. 'Note on a Method of Plotting Electron Distribution Curves for the F layer'. *Can. Jour. Res.*, 18, 98–103.

McLennan, J.C., 1928. 'The Aurora and its Spectrum'. *Proc. Roy. Soc. Lon.,* 120, 327–57.

McNicol, D., 1946. *Radio's Conquest of Space: The Experimental Rise in Radio Communication*. London: Chapman & Hall.

McNulty, J., 1988. 'Technology and Nation-Building in Canadian Broadcasting'. In R.M. Lorimer & D.C. Wilson, eds., *Communication Canada: Issues in Broadcasting and New Technologies*. Toronto: Kagan & Woo Ltd., 175–98.

Meaney, N., 1976. *The Search for Security in the Pacific, 1901–14*. Sydney: Sydney University Press.

Meissner, A., 1924. 'Die Ausbreitung der elektrischen Wellen über die Erde'. *Jahrbuch der drathlosen Telegraphie,* 24, 85–92.

Mellor, D.P., 1958. *The Role of Science and Industry: Australia in the War, 1939–45.* Canberra: Australian War Memorial.

Melrose, D.B. & Minett, H.C., 1998. 'Jack Hobart Piddington'. *HRAS*, 12 (2), 229–46.

Menzel, D., 1953. *Our Sun.* Massachussets: Harvard University Press.

Merchant, E.E., 1916. 'The Heaviside Layer'. *Proc. IRE,* 4, 511–21.

Merchant, E.W., 1915. 'Conditions Affecting the Variations in Strength of Wireless Signals'. *Jour. IEE, 53*, 329–44.

Mesny, R., 1926. 'Propagation des Ondes Courtes'. *Onde Électrique,* 5, 436–59.

Metropolitan-Vickers Electrical Co., 1921. *Education for Industry. A Description of the Apprenticeship Courses and Methods of Training of the Metropolitan-Vickers Electrical Co.* Manchester.

Middleton, W.E.K., 1979. *Physics at the National Research Council of Canada, 1929–1952.* Waterloo: Wilfrid Laurier University Press.

——, 1981. *Radar Development in Canada: The Radio Branch of the National Research Council of Canada, 1939–1946.* Waterloo, Ont.: Wilfrid Laurier University Press.

Miles, W.G.H., 1926. 'Short Waves for Long Ranges: A Review'. *EW &WE,* 3, 22–33.

Millar, T.B., 1978. *Australia in Peace and War: External Relations 1788–1977.* Canberra: Australian National University Press.

Millington, G., 1932. 'Ionization Charts of the Upper Atmosphere'. *Proc. Phys. Soc. Lond.,* 44, 580–93; 47 (1935), 263–76.

——, 1938. 'The Relation between Ionospheric Transmission Phenomena at Oblique Incidence and Those at Vertical Incidence'. *Proc. Phys. Soc. Lond.,* 50, 801–25.

——, 1948. *Fundamental Principles of Ionospheric Transmission.* London: His Majesty's Stationery Office, Sep 1943, DSIR and Admiralty.

Millis, C.T., 1932. *Education for Trades and Industries.* London: E. Arnold.

Mills, J.S., 1924. *The Press and Communications of the Empire.* London.

Milne, E.A., 1926. On the Possibility of the Emission of High-Speed Atoms from the Sun and Stars. *Mon. Not. Roy. Astr. Soc.*, 86, 459–473.

——, 1939–41. Love, Augustus Edward Hough. *Ob. Not. Fell. Roy. Soc.*, 3, 467–482.

Mimno, H.R., 1937. The Physics of the Ionosphere. *Reviews of Modern Physics,* 9, 1–44.

Mitra, A.P., 1984. *Fifty Years of Radio Science in India.* New Delhi: Indian National Science Academy.

Mokyr, J., 1991. *The Level of Riches: Technological Creativity and Economic Progress.* New York: Oxford University Press.

Moreau, L., 1991. 'The Military Communications Explosion, 1914–1918'. *The Antique Wireless Association Review*, 6, 135–54.

Morecroft, J.N., 1924. 'The Growing Importance of Short Waves'. *Radio Broadcast*, Aug 1924, 296.

Morrell, J., 1992. 'Research in Physics at the Clarendon Laboratory, Oxford, 1919–1939'. *HSPS*, 22, 263–307.

Morrell, J. & Thackray, A., 1981. *Gentlemen of Science: Early Years of the British Association for the Advancement of Science.* Oxford: Clarendon Press.

Morris, P.R., 1993–94. 'A Review of the Development of the Thermionic Valve Industry'. *Trans. Newcomen Soc.,* 65, 57–73.

Morse, A.H., 1925a. *Radio: Beam and Broadcast, Its Story and Patents* London: E. Benn Ltd.

——, 1925b. 'The Short Wave: Its Discovery, Abandonment and Revival'. *The Wireless World*, 17, 389–92.

Moseley, R., 1977. 'Tadpoles and Frogs: Some Aspects of the Professionalization of British Physics, 1870–1939'. *Soc. Stud. Sc.*, 7, 423–46.

Mosini, V., 1996. 'Realism vs. Instrumentalism in Chemistry: The Case of the Resonance Theory'. *Rivista della Storia della Scienza*, 4 (2), 145–68.

Mowery, D., 1986. 'Industrial Research in Britain, 1900–1950'. In R. Elbaum & W. Lazonick, eds. *The Decline of the British Economy.* Oxford: Clarendon Press.

Moyal, A., 1983. 'Telecommunications in Australia: An Historical Perspective, 1854–1930'. *Prometheus*, 1 (1), 23–41.

——, 1984. *Clear Across Australia. A History of Telecommunications.* Melbourne, Victoria: Thomas Nelson.

——, 1986. *A Bright and Savage Land: Science in Colonial Australia.* Sydney: William Collins.

——, 1987. 'The History of Telecommunication in Australia: Aspects of the Technological Experience, 1854–1930'. In N. Reingold & M. Rothenberg eds. *Scientific Colonialism: A Cross Cultural Comparison.* Washington, D.C.: Smithsonian Institution Press, 35–54.

Moyal, A., 2002. 'Sydney Herbert Witt, 1892–1973. Electrical Engineer'. *ADB*, 16, 574–5.

Muirhead, E., 1996. *A Man Ahead of his Times: T.H. Laby's Contributions to Australian Science*. Melbourne: University of Melbourne.

Munns, D., 1997. 'Linear Accelerators, Radio Astronomy, and Australia's Search for International Prestige'. *HSPS*, 27, 299–317.

Munro, G.H., 1928. 'The Reflecting Layer of the Upper Atmosphere. An Estimation of the Height for Wireless Waves of 600 Metres Wavelength in New Zealand'. *EW & WE*, 5, 242–4.

Munro, G.H. & Huxley, L.G.H., 1932. 'Shipboard Observations with a Cathode-ray Direction-finder between England and Australia'. *Jour. Ins. Ele. Eng.*, 71, 488–96.

Murray, F.H. & Barton Hoag, J., 1937. 'Heights of Reflection of Radio waves in the Ionosphere'. *Physical Review,* 51, 333–41.

Murray, G., 1926. 'Empire Wireless Possibilities'. *Unit. Emp. RES. J.*, 17, 167–9.

Murray-Smith, S., 1966. *A History of Technical Education in Australia, with Special Reference to the Period before 1914*. Melbourne, PhD diss.

Murray-Smith, S. & Dare, A.J., 1987. *The Tech. A Centenary History of the Royal Melbourne Institute of Technology*. Melbourne: Hyland House.

Myers, D.M., 1983. 'Sir John Percival Vissing Madsen, 1879–1969. Physicist and Engineer'. *ADB*, 10, 376–7.

Myers, J.M., 1925. *Wireless in Australia*. Sydney: AWA.

Nahin, P.J., 1987. 'Oliver Heaviside, Sage in Solitude: The Life, Work, and Times of an Electrical Genius of the Victorian Age'. New York: IEEE Press.

Naismith, R., 1961. 'Early Days at Ditton Park'. *R.R.O. Newsletter*, 15th Sept. 5, 1.

Nalder, R.H.F., 1958. *The Royal Corps of Signals. A History of its Antecedents and Development circa 1800–1995.* London: Royal Signal Institution.

Nash, H., 1990. 'William Shand Watt, 1876–1958: Mining Engineer and Meteorologist'. *ADB*, 1990, 12, 416–17.

National Research Council (1930–41). *Annual Report of the National Research Council Containing the Report of the President and Financial Statement*. Ottawa, vols. 14–23.

——, 1937. *Annual Report of the National Research Council Containing the Report of the President and Financial Statement for the Fiscal Year 1936–37.* Ottawa.

Neal, A.L., 1941. 'Development of Radio Communication in Canada'. *Canadian Geographical Journal*, 22, 164–91.

Neill, N., 1991. *Technically… and Further: Sydney Technical College 1891–1991.* Sydney: Hale & Ironmonger.

Newton-Smith, W.H., 1981. *The Rationality of Science.* London: Routledge.

Nicholson, J.W., 1910. 'On the Bending of Electric Waves Round the Earth'. *Phil. Mag.*, 19, 276–8; 20, 157–72.

Nolan, M., 1989. 'An Infant Industry: Canadian Private Radio, 1919–36'. *Can. His. Rev.*, 70, 496–518.

Oliver, W.H. & Williams, B.R., eds., 1981. *The Oxford History of New Zealand.* Wellington: Oxford University Press.

Orde, A., 1978. *Great Britain and International Security 1920–1926.* London: Royal Historical Society.

——, 1996. *The Eclipse of Great Britain: The United States and British Imperial Decline, 1895–1956.* London: Macmillan.

Oreskes, N. & Doel, R.E., 2002. 'The Physics and Chemistry of the Earth'. In M.J. Nye, ed. *The Cambridge History of Science: The Modern Physical and Mathematical Science.* Cambridge: Cambridge University Press, 5, 538–57.

Orme, M.R., 1920. 'Radiotelegraphy and Aviation'. In *The Yearbook of Wireless Telegraphy and Telephony.* London: Wireless Press, 988–94.

Otto Sibum, H., 2004. 'What Kind of Science is Experimental Physics?'. *Science*, 306, 60–1.

Packman, M.E., 1915. 'The Training of the Radio Operator'. *Proc. IRE*, 3, 311–33; 1916, 4, 41–6.

Parkinson, D.W. 'Geomagnetism. Theories since 1900'. In G.A. Good, ed. *Sciences of the Earth. An Encyclopedia of Events, People and Phenomena.* New York: Garland Publishing, 357–65.

Pawley, E., 1972. *BBC Engineering, 1922–1972.* London: BBC Publications.

Peck, J.L.H., 1946. 'Out of this World: The Story of the Ionosphere'. *Harper's Magazine,* 192, 502–9.

Pedersen, P.O., 1927. *The Propagation of Radio Waves along the Surface of the Earth and in the Atmosphere.* Copenhagen: Danmarks Naturvioenskabeliage Samfund.

Pedgley, D.E., 1995. 'Pen Portraits of Presidents-Sir George Clarke Simpson'. *Weather*, 50, 347–9.

Peers, F.W., 1966. 'The Nationalist Dilemma in Canadian Broadcasting'. In P. Russell, ed., *Nationalism in Canada.* Toronto: McGraw Hill, 252–71.

Peers, F.W., 1969. *The Politics of Canadian Broadcasting, 1920–1951.* Toronto: Toronto University Press.

Pegg, M., 1983. *Broadcasting and Society, 1918–1939.* London: Croom Helm.

Pestre, D., 1997. 'Studies of the Ionosphere and Forecasts for Radiocommunications. Physicists and Engineers, the Military and National Laboratories in France (and Germany) after 1945'. *History and Technology*, 13, 183–205.

Phillipson, D.J.C., 1991. 'The National Research Council of Canada: Its Historiography, its Chronology, its Bibliography'. *Scientia Canadensis*, 15, 177–200.

Phipps, S.P., 1991. 'The Commercial Development of Short Wave Radio in the United States, 1920–1926'. *Historical Journal of Film, Radio and Television,* 11, 215–27.

Picken, D.K., 1948. 'Thomas Howell Laby, 1880–1946'. *Ob. Not. Roy. Soc*, 5, 733–55.

Pickering, A., 1984. 'Against Putting the Phenomena First: The Discovery of the Weak Neutral Current'. *HSPS,* 15, 85–117.

Pickworth, G., 1993. 'Germany's Imperial Wireless System'. *Electronics World & Wireless World*, 99, 427–32.

Piddington, J.H. & Oliphant, M.L., 1971. 'David Forbes Martyn, 1906–1970'. *RAAS*, 2, 2. 47–60.

Piggott, W.R., 1994. 'Some Reminiscences of Work with Sir Edward Appleton'. *Jour. Atm. Terr. Phys.*, 566, 727–31.

Plummer, A., 1937. *New British Industries in the Twentieth Century.* London: Isaac Pitman.

Pocock, H.S., 1971. 'Sixty Years'. *Wireless World*, 77, 153–5.

Pocock, R.F., 1986. 'Radio Telegraphy in the Royal Navy, 1887–1900'. In *Papers Presented at the 13th IEE Weekend Meeting on the History of Electrical Engineering.* London, 12/1–12/6.

——, 1988. *The Early British Radio Industry.* Manchester: Manchester University Press.

Pocock, R.F. & Garratt, G.R.M., 1972. *The Origins of Maritime Radio: The story of the Introduction of Wireless Telegraphy in the Royal Navy between 1896 and 1900.* London: H.M.S.O.

Post Office, 1939. 'History of the Engineering Department'. *Post Office Green Papers*, 46, 3–41.

Prang, M., 1965. 'The Origins of Public Broadcasting in Canada'. *Can. His. Rev.* 46, 1–31.

Pratt, B.W., 1977. 'Electrical Equipment and Appliances'. In *The Australian Encyclopaedia.* 3rd ed. Sydney: Grolier Society of Australia.

Prostak, E.J., 1983. *Up in the Air: The Debate over Radio Use during the 1920s.* Kansas: University of Kansas, Ph.D. diss.

Pulley, O.O., 1934a. 'Technique of Height Measurement of the Ionosphere by the Pulse Method'. *Nature*, 133, 576–7.

——, 1934b. 'A Self Synchronized System for Ionospheric Investigation by the Pulse Method'. *Proc. Phys. Soc. Lond.*, 46, 853–71.

Pyatt, E., 1983. *The National Physical Laboratory: A History.* Bristol: Hilger.

Ramsbottom, C.E., 1995. 'The Era of the Home Wireless Constructor'. In *100 Years of Radio*, IEE Conference Publication 411. London: IEE, 114–18.

Ratcliffe, J.A., 1959. 'Thomas Lydwell Eckersley', *Biog. M. Fell. Roy. Soc.*, 5, 69–74.

——, 1966a. 'Edward Victor Appleton, 1892–1965'. *Biog. M. Fell. Roy. Soc.*, 12, 1–21.

——, 1966b. 'Appleton as a Radio Scientist'. *Electronics and Power*, 12, 34–6.

——, 1970. *Sun, Earth and Radio. An Introduction to the Ionosphere and Magnetosphere.* London: World University.

——, 1971. 'William Henry Eccles, 1875–1966'. *Biog. M. Fell. Roy. Soc.*, 17, 195–214.

——, 1974. 'Scientists' Reactions to Marconi's Transatlantic Radio Experiment'. *Proc. IEE,* 121, 1,033–38.

——, 1975a. 'Robert Alexander Watson-Watt, 1892–1973'. *Biog. M. Fell. Roy. Soc.*, 21, 549–68.

——, 1975b. 'Physics in a University Laboratory before and after World War II'. *Proc. Roy. Soc. Lond.*, A342, 457–64.

——, 1978. 'Wireless and the Upper Atmosphere, 1900–1935'. *Contemporary Physics*, 19, 495–504.

Rawer, K., 1958. *The Ionosphere: Its Significance for Geophysics and Radio Communications.* London, transl. German (1952) by L. Katz.

Rayleigh, L., 1941. 'Joseph John Thomson'. *Ob. Not. Fell. Roy. Soc.*, 3, 587–609.

Read, D., 1998a. 'The Radio Communication Company'. *Bull. BVWS,* 23 (3), 4–9.

—— 1998b. 'Metropolitan-Vickers Electrical Company-Trafford Park, Manchester'. *Bull. BVWS*, 23, 16–21.

Reade, L., 1963. *Marconi and the Discovery of Wireless.* London: Faber & Faber.

Reader, W.J., 1987. *A History of the Institution of Electrical Engineers, 1871–1971.* London: Peter Peregrinus Ltd.

Redhead, M., 1980. 'Models in Physics'. *Brit. Jour. Phil. Sc.*, 31, 145–63.

Reich, L.S., 1977. 'Research, Patents, and the Struggle to Control Radio: A Study of Big Business and the Uses of Industrial Research'. *Business History Review*, 51, 208–34.

——,1985. *The Making of American Industrial Research: Science and Business at GE and Bell, 1876–1926.* Cambridge: Cambridge University Press.

Report of the Committee of the Privy Council for Scientific and Industrial Research for the Year 1920–21. London, 1921.

Richardson, H.W., 1961. 'The New Industries Between the Wars'. *Oxford Economic Papers*, 13, 360–83.

Rigby, C.A., 1944. *The War on the Short Waves.* London: Lloyd Cole.

Rishbeth, H. & Garriot, O.K., 1969. *Introduction to Ionospheric Physics.* New York: Academy.

Rishbeth, H., 1994. 'What Became of Appleton's Ionosphere?'. *Jour. Atm. Terr. Phys.*, 56 (6), 713–26.

Ristow, A., 1927. *Die Funkentelegraphie, ihre internationale Entwicklung und Bedeutung.* Berlin: E. Ebering.

Roach, J.P.C., ed., 1959. *Victoria County History of Cambridgeshire.* Oxford: Oxford University Press.

Roberts, J.H.T., 1923. 'The Development of High-Power Silica Valves'. *Modern Wireless*, 1, 535–9.

Robinson, H. 1964. *A History of the Post Office in New Zealand.* Wellington: Government Printer.

Rogers, W.S., 1924. 'Air as Raw Material'. *The Annals of the American Academy*, 112, 251–5.

Roget, S.R., 1924. *A Dictionary of Electrical Terms, Including Telegraphy, Telephony and Wireless.* London: Isaac Pitman & Sons.

Rose, D.C., 1933. 'Radio Observations on the Upper Ionized Layer of the Atmosphere at the Time of the Total Eclipse of August 31st 1932'. *Can. Jour. Res*, 8, 15–28.

Ross, A.T., 1987. 'The Politics of Secondary Industry Research in Australia, 1926–1939'. *HRAS*, 7 (4), 373–92.

——, 1990. 'Australian Overseas Trade and National Development Policy, 1932–1939: A story of Colonial Larrikins or Australian Statesmen?'. *Aus. Jour. Pol. Hist.*, 36 (2), 184–204.

Ross, J.F., 1978. *A History of Radio in South Australia, 1897–1977.* Adelaide: Lutheran Publishing House.

——, 1998. *Radio Broadcasting Technology: 75 Years of Development in Australia, 1923–1998.* New South Wales: Port Macquarie.

Round, H.J., 1920. 'Direction and Position Finding'. *Jour. I.E.E.*, 58, 224–47.

Round, H.J., Eckersley, T.L., Tremellen, K., & Lunnon, F.C., 1925. 'Report on Measurements Made on Signal Strength in 1922–23 in Australia'. *Jour. IEE*, 63, 933–1,011.

Rouse Ball, W.W., 1912. 'The Cambridge School of Mathematics'. *The Mathematical Gazette*, 6, 311–23.

Royal Canadian Corps of Signals, 1960. *A Short History of the Northwest Territories and Yukon Radio System, RC Signs.* Edmonton.

Rudd, J.B., 1938. 'Equipment for the Measurement of Loudspeaker Response'. *AWA Tech. Rev.*, 3 (4), 143–56.

Rudwick, M., 1976. 'The Emergence of a Visual Language for Geological Science, 1760–1840'. *History of Science,* 14, 149–95.

Rukop, H., 1926. Recent Developments in Short-Wave Wireless Telegraphy. *EW & WE,* 3, 606–612.

Russell, J., 1992. 'Research in Physics at the Clarendon Laboratory, Oxford, 1919–1939'. *HSPBS,* 22, 263–307.

Ryan, H.N., 1923. 'General Efficiency of Reception on Short Waves'. *EW & WE*, 1, 140–8.

Sacazes, J.M., 1928. 'Les Ondes Courtes et la Télégraphie sans Fil Souterrain'. *QST*, 49, 57–9.

Sacklowski, A., 1927. 'Die Ausbreitung der Elektromagnetischen Wallen'. *Elektrische Nachrichten—Technik*, 4 Jan 1927, 31–74.

Sanderson, M., 1972a. *The Universities and British Industry, 1850–1970.* London: Routledge & Kegan Paul.

——, 1972b. 'Research and the Firm in British Industry, 1919–39'. *Science Studies*, 2, 107–51.

——, 1999. *Education and Economic Decline in Britain, 1870 to the 1990s.* Cambridge: Cambridge University Press.

Sarnoff, D., 1928. 'The Development of the Radio Art and Radio Industry since 1920'. In E.E. Bucher et al. *The Radio Industry: The Story of its Development.* Chicago: A.W. Shaw Co., 97–113.

Saskatoon Amateur Radio Club, 1968. *From Spark to Space: The Story of Amateur Radio in Canada.* Saskatoon.

Saul, S.B., 1979. 'Research and Development in British Industry from the End of the Nineteenth Century to the 1960s'. In T.C. Smout, ed., *The Search for Wealth and Stability*. London: Macmillan, 114–38.

Savours, A. & McConnell, A., 1982. 'The History of the Rossbank Observatory, Tasmania'. *Ann. Sci.*, 39, 527–64.

Sayer, R.S., 1950. 'The Springs of Technical Progress in Britain, 1919–1939'. *The Economic Journal*, 60, 275–91.

Schaffer, Simon, 1986. 'Scientific Discoveries and the End of Natural Philosophy'. *Soc. Stud. Sc.*, 16, 387–420.

Schedvin, C.B., 1970. *Australia and the Great Depression: A Study of Economic Development and Policy in the 1920s and 1930s*. Sydney: Sydney University Press.

——, 1982. 'The Culture of CSIRO'. *Aus. Cul. Hist.*, 2, 76–89.

——, 1987. *Shaping Science and Industry: A History of Australia's Council for Scientific and Industrial Research, 1926–1949*. Sydney: Unwin.

Schmidt, M., 1988. *Pioneers of Ozone Research: A Historical Survey*. Katlenburg-Lindau: Max Planck Institute for Aeronomy.

Schonfield, H.J., ed., 1933. *The Book of British Industry*. London: Denis Archer.

Schröter, F., 1938. 'Television Problems and their Practical Solution'. *AWA Tech. Rev.*, 36. 283–96.

Schuster, A., 1889. 'The Diurnal Variation of Terrestrial Magnetism'. *Phil. Trans. Roy. Soc. Lond.,* A180, 467–512.

Schuster, A., 1908. 'The Diurnal Variation of Terrestrial Magnetism'. *Phil. Trans. Roy. Soc. Lond.,* A208, 163–204.

Schuster, A., 1922. 'Correspondence: A Short Story in Wireless'. *The Electrician*, 89, 325.

Scott, Y., 1962. *History of Marine Department, 1913–62.* Sydney: AWA.

Scrase, F.J., 1969. 'Some Reminiscence of Kew Observatory in the 20s'. *Met. Mag.*, 98, 180–7.

Serle, G., 1973. *From Deserts the Prophets Come: The Creative Spirit in Australia, 1788–1972.* Melbourne: Heinemann.

Shoup, G.S., 1928. 'Wireless Communication in the British Empire'. *Trade Information Bulletin*, 551, 1–28.

Siegfried, A., 1914. *Democracy in New Zealand.* London: G. Bell & Sons.

Simpson, G.C., 1928. 'Charles Chree, 1860–1928'. *Proc. Roy. Soc. Lon.*, 122, vii–xiv.

——, 1935. 'Sir Arthur Schuster, 1851–1934', *Ob. Not. Fell. Roy. Soc.*, 1, 409–23.

Sinclair, K., 1972. 'Fruit Fly, Fireblight and Powdery Scab: Australia-New Zealand Trade Relations, 1919–39'. *The Journal of Imperial and Commonwealth History*, 1 (1), 27–48.

Sinclair, W.A., 1976. *The Process of Economic Development in Australia*. Melbourne: Longman Cheshire.

Slotten, H.R., 1995. 'Radio Engineers, the Federal Radio Commission, and the Social Shaping of Broadcast Technology'. *Technology and Culture*, 36, 950–86.

Smith, A.J., 1928. 'Wireless Export Trade. The British Position in International Competition'. *The Electrical Review*, 102, 589–90.

Smith, F., 1933. 'Scientific Research Essential to Prosperity: Industry More Than Ever Dependent on Chemists, Physicists, etc'. *JC*, 12, 19–22.

Smith, N., 1937. 'Extension of Normal-Incidence Ionosphere Measurements to Oblique-Incidence Radio Transmission'. *Jour. Res. NBS*, 19, 89–94.

Smith, N., Kirby, S.S., & Gilliland, T.R., 1938. 'The Application of Graphs of Maximum Usable Frequency to Communication Problems'. In *Papers Presented to the General Assembly of the International Scientific Radio Union (URSI) Held in Venice, Italy, in September 1938*. Brussels, 5, 127–33

Smith, N.S., 1948. 'High-Frequency Broadcasting in Australia'. *Proceedings of the Institute of Radio Engineers, Australia*, 9, 10. 4–20.

Smith-Rose, R.L., 1918. 'The Evolution of the Thermionic Valve'. *The Wireless World*, 66 (3), 142–7; 66 (4), 229–35, 281–8.

——, 1923. 'Directive Radio Telegraphy and Telephony'. *EW & WE*, 1, 119–25.

——, 1926. 'The Cause and Elimination of Night Errors in Radio Direction-Finding'. *Jour. IEE*, 64, 831–43.

——, 1927. *A Study of Radio Direction Finding*. *Rad. Res. Spec. Rep., 5*. London.

——, 1928a. *An Investigation of a Rotating Radio Beacon, Rad. Res. Spec. Rep., 6*. London: H.M.S.O.

——, 1928b. 'Radio Direction-Finding by Transmission and Reception, with Particular Reference to its Application to Marine Navigation'. In *Papers of the Assembly of the U.R.S.I. Held in Washington, October 1927*. Washington, 54–84.

——, 1931. *The Orfordness Rotating Beacon and Marine Navigation, Rad. Res. Spec. Rep., 10*. London: H.M.S.O.

——, 1934. 'Report of the British National Committee to Commission II on Investigations of the Propagation of Waves Carried out in Great Britain from April 1931 to June 1934'. In *Papers Ass. URSI, London, September 1934*. Brussels, 4, 153–65.

Smith-Rose, R.L. & Chapman, S.R., 1928. 'Some Experiments on the Application of a Rotating Beacon Transmitter to Marine Navigation'. *Jour. IEE*, 66, 256–69.

Smith-Rose, R.L. & Hopkins, H.G., 1938. 'Radio Direction-Finding on Wavelengths between 6 and 10 Metres Frequencies 50–30 Mc./Sec.'. *Jour. IEE*, 83, 87–97.

Smythe, D., 1981. *Dependency Road: Communications, Capitalism, Consciousness and Canada*. Norwood, NJ: Ablex.

Solla Prize, D.J. de, 1984. 'Of Sealing Wax and String'. *Natural History*, 93, 49–57.

Sommerville, R.S., 1926. 'Is Canada Becoming Americanized?'. *Empire Review*, 43, 537–40.

Soward, F.H., 1938. 'What Radio Programs School Children Hear'. In H.F. Angus, ed., *Canada and her Great Neighbor: Sociological Surveys of Opinions and Attitudes in Canada concerning the United States*. New Haven.

Spencer, E.H., 1939. *A Report on Technical Education in Australia and New Zealand*. New York.

Spry, G., 1929. 'One Nation, Two Cultures'. *Canadian Nation*, 1, 21.

——, 1931. 'The Case for Nationalized Broadcasting'. *Queen's Quarterly*, 38, 151–69.

——, 1935. 'Radio Broadcasting and Aspects of Canadian-American Relations'. In W.W. McLaren et al., eds. *Proceedings of the Conference on Canadian-American Affairs Held at the St Lawrence University, Canton, New York, June 17–22, 1935*. New York, 106–27.

Starr, A.T., 1935. *Definitions and Formulae for Students: Radio Engineering*. Bath: Pitman Press.

Stewart, B., 1882. 'Terrestrial Magnetism'. In *Encyclopaedia Britannica*. London, 16, 159–84.

Stokes, J.W., 1982. *70 Years of Radio Tubes and Valves*. New York: Vestal Press.

Stranger, R., 1933. *Dictionary of Wireless Terms*. London: G. Newman.

Strangeways, H.J., 1994. 'Appleton's Magneto-Ionic Theory and the Lorentz Polarisation Term'. *Jour. Atm. Terr. Phys.*, 56 (6), 705–12.

Strutt, C.R., 1964. 'The Optics Research of Robert John Strutt, Fourth Baron Rayleigh'. *Applied Optics*, 3, 1,113–15.

Strutt, R.J., 1918. 'Ultra-Violet Transparency of the Lower Atmosphere, and its Relative Poverty in Ozone'. *Proc. Roy. Soc.*, 94, 260–8.

Sturmey, S. G., 1958. *The Economic Development of Radio*. London: Duckworth.

——, 1960. 'Patents and Progress in Radio'. *The Manchester School*, 28, 19–36.

Sullivan, J., 1998. 'Jack, Robert 1877–1957'. *DNZB*, 4.

Sullivan, W.T. ed., 1984. *The Early Years of Radioastronomy*. Cambridge: Cambridge University Press.

——, 1988. 'Early Years of Australian Radio Astronomy'. In R. Home, ed. *Australian Science in the Making*. Cambridge: Cambridge University Press, 308–44.

Süsskind, Ch., 1974. 'Guglielmo Marconi, 1874–1937'. *Endeavour*, 33, 67–72.

Sviedrys, R., 1976. 'The Rise of Physics Laboratories in Britain'. *HSPS*, 7, 405–36.

Swann, W.F.G., 1916. 'On the Ionization of the Upper Atmosphere'. *Terrestrial Magnetism and Atmospheric Electricity*, 21, 1–8.

Tabarroni, G. et al., 1974. *Marconi: Cento Anni dalla Nascita*. Torino: ERI.

Tarplee, P., 1996. *Abinger and the Royal Greenwich Observatory: The Recording of Magnetism and Time*. Guildford: Surrey Industrial Hist. Gr.

Taylor, A.H., 1933. 'High-Frequency Transmission and Reception'. In Keith Henney, ed., *The Radio Engineering Handbook*. London: McGraw-Hill Book Co., 433–52.

——, 1948. *Radio Reminiscences: A Half Century*. Navy's Research Laboratory.

Taylor, J.E., 1903. 'Characteristics of Electric Earth-Current Disturbances, and their Origin'. *Phys. Roy. Soc. Lond.*, 18, 225–7.

Terman, F.E., 1938. *Fundamentals of Radio*. New York, London: McGraw-Hill.

——, 1943. *Radio Engineer's Handbook*. New York, London: McGraw-Hill.

Terrall, M., 1998. 'Heroic Narratives of Quest and Discovery'. *Configurations: A Journal of Literature, Science, and Technology*, 6, 223–42.

The British Institution of Radio Engineers, 1944. *Post-War Development in Radio Engineering*. London.

The Canadian Beam Service, 1926. *The Electrician*, 97, 474, 498.

The Countess Eileen de Armil, 1931. 'Wireless Bonds, Why Not an Empire Broadcasting?'. *Unit. Emp. RES. J.*, 22, 544–5.

The Inauguration of the Canadian Beam Service, 1926. *Experimental Wireless & The Wireless Engineer*, 3, 715–16.

The Wireless World, 1926. *Dictionary of Wireless Technical Terms, Compiled by S.O. Pearson*. London: Iliffe and Sons.

Thistle, M., 1966. *The Inner Ring: The Early History of the National Research Council.* Toronto: Toronto University Press.

Thomas, M., 1985. Manufacturing and Economic Recovery in Australia, 1932–1937. *Working Papers in Economic History*, Australian National University, 46, [also published in R. Gregory & N. Butlin, eds., 1989. *Recovery from the Depression: Australia and the World Economy in the 1930s.* Cambridge: Cambridge University Press.]

Thompson, J.H. & Randall, S.J., 1997. *Canada and the United States. Ambivalent Allies.* 2nd ed. London: University of Georgia Press.

Thompson, R., 1980. *Australian Imperialism in the Pacific, 1820–1920.* Melbourne: Melbourne University Press.

Thoms, D., 1990. 'Technical Education and the Transformation of Coventry's Industrial Economy, 1900–1939'. In P. Summerfield & E.J. Evans, eds. *Technical Education and the State since 1850: Historical and Contemporary Perspectives.* Manchester: Manchester University Press, 37–54.

Thoms, D. & Donnelly, T., 1985. *The Motor Car Industry in Coventry since the 1890s.* London: MacMillan.

Thomson, J.J., 1931. 'Training and Careers in Physics'. *JC*, 10, 18–20.

Thomson, M., 1925. 'Scientific Research in New Zealand'. *N.Z. Jour. Sc. Tech.*, 8 (1), 41–61.

Thon, G.P., 1958. 'Frederick Alexander Lindemann, Viscount Cherwell, 1886–1957'. *Biog. M. Fell. Roy. Soc.*, 4, 45–71.

Thrower, K.R., 1992. *History of the British Radio Valve to 1940.* London: MMA.

Touzet, A., 1918. 'Le Réseau Radiotélégraphique Indochinois'. *Revue Indochinoise*, 245, 7–21.

Tremellen, K.W., 1938. 'Effect of the Eleven Years Sunspot Cycle on Short Wave Communication'. *The Marconi Review*, 70, 43–4.

——, 1939. 'The Ionosphere'. *The Marconi Review,* 72, 1–14.

Tribolet, L.B., 1929. *The International Aspects of Electrical Communications in the Pacific Area.* Baltimore: John Hopkins.

Turner, L.B., 1926. 'Notes on Wireless Matters'. *The Electrician,* 9 Jul 1926, 42–3.

——, 1931. *Wireless.* Cambridge: Cambridge University Press.

Tuska, C.D., 1944. 'Historical Notes on the Determination of Distance by Timed Radio Waves'. *Jour. Fran. Inst.,* 237, 1–20, 83–102.

Tuve, M.A., 1932. 'The Geophysical Significance of Radio Measurements of the Ionized Layers'. *Trans. Am. Geo. Un.,* 13, 160–7.

——, 1947. 'Review of Magnetic Survey and Observatory Program of the Department of Terrestrial Magnetism, 1904–1946'. In *Carnegie Institution of Washington Year Book, 1946–1947*. Washington, D.C., 46, 43–53.

Tyne, G.F.J., 1977. *Saga of the Vacuum Tube*. Indianapolis: H.W. Sams Co.

Varcoe, I., 1970. 'Scientists, Government and Organised Research in Great Britain, 1914–16: The Early History of the DSIR'. *Minerva*, 8, 196–216.

Vardalas, J., 2001. *The Computer Revolution in Canada: Building National Technological Competence*. Cambridge, Mass.: MIT Press.

Vestine, E.H., 1967. 'Geomagnetism and Solar Physics'. In S.I. Akasofu et al., eds., *Sydney Chapman, Eighty: From his Friends*. Boulder: University of Colorado, 19–23.

Villard, O.G., 1976. 'The Ionospheric Sounder and its Place in the History of Radio Science'. *Radio Science*, 11 (11), 847–60.

Vincent-Smith, T., 1919. 'Wireless in the RFC [Royal Flying Corps] During the War'. *The Electrician*, 83, 445–7.

Vipond, M., 1986. 'CKY Winnipeg in the 1920's: Canada's only Experiment in Government Monopoly Broadcasting'. *Manitoba History*, 12, 2–13.

——, 1992. *Listening In The First Decade of Canadian Broadcasting, 1922–1932*. Montreal: McGill-Queen s University Press.

Vyvyan, R.N., 1974. *Marconi and Wireless*. Yorkshire: E.P. Publishing, originally published as *Wireless Over Thirty Years*. London: Routledge & Kegan Paul, 1933.

Wait, G.R., 1923. 'The Work and Equipment of the Watheroo Magnetic Observatory'. In G. Lightfoot, ed. *Proceedings of the Pan Pacific Congress*, 1. Melbourne, 505–9.

Walker, A., 1992. *A Skyful of Freedom: 60 Years of the BBC World Service*. London: Broadside Books.

Walker, D., 1997–98. 'Climate, Civilization and Character in Australia, 1880–1940'. *Aus. Cul. Hist.*, 16, 77–95.

Walshaw, C.D., 1990. 'The Early History of Atmospheric Ozone'. In J.J. Roche, ed., *Physicists Look Back: Studies in the History of Physics*. Bristol: Adam Hilger, 313–26.

Wark, I.W., 1977. '1851 Science Research Scholarship Awards to Australians'. *RAAS*, 3, 47–52.

Warwick, A., 1993. 'Cambridge Mathematics and Cavendish Physics: Cunningham, Campbell and Einstein's Relativity, 1905–1911. Part II: Comparing Traditions in Cambridge Physics'. *SHPS*, 24, 1–25.

Watson, George N., 1918. 'The Diffraction of Electric Waves by the Earth'. *Proc. Roy. Soc. Lond.*, 95, 83–99.

Watson Watt, R.A., 1918–1919. 'The Transmission of Electric Waves Round the Earth'. *Proc. Roy. Soc. Lond.*, 95, 546–63.

——, 1923. 'Directional Observations on Atmospherics, 1916–1920'. *Phil. Mag.*, 45, 1,010–26.

——, 1925. 'Atmospherics. Discussion on Ionization in the Atmosphere'. *Proc. Phys. Soc. Lond.*, 37, 23D-31D.

——, 1929. 'Weather and Wireless'. *Quart. Jover. Roy. Met. Soc.*, 55, 273–301.

——, 1936. 'Polarisation Errors in Direction Finders'. *Wireless Engineering*, 13, 3–6.

——, 1957. *Three Steps to Victory: A Personal Account by Radar's Greatest Pioneer.* London: Odhams Press.

Watson Watt, R.A. et al., 1933. *Applications of the Cathode Ray Oscillograph in Radio Research.* London: H.M.S.O.

Wayne, R.P., 1970. *Photochemistry.* London: Butterworths.

——, 1985. *Chemistry of Atmosphere.* Oxford: Clarendon Press.

Waynick, A.H., 1975. 'The Early History of Ionospheric Investigations in the United States'. *Phil. Trans. Roy. Soc. Lond.*, 280, 11–25.

Webb, E.K., ed., 1997. *Windows on Meteorology: Australian Perspective.* Melbourne: CSIRO.

Wedlake, G.E.C., 1973. *SOS: The Story of Radio-Communication.* New York: Crane, Russak & Co.

Weir, E.A., 1965. *The Struggle for National Broadcasting in Canada.* Toronto: McClelland & Stewart.

White, F.W.G., 1930. 'A Standard Frequency-Meter'. *N.Z. Jour. Sc. Tech.*, 11, 328–33.

——, 1939. 'Early Observations of Aurora Australis'. *N.Z. Jour. Sc. Tech.*, 20, 267b-71b.

——, 1970. 'John Percival Vissing Madsen 1879–1969'. *RAAS*, 2, 1. 51–65.

——, 1975. 'Early Work in Australia, New Zealand and at the Halley Stewart Laboratory, London'. *Phil. Tran. Roy. Soc. Lond.*, 280, 35–46.

White, F.W.G., Geddes, M., & Skey, H.F., 1938. 'Radio Fade-Outs, Auroras and Magnetic Storms'. *Nature*, 142, 289.

——, 1939. 'The Antarctic Zone of Maximum Auroral Frequency'. *Terr. Mag. & Atm. Elect.*, 4, 367–77.

White, F.W.G. & Huxley, L.G.H., 1974. 'Radio Research, Australia 1927–1939'. *RAAS*, 3, 1. 7–30.

White, L., 1947. *The American Radio*. Chicago: Chicago University Press.

White, R., 1981. *Inventing Australia: Images and Identity, 1688–1980*. Sydney: Allen and Unwin.

Whittaker, E.T., 1935. 'Macdonald, Hector Munro'. *Ob. Not. Fell. Roy. Soc.*, 11, 551–58.

——, 1966. 'Watson, George Neville'. *Biog. M. Fell. Roy. Soc.*, 12, 521–30.

Whitten, R.C. & Popoff, I.G., 1965. *Physics of the Lower Ionosphere*. New Jersey: Prentice-Hall.

——, 1971. *Fundamentals of Aeronomy*. New York: J. Wiley & Sons.

Wicken, O., 1997. 'Space Science and Technology in the Cold War: The Ionosphere, the Military, and Politics in Norway'. *History and Technology*, 13, 207–29.

Wickenden, W.E., 1929. *A Comparative Study of Engineering Education in the United States and in Europe*. New York: The Society for the Promotion of Engineering Education.

Wiener, M., 1981. *English Culture and the Decline of the Industrial Spirit*. Cambridge: Cambridge University Press.

Wild, J.P., 1972. 'The Beginnings of Radio Astronomy in Australia'. *RAAS*, 2 (3), 52–61.

Wilkes, M.V., 1975. 'Early History of Ionospheric Investigations'. *Phil. Tran. Roy. Soc. Lond.*, 280, 54–5.

——, 1997. 'Sir Edward Appleton and Early Ionosphere Research'. *Notes Rec. Roy. Soc. Lond.,* 51 (2), 281–90.

Wilkins, M., 1974. *The Maturing of Multinational Enterprise: American Business Abroad from 1914 to 1970*. Cambridge, Mass.: Harvard University Press.

Williams, N., 1989. 'Ernest T. Fisk: Pioneer, Visionary and Entrepreneur-1'. *Electronics Australia*, June, 40–3.

Wilson, A.C., 1994. *Wire and Wireless: A History of Telecommunications in New Zealand, 1890–1987*. Palmerston North: Dunmore Press.

Wilson, D.B., 1982. 'Experimentalists among the Mathematicians: Physics in the Cambridge Natural Sciences Tripos, 1851–1900'. *HSPS*, 12, 325–71.

Wilson, W., 1956. 'Nicholson, John William'. *Biog. M. Fell. Roy. Soc.*, 2, 209–14.

Wireless Development in Canada, 1930. *Engineer*, 150, 561–3.

Wireless World and Radio Review, 1922–23, 11 279.

Withley, J.H., 1933. 'Empire Broadcasting and the New Service'. *Unit. Emp. RES. J.*, 24, 85–95.

Witt, S.H., 1938. 'The Research Laboratories of the Postmaster-General's Department'. *Proc. Wor. Rad. Con.* Sydney, 3–7.

Wood, H.B., 1936. 'The Design of an Automatic Variable-Frequency Radio Transmitter with Automatically Tuned Receiver, for Use in the Investigation of Radio Propagation in the Ionosphere'. *Jover. Inst. Eng. Aust.*, 8, 403–414.

Wood, J., 1992. *History of International Broadcasting*. London: Peter Peregrinus.

Wood, S., 1989. 'The Valve: Industrial Aspects before 1925'. *Bull. BVWS*, 14, 12–17.

Woolgar, S., 1976. 'Writing an Intellectual History of Scientific Developments: The Use of Discovery Accounts'. *Soc. Stud. Sc.,* 6, 395–422.

Wright, J.W. & Smith, G.H., 1967. 'Introductory Paper Review of Current Methods for Obtaining Electron-Density Profiles from Ionograms'. *Radio Science,* 2, 1,119–24.

Wrong, H., 1975. 'The Canada-United States Relationship, 1927–1951. *International Journal*, 31, 529–45.

Yang, H., 1928. 'Beam Transmission of Ultra Short Waves'. *Proc. IRE*, 16, 715–41; repr. by *Proc. IEEE*, 1984, 72, 635–45.

Yavetz, I., 1995. *From Obscurity to Enigma: The Work of Oliver Heaviside, 1872–1889.* Berlin: Birkhäuser Verlag.

Yeang, C.P., 2003. 'The Study of Long-Distance Radio-Wave Propagation, 1900–1919'. *HSPS*, 33, 2. 369–404.

——, 2004. 'Scientific Fact or Engineering Specification? The U.S. Navy's Experiments on Wireless Telegraphy *Circa* 1910'. *Technology and Culture,* 45, 1–29.

Young, P., 1983. *Power of Speech: A History of Standard Telephones and Cables, 1883–1983.* London: George Allen & Unwin.

Zimmerman, D.K., 1986. 'The Organization of Science for War: The Management of Canadian Radar Development, 1939–45'. *Scientia Canadensis,* 10, 93–108.

——, 1988. 'Radar and Research Enterprises Limited: A Study of Wartime Industrial Failure'. *Ontario History*, 80, 121–42.

INDEX